Subcellular Biochemistry

Volume 20
Mycoplasma Cell Membranes

SUBCELLULAR BIOCHEMISTRY

SERIES EDITOR

J. R. HARRIS, Institute of Zoology, University of Mainz, Mainz, Germany

ASSISTANT EDITORS

H. J. HILDERSON, University of Antwerp, Antwerp, Belgium
D. A. WALL, SmithKline Beecham Pharmaceuticals, King of Prussia, Pennsylvania, U.S.A.

A Continuation Order Plan is available for this series. A continuation order will bring delivery of each new volume immediately upon publication. Volumes are billed only upon actual shipment. For further information please contact the publisher.

Subcellular Biochemistry

Volume 20
Mycoplasma Cell Membranes

Edited by

Shlomo Rottem

and

Itzhak Kahane

Department of Membrane and Ultrastructure Research
The Hebrew University–Hadassah Medical School
Jerusalem, Israel

SPRINGER SCIENCE+BUSINESS MEDIA, LLC

The Library of Congress cataloged the first volume of this title as follows:

Sub-cellular biochemistry.

  London, New York, Plenum Press.
    v. illus. 23 cm. quarterly.
  Began with Sept. 1971 issue. Cf. New serial titles.
  1. Cytochemistry—Periodicals. 2. Cell organelles—Periodicals.
QH611.S84 574.8'76   73-643479

ISBN 978-0-306-44394-7   ISBN 978-1-4615-2924-8 (eBook)
DOI 10.1007/978-1-4615-2924-8

This series is a continuation of the journal *Sub-Cellular Biochemistry*,
Volumes 1 to 4 of which were published quarterly from 1972 to 1975

© 1993 Springer Science+Business Media New York
Originally published by Plenum Press in 1993

Contributors

Joel B. Baseman   Department of Microbiology, The University of Texas Health Science Center at San Antonio, San Antonio, Texas 78284

Robert Bittman   Department of Chemistry and Biochemistry, Queens College of The City University of New York, Flushing, New York 11367

J. M. Bové   Laboratory of Cellular and Molecular Biology, INRA and the University of Bordeaux II, 33883 Villenave d'Ornon Cedex, France

Vincent P. Cirillo   Department of Biochemistry and Cell Biology, State University of New York, Stony Brook, New York 11794

Jean Dahl   Department of Pathology, Harvard Medical School, Boston, Massachusetts 02115

X. Foissac   Laboratory of Cellular and Molecular Biology, INRA and the University of Bordeaux II, 33883 Villenave d'Ornon Cedex, France

Shulamith Horowitz   Mycoplasma Laboratory, Department of Microbiology and Immunology, Ben Gurion University of the Negev, Beersheva 84105, Israel

Itzhak Kahane   Department of Membrane and Ultrastructure Research, The Hebrew University–Hadassah Medical School, Jerusalem 91010, Israel

Göran Lindblom   Department of Physical Chemistry, University of Umeå, S-901 87 Umeå, Sweden

Ronald N. McElhaney   Department of Biochemistry, University of Alberta, Edmonton, Alberta, Canada T6G 2H7

F. Chris Minion   Veterinary Medical Research Institute, Iowa State University, Ames, Iowa 50011

Shmuel Razin   Department of Membrane and Ultrastructure Research, The Hebrew University–Hadassah Medical School, Jerusalem 91010, Israel

Leif Rilfors   Department of Physical Chemistry, University of Umeå, S-901 87 Umeå, Sweden

Ricardo F. Rosenbusch   Veterinary Medical Research Institute, Iowa State University, Ames, Iowa 50011

Shlomo Rottem   Department of Membrane and Ultrastructure Research, The Hebrew University–Hadassah Medical School, Jerusalem 91010, Israel

Colette Saillard   Laboratory of Cellular and Molecular Biology, INRA and the University of Bordeaux II, 33883 Villenave d'Ornon Cedex, France

Mitchell H. Shirvan   Department of Pediatrics, Hadassah University Hospital, Jerusalem 91010, Israel

Åke Wieslander   Department of Biochemistry, University of Umeå, S-901 87 Umeå, Sweden

Preface

The mycoplasmas, a trivial name used to denote organisms included in the class Mollicutes, are a group of prokaryotic organisms comprising more than 120 species distinguished from ordinary bacteria by their small size and the total lack of cell walls. The absence of a cell wall in mycoplasmas is a characteristic of outstanding importance to which the mycoplasmas owe many of their peculiarities, for example, their morphological instability, osmotic sensitivity, unique ion pumping systems, resistance to antibiotics that interfere with cell wall biosynthesis, and susceptibility to lysis by detergents and alcohols.

The fact that the mycoplasma cells contain only one membrane type, the plasma membrane, constitutes one of their most useful properties for membrane studies; once the membrane is isolated, it is uncontaminated with other membrane types. Another advantage in using mycoplasmas as models for membrane studies stems from the fact that their membrane lipid composition can be altered in a controlled manner. This characteristic results from the partial or total inability of the mycoplasmas to synthesize long-chain fatty acids and cholesterol, making mycoplasmas dependent on the supply of fatty acids from the growth medium. The ability to introduce controlled alterations in the fatty acid composition and cholesterol content of mycoplasma membranes has been utilized in studying the molecular organization and physical properties of biological membranes.

Many developments have occurred over the last decade in our methods of studying the abundance, sequence, structural organization, and function of mycoplasma membrane proteins. These studies took into consideration the structural simplicity of mycoplasmas, the relatively low coding capacity of their genome, and the ecological niches of these organisms (many of them live in close association with eukaryotic host cells as membrane surface parasites). The recent developments in identifying and characterizing mycoplasma membrane components that act as adhesins will shape and direct future efforts toward understanding the mechanism of adherence to infected target tissue, the biosynthesis of the adhesins, and their molecular organization.

The present volume is the first attempt to bring together in a comprehensive and comparative way the vast knowledge accumulated on the mycoplasma cell membranes. We hope that this volume, written by experts in their fields, will provide an up-to-date, exhaustive treatment of the subject and an impetus for mycoplasma membrane research in the future.

Shlomo Rottem
Itzhak   Kahane

Jerusalem, Israel

Contents

Chapter 3

**Physical Studies of Lipid Organization and Dynamics
in Mycoplasma Membranes**

Ronald N. McElhaney

Chapter 4

**Regulation and Physicochemical Properties of the Polar Lipids
in *Acholeplasma laidlawii***

Leif Rilfors, Åke Wieslander, and Göran Lindblom

Chapter 5
The Role of Cholesterol in Mycoplasma Membranes
Jean Dahl

Chapter 9

The Cytadhesins of *Mycoplasma pneumoniae* and *M. genitalium*

Joel B. Baseman

Chapter 10

Ion Pumps and Volume Regulation in Mycoplasmas

Mitchell H. Shirvan and Shlomo Rottem

Chapter 11
Transport Systems in Mycoplasmas
Vincent P. Cirillo

Chapter 1

Mycoplasma Membranes as Models in Membrane Research

Shmuel Razin

The mycoplasmas, lacking a cell wall and intracytoplasmic membranes, have only one type of membrane, the plasma membrane. The ease with which this membrane can be isolated and the fact that controlled alterations in its composition are introducible have made mycoplasma membranes most effective and popular tools in biomembrane research. The aim of this chapter is to introduce the reader to the mycoplasmas and provide a historical background of the various phases in mycoplasma membrane research. Focus will be on the advantages of mycoplasma membranes as models in studying structure and function of biological membranes, and on the development of concepts that have arisen since the beginning of mycoplasma membrane research, about 30 years ago.

1. AN INTRODUCTION TO MOLLICUTES

The first cultivation of a mycoplasma, the bovine pleuropneumonia agent, was reported almost 100 years ago (Nocard and Roux, 1898). Yet, despite our long acquaintance with mycoplasmas, their nature, relationship to other organ-

Shmuel Razin   Department of Membrane and Ultrastructure Research, The Hebrew University–Hadassah Medical School, Jerusalem 91010, Israel.
Subcellular Biochemistry, Volume 20: Mycoplasma Cell Membranes, edited by Shlomo Rottem and Itzhak Kahane. Plenum Press, New York, 1993.

isms, and taxonomic status were a continuing enigma to microbiologists. Owing to their minute size and ability to pass through filters which block the passage of bacteria, the mycoplasmas were for a long time considered as viruses. Then, following the discovery of bacterial L-forms which resemble mycoplasmas in morphology and in the peculiar "fried egg" colony shape (Klieneberger, 1935), mycoplasmas were confused with the L-forms, which are bacteria that lost partially or entirely their cell walls. The mycoplasma literature in the 1950s and 1960s was full of papers supporting or opposing the definition of mycoplasmas as bacterial L-forms. The controversy came to an end in the late 1960s when the first genomic analysis data obtained by DNA hybridization ruled out any relationship of mycoplasmas to stable L-forms of present-day walled bacteria (Razin, 1969a). Nevertheless, it should be stressed that associating mycoplasmas with L-forms is not entirely wrong when viewed from the long-range evolutionary perspective. Mycoplasmas are currently considered to have evolved from gram-positive walled bacteria by degenerative evolution so that their evolutionary history appears to include the loss of a cell wall, a step reminiscent of the induction of L-forms (Woese, 1987; Weisburg et al., 1989). The major difference, however, between mycoplasmas and L-forms is that the loss of the cell wall was apparently only one step in the lengthy process of mycoplasma evolution involving many more steps resulting in marked diminution of the genome, while the present-day L-forms are actually laboratory artifacts produced by partial or complete cell wall removal, with minimal changes in the genomic setup of the parent bacterium (Razin, 1992).

Morphologically, mycoplasmas vary in shape from spherical or pear-shaped structures (0.3–0.8 μm in diameter) to branched or helical filaments. Genome replication precedes, but is not necessarily synchronized with, cell division. Thus, budding forms, filaments, and chains of beads may be observed in mycoplasma cultures. The single most important characteristic which distinguishes mycoplasmas from all other prokaryotes is their complete lack of a cell wall. This unique property has led to their placement in a separate division, Tenericutes (wall-less bacteria), forming one of the four divisions of the kingdom Prokaryotae. The other three divisions are: Firmicutes, the gram-positive bacteria; Gracilicutes, the gram-negative bacteria; and Mendosicutes, the archaebacteria (Murray, 1984). The division Tenericutes is comprised presently of one class, Mollicutes, the name being derived from the Latin, *mollis*, soft, and *cutis*, skin, to denote the lack of a rigid cell wall from mycoplasmas. The current classification of Mollicutes is presented in Table I.

While the trivial term mycoplasmas has been used to denote any species included in Mollicutes, the trivial names acholeplasmas, ureaplasmas, anaeroplasmas, and spiroplasmas are commonly used for members of the corresponding genus. In order to keep the term mycoplasmas only for organisms included in the genus *Mycoplasma*, the trivial term mollicute(s) was introduced to describe

Table I

Taxonomy and Properties of Mycoplasmas (Class Mollicutes)[a]

Classification	Current No. of recognized species	Genome size[b] (in kbp)	Mole % G + C of DNA	Cholesterol requirement	Distinctive properties	Habitat
Order I: *Mycoplasmatales*						
Family I: *Mycoplasmataceae*						
Genus I: *Mycoplasma*	92	580–1300	23–41	+	—	Man, animals, plants, insects
Genus II: *Ureaplasma*	5	730–1160	27–30	+	Urease positive	Man, animals
Family II: *Spiroplasmataceae*						
Genus I: *Spiroplasma*	11	1350–1700	25–31	+	Helical filaments	Arthropods (including insects), plants
Order II: *Acholeplasmatales*						
Family I: *Acholeplasmataceae*						
Genus I: *Acholeplasma*	12	about 1600	27–36	−	—	Animals, plants, insects
Order III: *Anaeroplasmatales*						
Family I: *Anaeroplasmataceae*						
Genus I: *Anaeroplasma*	4	about 1600[c]	29–33	+	Obligate anaerobes (oxygen sensitive)	Bovine–ovine rumen
Genus II: *Asteroleplasma*	1	about 1600[c]	40	−		Bovine–ovine rumen
Uncultivated, unclassified		600–1185	23–29	Not established	Uncultivated as yet	Plants, insects
Mycoplasma-like organisms (MLOs)						

[a]Updated and modified from Razin (1978a).
[b]According to recent data obtained by pulse-field gel electrophoresis.
[c]Data obtained by renaturation kinetics.

all members of the class. Nevertheless, the term mycoplasmas is still widely used in the broader sense and will be used as such in this chapter.

The total lack of a cell wall explains many of the unique properties of the mycoplasmas, such as sensitivity to osmotic shock and detergents, resistance to penicillin, and a peculiar fried-egg colony shape (Razin and Oliver, 1961). Thin sections of mycoplasmas reveal that the cells are built essentially of three organelles: the cell membrane, ribosomes, and the characteristic prokaryotic genome. The genome size of most mycoplasmas is the smallest known for self-replicating organisms. Its size may be about 600–700 kbp, so that the estimated number of genes in these mycoplasmas does not exceed 500, about one-fifth the number of genes in *E. coli* (Razin, 1985a; Muto, 1987; Neimark and Lange, 1990). This is expressed by a small number of cell proteins (Kawauchi *et al.*, 1982) and by the lack of many enzymatic activities and metabolic pathways, in line with the parasitic mode of life and fastidious nature of mycoplasmas. Detailed description of the cell biology, ecology, and taxonomy of Mollicutes can be found in several recent reviews (Razin, 1991, 1992). For reviews on the molecular biology and genetics of mycoplasmas, the reader is referred to Razin (1985a) and Dybvig (1990). Extensive treatment of all aspects of mycoplasmology is provided by the five volumes of the series *The Mycoplasmas* (Barile and Razin, 1979; Tully and Whitcomb, 1979; Whitcomb and Tully, 1979, 1989; Razin and Barile, 1985). The subject of mycoplasma membranes has been periodically reviewed by the author of this chapter since 1963 (Razin, 1963a, 1967, 1969b, 1973, 1975, 1978a, 1981). Specific aspects concerning mycoplasma membrane lipids were reviewed by Rottem (1980) and by McElhaney (1984, 1989).

2. CHARACTERIZATION OF MYCOPLASMA MEMBRANES

2.1. Proof for Lack of a Cell Wall in Mycoplasmas

The first indirect evidence for the lack of a cell wall in mycoplasmas was obtained by Kandler and Zehender (1957) and by Plackett (1959) reporting the absence of the specific components of the bacterial cell wall peptidoglycan, muramic and diaminopimelic acids, from a variety of mycoplasmas. The lack of peptidoglycan was also corroborated by the resistance of mycoplasmas to lysis by lysozyme (Plackett, 1959; Razin and Argaman, 1963 Razin, 1963a) and to growth inhibition by penicillin and other antibiotics that selectively inhibit peptidoglycan synthesis (Klieneberger-Nobel, 1962). Direct proof for the absence of cell walls in mycoplasmas was obtained at about the same time by electron microscopy of thin sections of mycoplasma cells (van Iterson and Ruys, 1960; Domermuth *et al.*, 1964). The osmium-fixed sections showed the mycoplasma cells to be limited by a single trilaminar-shaped membrane, about 80–100 Å thick, identical to the unit membrane shape characteristic of biological membranes. Moreover, these early sections already showed that mycoplasma cells have no intracytoplasmic membrane structures, such as mesosomes, leading to the conclusion that mycoplasmas have only one type of membrane—the plasma membrane. This has proved to represent perhaps the greatest advantage of mycoplasmas as models for membrane study; once isolated, one can be sure that the mycoplasma plasma membrane is uncontaminated with other membrane types.

2.2. Development of Procedures for Mycoplasma Membrane Isolation

Not only do mycoplasmas have only one membrane type, but for most mycoplasmas this membrane can be easily isolated by the gentle technique of osmotic lysis. We were led to this finding in the early 1960s, being impressed by the marked resemblance of thin-sectioned mycoplasmas to thin-sectioned bacterial protoplasts and L-forms. Comparative studies carried out by us showed that mycoplasmas resemble bacterial protoplasts in their sensitivity to lysis by osmotic shock and detergents, though some mycoplasmas showed higher resistance than the protoplasts to osmotic lysis (Razin and Argaman, 1962, 1963; Razin, 1963b, 1964). Fortunately, *Acholeplasma laidlawii,* which was then the object for our nutritional studies (Razin and Knight, 1960a,b; Razin and Cohen, 1963), was found to be very sensitive to osmotic lysis. This fact, as well as later findings concerning the ease of manipulating its membrane lipid composition, explain why so much of the mycoplasma membrane work was carried out on this organism. The fact that some species, like *Mycoplasma gallisepticum* and *M. pneumoniae,* resisted ordinary osmotic lysis prompted the development of other methods for cell rupture, such as intensifying osmotic shock by preloading the mycoplasmas with glycerol (Rottem *et al.,* 1968) or by induction of lysis of nonenergized dicyclohexylcarbodiimide (DCCD)-treated cells in isoosmotic NaCl-Tris buffer at pH 8.5 (Shirvan *et al.,* 1982). Another development depended on induction of cell lysis by digitonin, which perturbs mycoplasma membrane structure by complexing with membrane cholesterol (Rottem and Razin, 1972). Mechanical presses, sonic or ultrasonic oscillators (Pollack *et al.,* 1965b), and alternate freezing and thawing (Williams and Taylor-Robinson, 1967) have also been used frequently. The various methods for mycoplasma cell lysis and membrane isolation have been periodically reviewed and evaluated, the last time by Razin (1983).

2.3. Chemical Analysis of Isolated Membranes

The development of effective cell lysis procedures in the early 1960s facilitated the isolation of mycoplasma membranes in large quantities and in a rather pure state, enabling their chemical characterization. The first report on the gross chemical composition of mycoplasma membranes (Razin *et al.*, 1963) pointed out that the membrane is essentially built of protein (about 60–70% of the membrane mass) and lipid (about 20–30%) with little carbohydrate and small amounts of cytoplasmic contaminants (RNA and DNA). Results of this early analysis were later corroborated in numerous studies on a variety of mycoplasma membranes. Naturally, the initial gross analysis of the isolated membranes was soon followed by studies on specific membrane components, focusing first on membrane lipids, although they ranked second in quantity to membrane proteins. The main reason for this was the far simpler and better developed techniques for lipid than for protein analysis in the 1960s. Special attention was given from the beginning to cholesterol, as it is a peculiar component of mycoplasma membranes and generally not found in other prokaryotic membranes (Razin *et al.*, 1963; Argaman and Razin, 1965). These early studies constituted the prelude for the extensive use of mycoplasma membranes in studies of the role of cholesterol in the membrane (see Section 3.2.2).

Introduction of gas–liquid chromatography facilitated the analysis of fatty acid composition of membrane lipids (Razin *et al.*, 1966b), opening a new field of research on the effects of the fatty acid moieties of membrane lipids on organization, physical state, and function of the membrane (see Section 3.2.1). Mycoplasma membrane glycolipids became another target for research in the late 1960s and early 1970s because of their important role as antigenic components, particularly in *M. pneumoniae* membranes (Beckman and Kenny, 1968; Plackett *et al.*, 1969; Razin *et al.*, 1970, 1971b, 1972). Another class of lipids, peculiar to some mycoplasma membranes, are the lipoglycans, discovered in the 1970s (Smith *et al.*, 1976). Being composed of long oligosaccharide chains, linked covalently to a diglyceride, they resemble somewhat bacterial lipopolysaccharides. They attracted some attention as they exhibit a variety of biological activities, particularly on the immune system, serving also as membrane receptors for mycoplasma viruses (Smith, 1984).

Although the early chemical analysis of mycoplasma membranes showed proteins to constitute the major part of the membrane mass, little advance was made in terms of their characterization until the mid-1960s. The introduction of polyacrylamide gel electrophoresis (PAGE) had revolutionized protein analysis. Analysis of the hydrophobic membrane proteins had first to overcome their solubilization problem. The first solubilization mixture used by us for mycoplasma membranes was that described by Takayama *et al.* (1966). The membranes were dissolved in phenol–acetic acid–water (2:1:0.5, w/v/v) and the solution was run in polyacrylamide gels containing 5 M urea and 35% acetic acid. Although solubilization of the membranes was incomplete, so that some membrane protein was stuck on top of the gel, reproducible electrophoretic patterns of the membrane proteins were obtained. Most importantly, the patterns were highly specific for the different mycoplasma strains examined (Rottem and

Razin, 1967). This work laid the basis for the use of electrophoretic patterns of membrane proteins and, later, total cell proteins as "fingerprints" for strain identification (Razin, 1968). The improvements which followed were mostly technical, including the replacement of the phenol–acetic acid solubilizing mixture with a solution of sodium dodecyl sulfate (SDS) and inclusion of this detergent in the gel. Replacement of the cylindrical gels by slab gels greatly improved resolution of the protein bands, culminating in the two-dimensional SDS-PAGE method introduced by O'Farrell (1975) enabling the superior separation of membrane proteins and the application of the peptide patterns obtained to strain identification (Rodwell and Rodwell, 1978; Mouches and Bove, 1983).

2.4. Localization of Membrane Enzymes and Transport Systems

The development of mycoplasma cell lysis procedures in the early 1960s opened the way for enzyme localization studies. The relatively easy separation of mycoplasma membranes from cytoplasmic constituents by the gentle osmotic lysis or alternate freezing and thawing procedures has been of great advantage in this respect. Hence, starting in the mid-1960s (Pollack et al., 1965a; Rottem and Razin, 1966), there was a continuous flow of papers dealing with enzymatic activities localized in the membrane. Being the only membrane in the cell, the mycoplasma membrane is expected to be rich in enzymatic activities, as well as in specific carriers active in transport of nutrients through the membrane. Of special interest has been the early finding that in *Mycoplasma* (Pollack et al., 1965a) and in *Spiroplasma* species (Kahane et al., 1977), i.e., in most known mollicutes, the transfer of electrons from NADH to oxygen occurs in the cytoplasmic fraction, in sharp contrast to the localization of the NADH oxidase in *Acholeplasma* species and in other prokaryotes, where it is membrane-bound (Pollack et al., 1965a; Ne'eman and Razin, 1975; Larraga and Razin, 1976).

Another membrane-bound enzyme attracting much attention since the early days of mycoplasma membrane research has been the membrane-bound ATPase activity. While in other prokaryotes the larger part of the enzyme complex, that named F_1, can be easily detached from the membrane, the mycoplasma ATPase resisted detachment, behaving like an integral membrane protein complex (Rottem and Razin, 1966; Ne'eman and Razin, 1975). Only the recent cloning of the mycoplasma ATPase complex genes may provide the solution to this peculiar behavior.

Being dependent on many nutrients supplied by the host or growth medium, mycoplasmas carry a variety of active transport systems which basically resemble those of other prokaryotes. Thus, sugar transport in *A. laidlawii* is energized by the proton-motive force generated through the activity of the membrane-bound ATPase (Tarshis and Kapitanov, 1978), while fermentative *Mycoplasma* and *Spiroplasma* species utilize the highly efficient phosphoenolpyruvate-dependent sugar phosphotransferase system for sugar transport (Cirillo and Razin, 1973; Cirillo, 1979). Membrane vesicles prepared from wall-covered bacteria have been most useful in transport studies in that they enable the dissociation of the transport process *per se* from assimilation. Although one would expect the preparation of sealed membrane vesicles active in transport from the wall-less and osmotically sensitive mycoplasmas to be easy, results have usually been disappointing with no clear explanation for this failure (Razin, 1981).

3. ELUCIDATING MEMBRANE ORGANIZATION

3.1. The Membrane Reconstitution Approach

Membrane reconstitution formed a major approach in the 1960s and 1970s to elucidation of the molecular organization of the protein and lipid components in biomembranes. Reconstitution studies, ideally, have to include the disaggregation of the membrane into its building blocks, the biochemical and biophysical characterization of the solubilization products and their reassembly to a membrane identical to the original as regards structure and function (Razin, 1972). If biomembranes actually consisted of homogeneous lipoprotein subunits, according to the theory prevailing in the mid-1960s (Green *et al.*, 1967), then such a task should not be too difficult. All that would have to be done would be to disaggregate the membrane into its component subunits, e.g., by detergent action. The subunits would then reassemble spontaneously, re-forming membranous structures following the removal of the dissociative agent, since the structure-determining information is supposed to be innate in the subunits themselves (Green *et al.*, 1967).

The *A. laidlawii* membrane was essentially the first plasma membrane subjected to the membrane reconstitution approach. Why this membrane was selected has an interesting story behind it. Space science was initiated in the late 1950s, and the newly established National Aeronautics and Space Administration (NASA) became interested in early and primitive forms of life. The search for the simplest, smallest, and most primitive organisms living on Earth brought Morowitz and Tourtellotte (1962) to mycoplasmas. Being comprised of the minimum set of organelles required for independent growth and reproduction—a plasma membrane, ribosomes, and a circular double-helix DNA molecule—these organisms appeared to be the most appropriate candidates for an extremely ambitious project entertained by Morowitz of assembling a living cell from its components. Starting with solubilization and reassembly of the cell membrane seemed to be the most logical step to initiate the venture. The ease with which large quantities of pure cell membranes of *A. laidlawii* could be obtained by osmotic lysis of cells, and the high sensitivity of the membranes to lysis by SDS led us to the first membrane solubilization and reconstitution experiments (Razin *et al.*, 1965). The results appeared to support the subunit hypothesis by showing a single symmetrical schlieren peak of about 3 S in the solution of *A. laidlawii* membranes in SDS. Moreover, following the removal of the detergent by slow dialysis in the cold against Mg^{2+}, the solubilized material reaggregated into membranous structures resembling the native membranes in thin sections and in chemical composition (Razin *et al.*, 1965). Soon after, however, doubts about the validity of the subunit interpretation were raised by Engelman *et al.* (1967), who showed that membrane protein could be separated from membrane lipid by prolonged centrifugation of the solubilized membrane material on sucrose gradients. Further studies (Rodwell *et al.*, 1967; Rottem *et al.*, 1968; Razin and Barash, 1969) showed conclusively that membrane material in SDS consists not of lipoprotein subunits, but of protein–SDS complexes and of lipid–SDS micelles separable by various techniques.

Another major question concerned the molecular organization of the reconstituted mycoplasma membranes. These membranes could not be distinguished from native membranes in their chemical composition, density, and electron microscopical appearance when negative staining and thin-sectioning were used (Razin *et al.*, 1965, 1969; Razin and Barash, 1969; Terry *et al.*, 1967). However, the subsequent application of electron spin resonance, x-ray diffraction, and freeze-fracturing electron microscopy brought enough evidence indicating that the reconstituted membranes differ in important organizational details from native mycoplasma membranes, in particular with respect to the organization of the proteins (Rottem *et al.*, 1970; J. C. Metcalfe *et al.*, 1971; S. M. Metcalfe *et al.*, 1971; Tillack *et al.*, 1970). The finding that both the convex and concave fracture faces of the reconstituted membranes were fairly smooth, lacking the characteristic particles seen in native membranes (Figure 1), suggested that the proteins in the reconstituted membrane occupied mostly a surface position, with very little protein embedded in the lipid bilayer (Tillack *et al.*, 1970; Razin, 1974a; Ne'eman and Razin, 1975).

In retrospect, it was naive to expect that a membrane as complex as that of a mycoplasma could be reassembled into a structure identical to that of the native membrane in a single step consisting of dialysis, particularly in light of the denaturing effects of the detergent used for membrane solubilization. Nevertheless, studies on mycoplasma membrane reconstitution not only advanced considerably our knowledge of mycoplasma membranes, but gave a significant boost to the subject of membrane reconstitution in general, particularly by developing methods and criteria for reconstitution (Razin, 1972; Razin and Rottem, 1974).

Mycoplasma membrane reconstitution studies resulted also in an outgrowth of an applied nature. The reconstitution process enabled the production of "hybrid" membranes from SDS-solubilized membrane lipids of one mycoplasma and SDS-solubilized membrane proteins of another mycoplasma (Razin and Kahane,

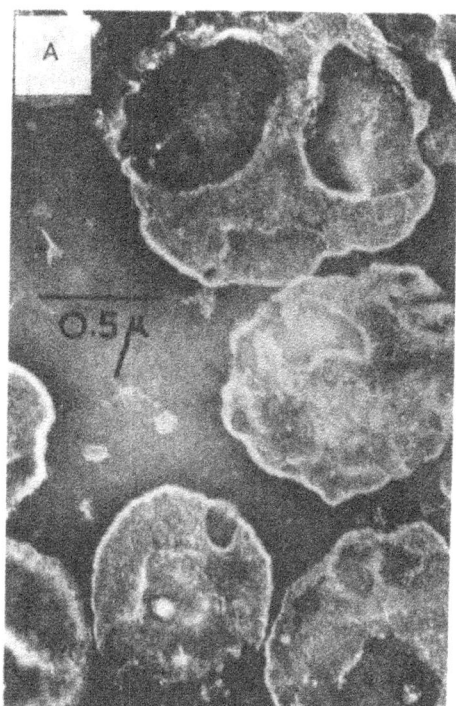

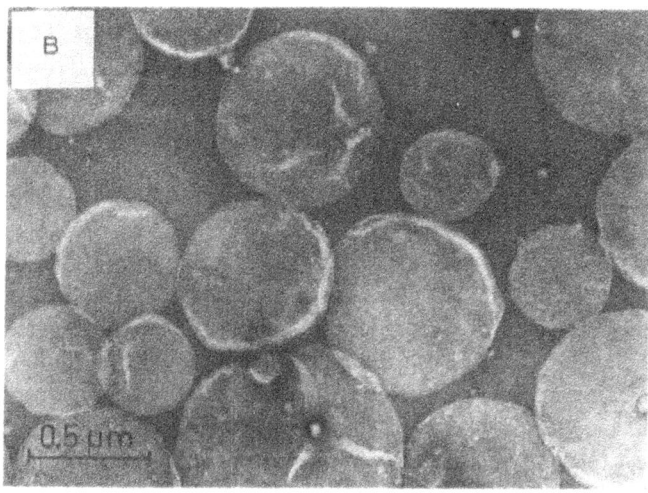

FIGURE 1. Native and reconstituted mycoplasma membranes. (A) Negatively stained isolated *A. laidlawii* membranes. The "holes" in the membranes are apparently the result of tears in the membranes caused by the osmotic shock. (B) Negatively stained reconstituted *A. laidlawii* membranes. (C) Thin sections of isolated *A. laidlawii* membranes. (D) Thin sections of reconstituted *A. laidlawii* membranes. (E) Freeze-fractured native *A. laidlawii* membranes. (F) Freeze-fractured reconstituted

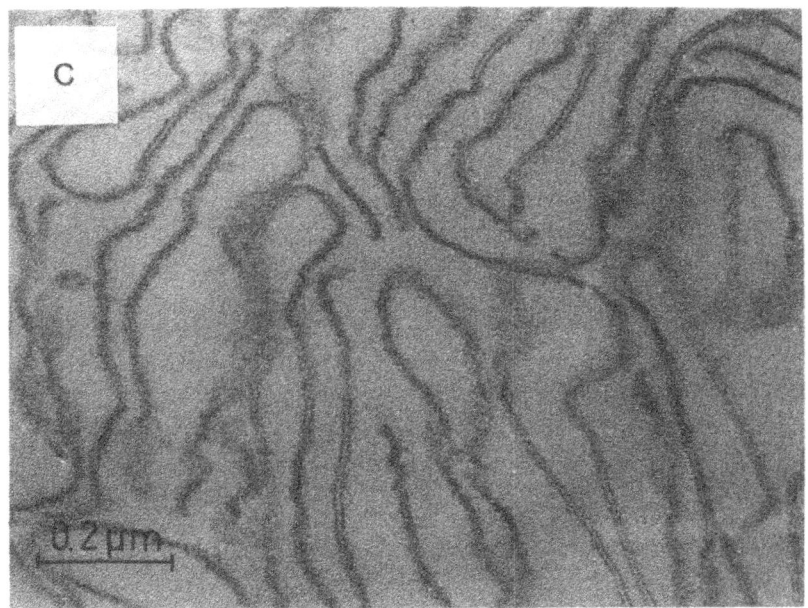

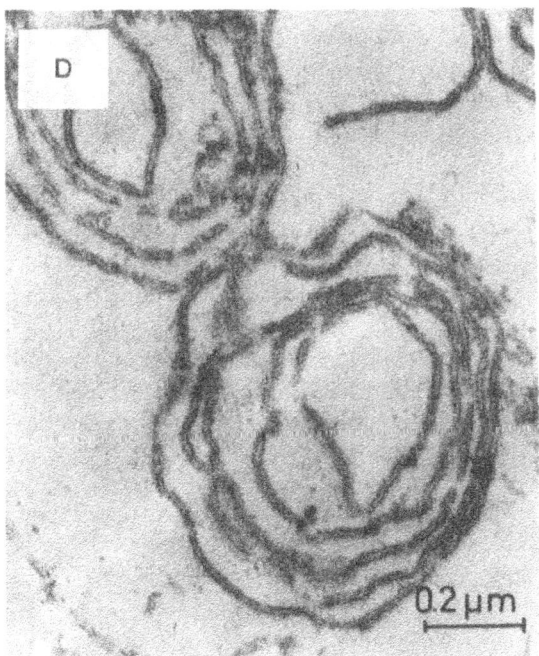

A. *laidlawii* membranes. The lack of intramembranous particles on the fracture faces is noticeable. (A and B from Rottem *et al.*, 1968; C from Razin, 1969b; D, E, and F from Razin, 1974a.)

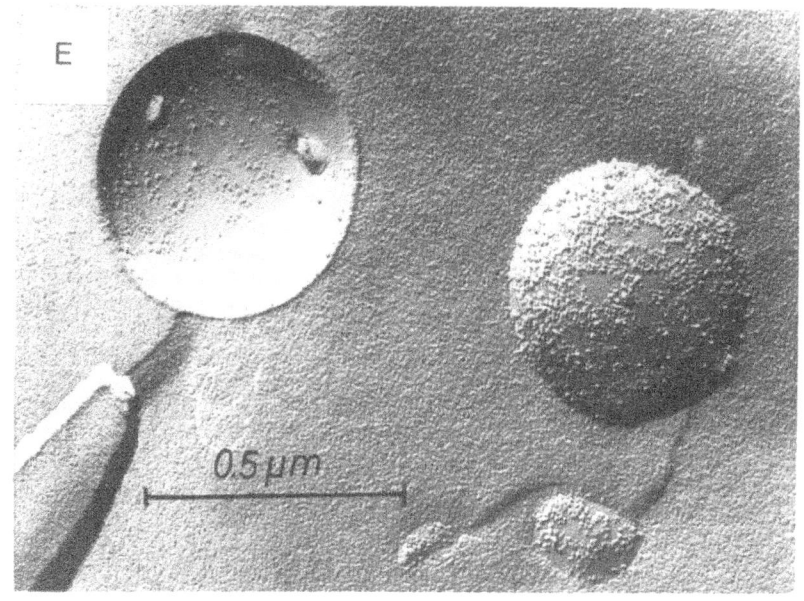

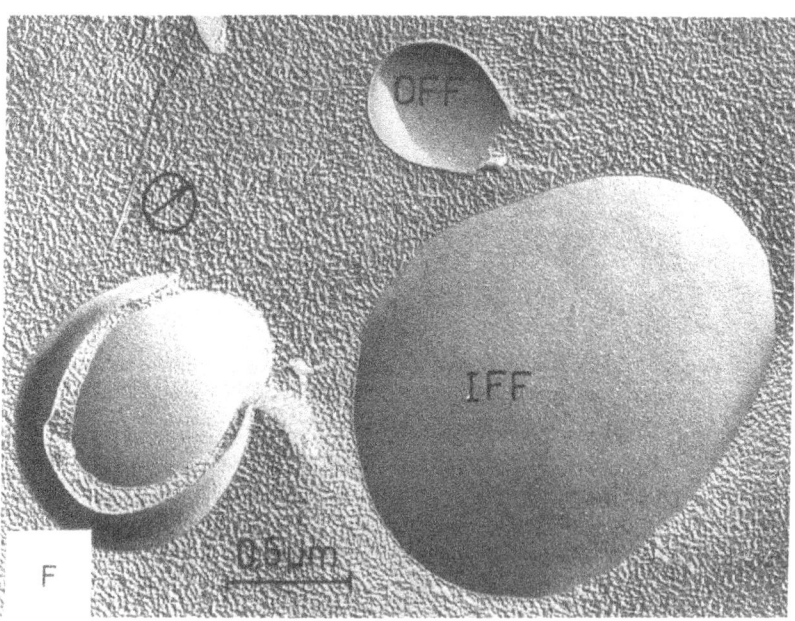

FIGURE 1. (*Continued*)

1969; Cole *et al.*, 1971). The finding that the reconstituted mycoplasma membranes retained their immunogenicity (Kahane and Razin, 1969) led to the use of hybrid membranes for the preparation of highly specific and potent antisera to serologically active membrane lipids. By themselves, the lipids are unable to elicit an antibody response. Hybrid membranes prepared from purified *M. pneumoniae* glycolipids (Razin *et al.*, 1970, 1971b) or cytolipin H from bovine spleen (Razin *et al.*, 1971a) and membrane proteins of *A. laidlawii* were most effective in eliciting the production, in rabbits, of antibodies to the lipid component. The main advantage of using reconstituted hybrid membranes for the production of antibodies to membrane lipids is that the lipid and protein components can be selected so that the specificity of the antibodies produced is controllable. This approach, developed in the early 1970s on mycoplasma membranes, has since been followed by others using different membrane systems.

3.2. The Membrane Lipid Manipulation Approach

The restricted genetic information in the small mycoplasma genome dictates limited biosynthetic abilities. Thus, mycoplasmas are partially or totally incapable of fatty acid synthesis, depending on the host or the growth medium for their supply. In addition, most mycoplasmas require cholesterol for growth, a unique requirement among prokaryotes. The fatty acid residues of membrane phospholipids and glycolipids and cholesterol constitute the major portion of the hydrophobic core of the biological membrane. The dependence of mycoplasmas on an exogenous supply of fatty acids and cholesterol has been one of their greatest advantages as models for membrane studies. The ability to introduce controlled alterations in mycoplasma membrane lipids, simply by controlling the composition and content of fatty acids and sterols in the growth medium, has been used most effectively in elucidating membrane lipid organization and function in the membrane.

3.2.1. Manipulation of Fatty Acid Composition

The early finding by Razin and Rottem (1963) that *A. laidlawii* requires unsaturated fatty acids for growth was the first step leading to the use of this mycoplasma as a model for studying membrane lipid organization. The second step was accomplished 3 years later when the effects of unsaturated and saturated fatty acids on growth and morphology of *A. laidlawii* were followed (Razin *et al.*, 1966a; Razin and Cosenza, 1966), culminating in the finding that it is possible to change rather dramatically the fatty acid composition of *A. laidlawii* membrane lipids by changing the fatty acid supplement in the growth medium (Razin *et al.*, 1966b). Membrane lipids of *A. laidlawii* grown with an oleic acid supplement contained up to 80% of this fatty acid, while membranes of cells

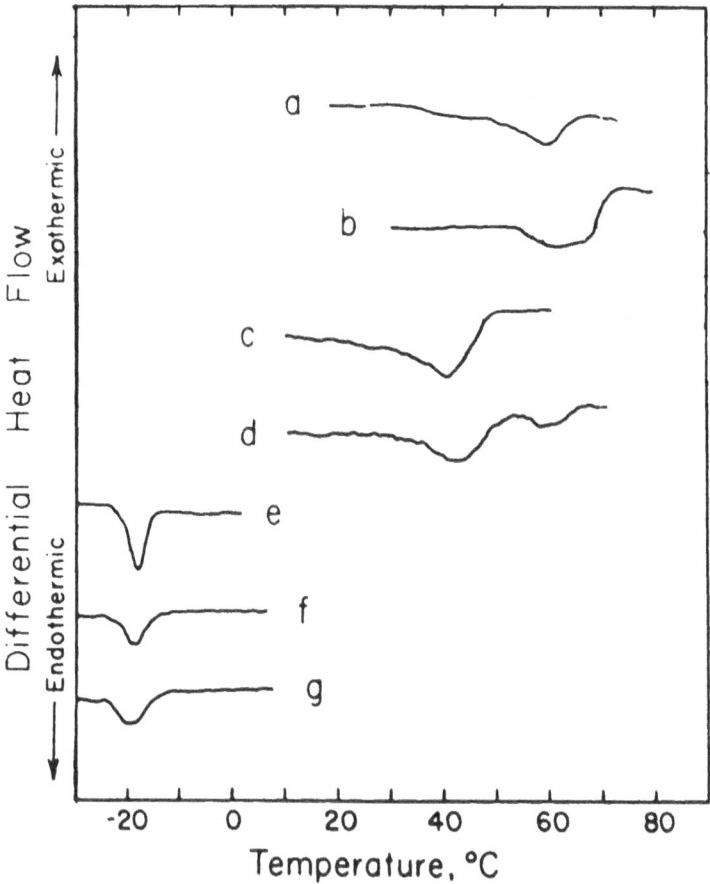

FIGURE 2. Differential scanning calorimetry of *A. laidlawii* membranes and extracted lipids. Curves a and b show, respectively, extracted lipids and membranes of cells grown in stearate-enriched medium; curves c and d show extracted lipids and membranes of cells grown in unsupplemented growth medium; and curves e, f, and g show extracted lipids, membranes, and whole cells from oleate-enriched medium. (From Steim *et al.*, 1969.)

grown with palmitic acid contained it as the dominant fatty acid. We had also noticed that *A. laidlawii* grew profusely with oleic or linoleic acids, producing filaments that later broke up to cocci. The same organism grew very poorly with palmitic or stearic acids, forming swollen cocci, exhibiting a marked tendency to lyse (Razin *et al.*, 1966a).

These observations laid the foundation for numerous studies having very significant implications to our understanding of biological membrane structure and function. There is no doubt that the study with the greatest impact has been that of Steim *et al.* (1969). This study made use of *A. laidlawii* membranes enriched with different fatty acids. Differential scanning calorimetry of these membranes (Figure 2) provided the strongest experimental evidence for the idea that the bulk of the biological membrane lipids form a bilayer, as in liposomes, laying to rest the lipoprotein subunit model (Figure 3). Moreover, the findings by Steim *et al.* (1969) and Reinert and Steim (1970) that the endothermic phase transition of *A. laidlawii* membrane lipids from a crystalline to a liquid-crystalline state occurred over the same temperature range in viable cells, isolated membranes, and in aqueous dispersions of extracted lipids, and the demonstration that transition temperatures decreased with increased unsaturation of the fatty acid residues, can be considered as the cornerstone for the wide field of membrane fluidity research. Clearly, the easy manipulation of the fatty acid composition of mycoplasma membrane lipids has served since 1965 as a most effective tool in studies on membrane fluidity, permeability, and disposition of membrane components. The voluminous literature on these aspects of mycoplasma membrane research has been extensively reviewed (Rottem, 1980, 1982; Razin, 1981; Melchior, 1982; McElhaney, 1984, 1989).

3.2.2. Manipulation of Membrane Sterols

Apart from the *Acholeplasma* and the single anaerobic *Asteroleplasma* species, all of the other Mollicutes require cholesterol for growth (Table I). The findings by many workers that none of the Mollicutes synthesizes or modifies sterols (Rodwell, 1963; Argaman and Razin, 1965; Rottem *et al.*, 1971) opened the way for changing the sterol type and content in the mycoplasma membrane and, in this way, examining the effects of sterols on membrane structure and function. The best test organisms for this purpose proved to be the caprine mycoplasmas *M. mycoides* subsp. *capri* and *M. capricolum*, as these mycoplasmas could be cultivated with little cholesterol, although growth was less than optimal under these conditions (Razin, 1967). In this way, the cholesterol content of the membrane could be reduced to less than 3% of the total membrane lipid, as compared with about 25% in membranes of the same organisms grown with optimal amounts of cholesterol (Rottem *et al.*, 1973b). The most remarkable difference between the cholesterol-poor and the cholesterol-rich membranes was that a thermotropic phase transition could be demonstrated only in the former (Rottem *et al.*, 1973a). Differential scanning calorimetry revealed an endothermic phase transition centered at about 25°C in the cholesterol-poor membranes, whereas no transition was observed in the cholesterol-rich ones. Other techniques, such as fluorescence polarization with diphenylhexatriene as a probe, Arrhenius plots of enzymatic and transport activities (Figure 4), and freeze-fracturing (Figure 5) further confirmed these findings. Chilling of the cholesterol-poor membranes to 4°C prior to the quick freezing caused the aggregation of the intramembranous particles, leaving more than two-thirds of the

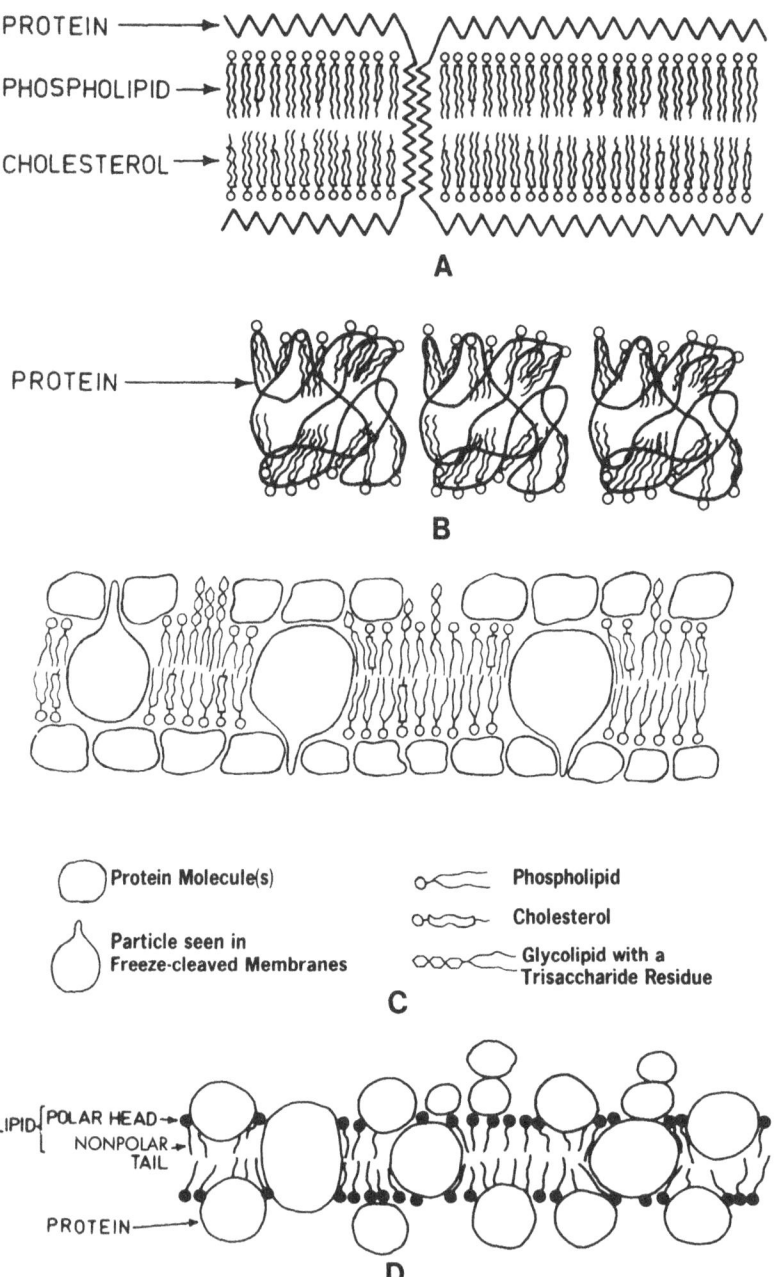

FIGURE 3. Evolution of the concepts concerning the molecular structure of biological membranes, as depicted in membrane models devised to present the possible molecular organization of mycoplasma membranes. (A) The Danielli–Davson "leaflet" model (Razin, 1969b). (B) The "lipoprotein subunit" model (Razin, 1969b). (C) A model depicting the possible organization of protein and lipid in the *M. pneumoniae* membrane (Razin *et al.*, 1972). (D) Another model showing possible arrangements of proteins and lipids in the membrane (Razin, 1972).

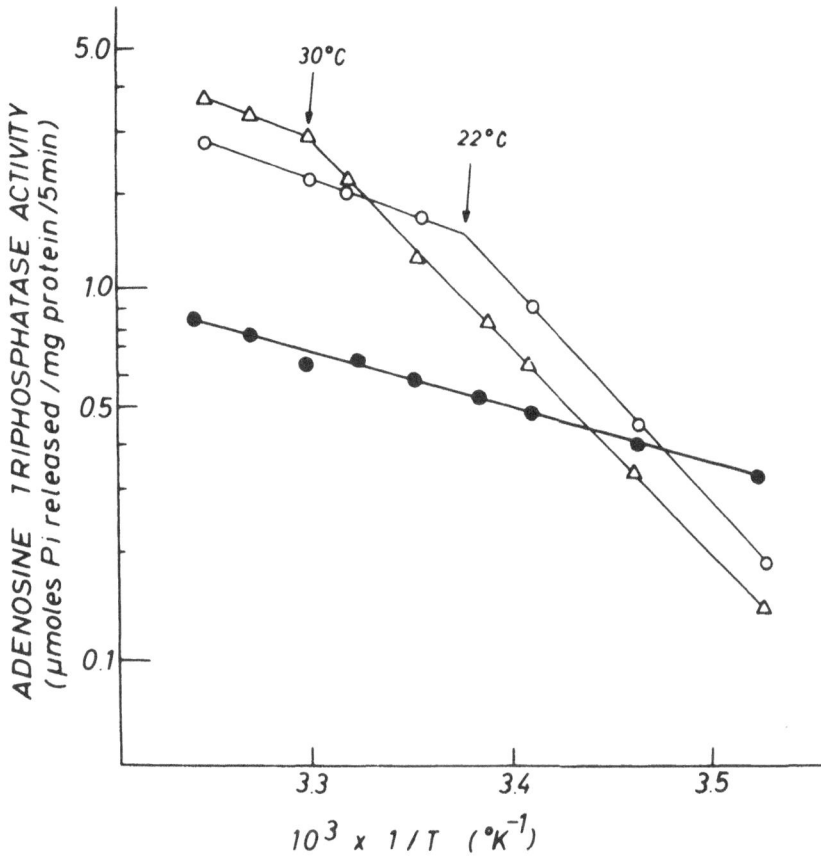

FIGURE 4. Arrhenius plots of ATPase activity of the cholesterol-poor *M. mycoides* subsp. *capri* membranes enriched with oleic acid (○) or elaidic acid (△). The Arrhenius plot of the ATPase activity of membranes from the cholesterol-rich parent strain (●) is included for comparison. A break in the plots, corresponding apparently to a phase transition of membrane lipids, can be observed only in the cholesterol-poor membranes. (From Rottem *et al.*, 1973a, as modified by Razin, 1975.)

fracture faces particle-free (Figure 5). Aggregation of the intramembranous parti-cles, believed to contain integral membrane proteins, is a manifestation of the gelation of the lipid domain (Verkleij *et al.*, 1972). No aggregation of particles was discernible in the cholesterol-rich membranes, even when quenched from 4°C (Figure 5). These experiments provided the first clear-cut evidence, with membranes of growing cells, that cholesterol regulates membrane fluidity during changes in growth temperature or following alterations in fatty acid composition of membrane lipids.

The inability of mycoplasmas to modify sterols could be exploited to test the

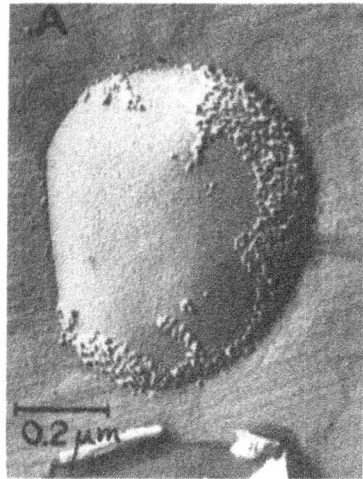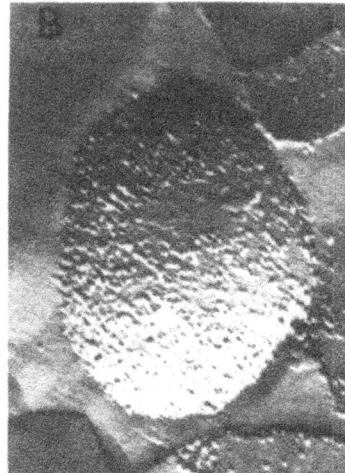

FIGURE 5. Fracture faces of freeze-cleaved *M. mycoides* subsp. *capri* membranes of cells grown with little cholesterol (A) and of cells grown with an adequate supply of cholesterol (B). The membranes were suspended in 20% glycerol and incubated at 4°C prior to freezing. The intra-membranous particles, representing membrane proteins, aggregated in the cholesterol-poor membranes because of the transition of membrane lipids to the gel state. No phase transition occurred in the cholesterol-rich membranes (Rottem *et al.*, 1973b).

structural features of the sterol molecule essential for performing its biological function. Experiments with artificial lipid membranes indicated that a planar tetracyclic ring system, an equatorial hydroxyl group at position C-3, and a branched aliphatic side chain at least eight carbon atoms long, are essential for this function (Demel and de Kruyff, 1976). These are exactly the structural features of sterols capable of promoting mycoplasma growth (Razin, 1975, 1982). However, later findings (Odriozola *et al.*, 1978; Lala *et al.*, 1979; C. E. Dahl *et al.*, 1980; J. S. Dahl *et al.*, 1980) that *M. capricolum* can grow with sterols that do not fulfill all three structural requirements, and the fact that the caprine mycoplasmas could not grow in the total absence of cholesterol, led us to suggest that cholesterol may have an additional function(s) to that of regulating bulk lipid fluidity. Since this role could be fulfilled by minute amounts of choles-terol, relative to the large amounts of cholesterol needed for regulating mem-brane fluidity, it was reasonable to assume that cholesterol may also play a catalytic role. Experimental support for this notion has, indeed, been obtained by Jean and Charles Dahl, showing that a small amount of membrane-associated cholesterol serves as a signal for membrane biogenesis and, in turn, macro-molecular synthesis and cell growth (Dahl and Dahl, 1984; Chapter 5, this volume).

The hypothesis put forward in the early 1970s that increased incorporation of cholesterol into the plasma membrane of the arterial intima cells triggers the formation of atherosclerotic lesions (Papahadjopoulos, 1974) prompted extensive studies on factors controlling cholesterol uptake in mycoplasma membranes. Among serum lipoproteins, those having the highest cholesterol/phospholipid ratios, such as human low-density lipoproteins, were found to be the best cholesterol donors (Slutzky *et al.*, 1976, 1977), in line with the weight given to low-density lipoproteins as risk factors in atherosclerosis. Experimental systems based on mycoplasma membranes and serum lipoproteins proved of value in studying the interaction of the lipoproteins with binding sites on the membranes, and the effects of cholesterol transfer on the integrity of the lipoprotein particle (Efrati *et al.*, 1982). Moreover, the availability of a variety of mycoplasma membranes, differing in their cholesterol binding capacity, opened the way for studying the effects of membrane components and their physical state on cholesterol incorporation. Thus, the ability to manipulate the phospholipid content of the mycoplasma membranes was utilized to show that cholesterol uptake depends on membrane phospholipid content. The marked decrease in membrane phospholipid content relative to that of the protein on aging of mycoplasma cultures was accompanied by a parallel decrease in membrane cholesterol, whereas an increase in membrane phospholipid content resulting from chloramphenicol treatment of cultures was accompanied by an increase in membrane cholesterol (Razin, 1974b). The lower cholesterol uptake capacity of acholeplasma membranes appeared to be associated with their high glycolipid content (Efrati *et al.*, 1986) and was influenced by the physical state of the lipid (Razin, 1978b). For previous reviews on the role of sterols in mycoplasma membranes, the reader is referred to Razin and Rottem (1978) and Razin (1982).

3.3. Transbilayer Disposition of Membrane Lipids and Proteins

Once it became clear that the bulk of membrane lipids forms a bilayer (Steim *et al.*, 1969), much of the efforts of membrane researchers were directed toward elucidation of the transbilayer distribution of membrane lipids and proteins. Again, mycoplasma membranes proved to serve as convenient models in this line of research. The lack of a cell wall facilitated disposition studies, as these depended primarily on the use of macromolecular reagents, such as enzymes and antibodies to membrane lipids and proteins incapable of penetrating the bacterial wall barrier. Extensive use was made of polycationic ferritin, lectins, and lactoperoxidase-mediated iodination, phospholipase, pronase, and trypsin digestion to selectively label or digest membrane lipids and proteins exposed on the outer surfaces of the cell membrane. The results of these numerous studies (for reviews, see Razin, 1981; Rottem, 1982) have shown a definite transbilayer asymmetry of the mycoplasma membrane with regard to the distribution of membrane phospholipids and glycolipids (Figure 6). The fact that cholesterol is

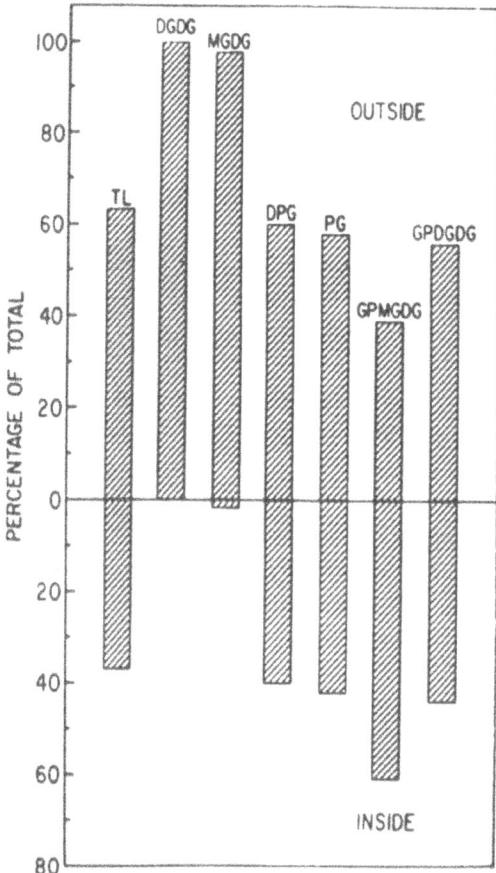

FIGURE 6. Transbilayer distribution of the major lipid species of *A. laidlawii* membranes, as revealed by the lactoperoxidase-mediated iodination technique. TL, total lipid; DGDG, diglucosyldiglyceride; DPG, diphosphatidylglycerol; MGDG, monoglucosyldiglyceride; PG, phosphatidylglycerol; GPMGDG, glycerophosphoryl monoglucosyldiglyceride; GPDGDG, glycerophosphoryl diglucosyldiglyceride. (From Gross and Rottem, 1979.)

taken up from the medium and thus must first be bound to the outer leaflet of the membrane lipid bilayer has also been utilized in studying the disposition and transmembrane mobility of cholesterol in biological membranes. Experiments carried out by Rottem and Bittman in the late 1970s have shown the ability of cholesterol to flip-flop from the outer to the inner half of the lipid bilayer, leading to its essentially equal distribution in the two halves of the bilayer (Bittman and Rottem, 1976; Rottem *et al.*, 1978).

The asymmetrical distribution of membrane proteins, demonstrated conclusively with mycoplasma membranes in the 1970s, led to the membrane model proposed by us (Figure 3, model D) as early as 1972 (Razin, 1972), at about the same time the fluid-mosaic membrane model of Singer and Nicolson was published (1972). The relative ease of changing membrane lipid fluidity in mycoplasmas considerably facilitated studies on the lateral mobility of membrane lipids and proteins within the plane of the membrane, and the vertical disposition of the proteins immersed in the lipid bilayer. Of special interest were the findings of the effects of aging of cultures and changes in the membrane electrochemical gradient on the vertical disposition and surface exposure of mycoplasma membrane proteins (Amar *et al.*, 1979), findings which could be relevant to the understanding of the interaction of pathogenic mycoplasmas with host cell surfaces, as will be discussed in the next section.

4. ROLE OF MEMBRANE COMPONENTS IN PATHOGENICITY

Most human and animal mycoplasmas can be considered typical surface parasites. They usually adhere to and colonize the epithelial linings of the respiratory and urogenital tract and rarely invade tissues. The surge of interest in mycoplasma adherence in the 1980s has been a consequence of the new trend in medical microbiology that emphasized the study of adherence of parasites to their host tissues, prompted by recognition of the fact that adhesion of the parasite to its target tissue constitutes the initial and essential step in the establishment of infection. Interference with adherence will thus prevent colonization and disease. Elucidation of the nature of the surface components of the parasite and host responsible for adhesion may lead to the development of means by which the adherence process can be blocked. The structural simplicity of mycoplasmas, and particularly their lack of a cell wall, has offered certain unique advantages to their use as models in adherence studies. The lack of a cell wall and any of the appendages, such as fimbriae, associated with adherence of other prokaryotes, had pointed to the fact that mycoplasma adhesins must constitute part of the cell membrane. Moreover, the lack of a cell wall facilitates the direct contact of the mycoplasma membrane with that of the host, creating a condition which, in principle, could lead to the fusion of the two membranes, or at least enable transfer or exchange of membrane components. These events, if they occur, may explain much of the damage to the epithelial linings colonized by the mycoplasmas and explain much of the flavor and attraction of the mycoplasma model in adherence research.

Although adherence of mycoplasmas to erythrocytes and other host cells has been known since the 1960s, the first indication for a specific mycoplasma membrane component associated with adhesion came from the work of Hu *et al.* (1977) pointing to a *M. pneumoniae* membrane protein, named P1, as a major adhesin. This work can be considered as a cornerstone in mycoplasma adherence research. The next major findings, using specific antibody labeling and reported almost simultaneously by Hu *et al.* (1982), Baseman *et al.* (1982), and Feldner *et al.* (1982), was that P1 concentrates at the surface of the tip structure of *M. pneumoniae,* acting, apparently, as an attachment organelle. Introduction of the molecular genetic tools has led to the third phase in the P1 research, comprising the cloning of the P1 gene and the complete molecular characterization of this major adhesin (Su *et al.,* 1987; Inamine *et al.,* 1988). There is no need to go any further in describing the significant developments that have taken place in the definition of the mycoplasma membrane components taking part in adhesion. This field was extensively and periodically reviewed (Razin, 1978a, 1985b, 1986; Razin and Yogev, 1989; Razin and Jacobs, 1992; Razin *et al.,* 1981; Kahane, 1984) and will be dealt with in Chapters 8 and 9 in this volume.

5. FUTURE PROSPECTS

The recent developments in the molecular definition of mycoplasma membrane components acting as adhesins illustrate the concepts and means shaping the current direction of mycoplasma membrane research. There is little doubt that research will continue to depend heavily on molecular genetic methodology. The complete sequencing of the minute mycoplasma genome is already under way. Taking into account that the mycoplasma membrane constitutes a significant part of the cell mass and that a major number of the mycoplasma cell proteins (Archer *et al.,* 1978) and all of the different species of lipids are located in the membrane, it is evident that a considerable number of the mycoplasma genes code for membrane components. The complete sequencing of the mycoplasma genome has been proposed by Morowitz (1984) as an essential step in a very ambitious, but experimentally feasible, project of testing the dogma of molecular biology. The idea is to assign coding space and completely define in molecular terms all of the mycoplasma cell components. If achieved, this would mean that the "logic of life" is finite, relatively simple, and subject to full exploration, a conclusion of far-reaching implications to our perception of the meaning of life.

6. REFERENCES

Amar, A., Rottem, S., and Razin, S., 1979, Is the vertical disposition of mycoplasma membrane proteins affected by membrane fluidity? *Biochim. Biophys. Acta* **552:**457–467.
Archer, D. B., Rodwell, A. W., and Rodwell, E. S., 1978, The nature and location of *Acholeplasma laidlawii* membrane protein investigated by two-dimensional gel electrophoresis, *Biochim. Biophys. Acta* **513:**268–283.

Argaman, M., and Razin, S., 1965, Cholesterol and cholesterol esters in Mycoplasma, *J. Gen. Microbiol.* **38**:153–168.

Barile, M. F., and Razin, S., (eds.), 1979, *The Mycoplasmas*, Volume I: *Cell Biology,* Academic Press, New York.

Baseman, J. B., Cole, R. M., Krause, D. C., and Leith, D. K., 1982, Molecular basis for cytadsorption of *Mycoplasma pneumoniae*, *J. Bacteriol.* **151**:1514–1522.

Beckman, B. L., and Kenny, G. E., 1968, Immunochemical analysis of serologically active lipids of *Mycoplasma pneumoniae*, *J. Bacteriol.* **96**:1171–1180.

Bittman, R., and Rottem, S., 1976, Distribution of cholesterol between the outer and inner halves of the lipid bilayer of mycoplasma cell membranes, *Biochem. Biophys. Res. Commun.* **71**:318–324.

Cirillo, V. P., 1979, Transport systems, in: *The Mycoplasmas*, Volume I (M. F. Barile and S. Razin, eds.), Academic Press, New York, pp. 323–349.

Cirillo, V. P., and Razin, S., 1973, Distribution of a phosphoenolpyruvate-dependent sugar phosphotransferase system in mycoplasmas, *J. Bacteriol.* **113**:212–217.

Cole, R. M., Popkin, T. J., Prescott, B., Chanock, R. M., and Razin, S., 1971, Electron microscopy of solubilized *Acholeplasma laidlawii* membrane proteins reaggregated with *Mycoplasma pneumoniae* glycolipids, *Biochim. Biophys. Acta* **233**:76–83.

Dahl, C. E., Dahl, J. S., and Bloch, K., 1980, Effects of alkyl-substituted precursors of cholesterol on artificial and natural membranes and on the viability of *Mycoplasma capricolum, Biochemistry* **19**:1462–1467.

Dahl, J. S., and Dahl, C. E., 1984, Effects of cholesterol on phospholipid, RNA, and protein synthesis in *Mycoplasma capricolum, Isr. J. Med. Sci.* **20**:807–811.

Dahl, J. S., Dahl, C. E., and Bloch, K., 1980, Sterols in membranes: Growth characteristics and membrane properties of *Mycoplasma capricolum* cultured on cholesterol and lanosterol, *Biochemistry* **19**:1467–1471.

Demel, R. A., and de Kruyff, B., 1976, The function of sterols in membranes, *Biochim. Biophys. Acta* **457**:109–132.

Domermuth, C. H., Nielsen, M. H., Freundt, E. A., and Birch-Anderson, A., 1964, Ultrastructure of *Mycoplasma* species, *J. Bacteriol.* **88**:727–744.

Dybvig, K., 1990, Mycoplasmal genetics, *Annu. Rev. Microbiol.* **44**:81–104.

Efrati, H., Oschry, Y., Eisenberg, S., and Razin, S., 1982, Preferential uptake of lipids by mycoplasma membranes from human plasma low-density lipoproteins, *Biochemistry* **21**:6477–6482.

Efrati, H., Wax, Y., and Rottem, S., 1986, Glycolipids—A major factor in determining the low cholesterol uptake capacity of *A. laidlawii*, *Arch. Biochem. Biophys.* **248**:282–288.

Engelman, D. M., Terry, T. M., and Morowitz, H. J., 1967, Characterization of the plasma membrane of *Mycoplasma laidlawii*. I. Sodium dodecyl sulfate solubilization, *Biochim. Biophys. Acta* **135**:381–390.

Feldner, J., Gobel, U., and Bredt, W., 1982, *Mycoplasma pneumoniae* adhesin localized to tip structure by monoclonal antibody, *Nature* **298**:765–767.

Green, D. E., Allman, D. W., Bachmann, E., Baum, H., Kopaczyk, K., Korman, E. F., Lipton, S., MacLennan, D. H., McConnell, D. G., Perdue, J. F., Rieske, J. S., and Tzagoloff, A., 1967, Formation of membranes by repeating units, *Arch. Biochem.* **119**:312–335.

Gross, Z., and Rottem, S., 1979, Lipid distribution in *Acholeplasma laidlawii* membrane: A study using the lactoperioxidase-mediated iodination, *Biochim. Biophys. Acta.* **555**:547–552.

Hu, P. C., Collier, A. M., and Baseman, J. B., 1977, Surface parasitism by *Mycoplasma pneumoniae* of respiratory epithelium, *J. Exp. Med.* **145**:1328–1343.

Hu, P. C., Cole, R. M., Huang, Y. S., Graham, J. A., Gardner, D. E., Collier, A. M., and Clyde,

W. A., Jr., 1982, *Mycoplasma pneumoniae* infection: Role of a surface protein in the attachment organelle, *Science* **216**:313–315.

Inamine, J. M., Denny, T. P., Loechel, S., Shaper, U., Huang, C.-H., Bott, K. F., Bott, and Hu, P. C., 1988, Nucleotide sequence of the P1-attachment-protein gene of *Mycoplasma pneumoniae, Gene* **64**:217–229.

Kahane, I., 1984, In vitro studies on the mechanism of adherence and pathogenicity of mycoplasmas, *Isr. J. Med. Sci.* **20**:874–877.

Kahane, I., and Razin, S., 1969, Immunological analysis of Mycoplasma membranes, *J. Bacteriol.* **100**:187–194.

Kahane, I., Greenstein, S., and Razin, S., 1977, Carbohydrate content and enzymic activities in the membrane of *Spiroplasma citri, J. Gen. Microbiol.* **101**:173–176.

Kandler, O., and Zehender, C., 1957, Uber das vorkormmen von α-ε diaminopimelinsaure bei verscheidenen L-Phasentypen von *Proteus vulgaris* und bei den pleuropneumonie-anhlichen organismen, *Z. Naturforsch.* **12B**:725–728.

Kawauchi, Y., Muto, A., and Osawa, S., 1982, The protein composition of *Mycoplasma capricolum* ribosomes, *Mol. Gen. Genet.* **188**:7–11.

Klieneberger, E., 1935, The natural occurrence of pleuropneumonia-like organisms in apparent symbiosis with *Streptobacillus moniliformis* and other bacteria, *J. Pathol. Bacteriol.* **40**:93–105.

Klieneberger-Nobel, E., 1962, *Pleuropneumonia-Like-Organisms (PPLO) Mycoplasmataceae*, Academic Press, New York.

Lala, A. K., Buttke, T. M., and Bloch, K., 1979, On the role of the sterol hydroxyl group in membranes, *J. Biol. Chem.* **254**:10582–10585.

Larraga, V., and Razin, S., 1976, Reduced nicotinamide adenine dinucleotide oxidase activity in membranes and cytoplasm of *Acholeplasma laidlawii* and *Mycoplasma mycoides* subsp. *capri, J. Bacteriol.* **128**:827–833.

McElhaney, R. N., 1984, The structure and function of the *Acholeplasma laidlawii* plasma membrane, *Biochim. Biophys. Acta* **779**:1–42.

McElhaney, R. N., 1989, The influence of membrane lipid composition and physical properties of membrane structure and function in *Acholeplasma laidlawii, Crit. Rev. Microbiol.* **17**:1–32.

Melchior, D. L., 1982, Lipid phase transition and regulation of membrane fluidity in prokaryotes, *Curr. Top. Membr. Transp.* **17**:263–316.

Metcalfe, J. C., Metcalfe, S. M., and Engelman, D. M., 1971, Structural comparisons of native and reaggregated membranes from *Mycoplasma laidlawii* and erythrocytes by X-ray diffraction and nuclear magnetic resonance techniques, *Biochim. Biophys. Acta* **241**:412–421.

Metcalfe, S. M., Metcalfe, J. C., and Engelman, D. M., 1971, Structural comparisons of native and reaggregated membranes from *Mycoplasma laidlawii* and erythrocytes using a fluorescence probe, *Biochim. Biophys. Acta* **241**:422–430.

Morowitz, H. J., 1984, The completeness of molecular biology, *Isr. J. Med. Sci.* **20**:750–753.

Morowitz, H. J., and Tourtellotte, M. E., 1962, The smallest living cells, *Sci. Am.* **206**:117–125.

Mouches, C., and Bove, J. M., 1983, Electrophoretic characterization of mycoplasma membrane proteins, in: *Methods in Mycoplasmology*, Volume I (S. Razin and J. G. Tully, eds.), Academic Press, New York, pp. 241–255.

Murray, R. G. E., 1984, The higher taxa, or a place for everything . . . ? in: *Bergey's Manual for Systematic Bacteriology*, Volume 1 (N. R. Krieg and J. G. Holt, eds.), Williams & Wilkins, Baltimore, pp. 31–34.

Muto, A., 1987, The genome structure of *Mycoplasma capricolum, Isr. J. Med. Sci.* **23**:334–341.

Ne'eman, Z., and Razin, S., 1975, Characterization of the mycoplasma membrane proteins. V.

Release and localization of membrane-bound enzymes in *Acholeplasma laidlawii, Biochim. Biophys. Acta* **375**:54–68.

Neimark, H. C., and Lange, C. S., 1990, Pulse-field electrophoresis indicates full-length mycoplasma chromosomes range widely in size, *Nucleic Acids Res.* **18**:5443–5448.

Nocard, E., and Roux, E. R., 1898, Le microbe de la peripneumonie, *Ann. Inst. Pasteur (Paris)* **12**:240–262.

Odriozola, J. M., Waitzkin, E., Smith, T. L., and Bloch, K., 1978, Sterol requirements of *Mycoplasma capricolum, Proc. Natl. Acad. Sci. USA* **75**:4107–4109.

O'Farrell, P. H., 1975, High resolution two-dimensional electrophoresis of proteins, *J. Biol. Chem.* **250**:4007–4021.

Papahadjopoulos, D., 1974, Cholesterol and cell membrane function: A hypothesis concerning the etiology of atherosclerosis, *J. Theor. Biol.* **43**:329–337.

Plackett, P., 1959, On the possible absence of "mucocomplex" from *Mycoplasma mycoides, Biochim. Biophys. Acta* **35**:260–262.

Plackett, P., Marmion, B. P., Shaw, E. J., and Lemcke, R. M., 1969, Immunochemical analysis of *Mycoplasma pneumoniae*. 3. Separation and chemical identification of serologically active lipids, *Aust. J. Exp. Med. Sci.* **47**:171–195.

Pollack, J. D., Razin, S., and Cleverdon, R. C., 1965a, Localization of enzymes in Mycoplasma, *J. Bacteriol.* **90**:617–622.

Pollack, J. D., Razin, S., Pollack, M. E., and Cleverdon, R. C., 1965b, Fractionation of Mycoplasma cells for enzyme localization, *Life Sci.* **4**:973–977.

Razin, S., 1963a, Structure, composition and properties of the PPLO cell envelope, in: *Recent Progress in Microbiology, VIII,* University of Toronto Press, Toronto, pp. 526–534.

Razin, S., 1963b, Osmotic lysis of Mycoplasma, *J. Gen. Microbiol.* **33**:471–475.

Razin, S., 1964, Factors influencing osmotic fragility of Mycoplasma, *J. Gen. Microbiol.* **36**:451–459.

Razin, S., 1967, The cell membrane of mycoplasma, *Ann. N.Y. Acad. Sci.* **143**:115–129.

Razin, S., 1968, Mycoplasma taxonomy studied by electrophoresis of cell proteins, *J. Bacteriol.* **96**:687–694.

Razin, S., 1969a, Structure and function in Mycoplasma, *Annu. Rev. Microbiol.* **23**:317–356.

Razin, S., 1969b, The Mycoplasma membrane, in: *Mycoplasmatales and the L-Phase of Bacteria* (L. Hayflick, ed.), Appleton–Century–Crofts, New York, pp. 317–348.

Razin, S., 1972, Reconstitution of biological membranes, *Biochim. Biophys. Acta* **265**:241–296.

Razin, S., 1973, Physiology of mycoplasmas, *Adv. Microb. Physiol.* **10**:1–80.

Razin, S., 1974a, Reconstitution of mycoplasma membranes, *J. Supramol. Struct.* **2**:670–681.

Razin, S., 1974b, Correlation of cholesterol to phospholipid content in membranes of growing mycoplasma, *FEBS Lett.* **47**:81–85.

Razin, S., 1975, The Mycoplasma membrane, *Prog. Surf. Membr. Sci.* **9**:257–312.

Razin, S., 1978a, The mycoplasmas, *Microbiol. Rev.* **42**:414–470.

Razin, S., 1978b, Cholesterol uptake is dependent on membrane fluidity in mycoplasmas, *Biochim. Biophys. Acta* **513**:401–404.

Razin, S., 1981, The Mycoplasma membrane, in: *Organization of Prokaryotic Cell Membranes,* Volume I (B. K. Ghosh, ed.), CRC Press, Boca Raton, Fla., pp. 165–250.

Razin, S., 1982, Sterols in mycoplasma membranes, in: *Microbial Membrane Lipids* (S. Razin and S. Rottem, eds.), Academic Press, New York, pp. 183–205.

Razin, S., 1983, Cell lysis and isolation of membranes, in: *Methods in Mycoplasmology,* Volume I (S. Razin and J. G. Tully, eds.), Academic Press, New York, pp. 225–233.

Razin, S., 1985a, Molecular biology and genetics of mycoplasmas (mollicutes), *Microbiol. Rev.* **49**:419–455.

Razin, S., 1985b, Mycoplasma adherence, in: *The Mycoplasmas*, Volume 14: *Mycoplasma Pathogenicity*, (S. Razin and M. F. Barile, eds.), Academic Press, New York, pp. 161–202.

Razin, S., 1986, Mycoplasmal adhesins and lectins, in: *Microbial Lectins and Agglutinins* (D. Mirelman, ed.), Wiley, New York, pp. 217–235.

Razin, S., 1991, The genera *Mycoplasma, Ureaplasma, Acholeplasma, Anaeroplasma* and *Asteroleplasma*, in: *The Prokaryotes*, Volume II, 2nd ed. (A. Balows, H. G. Truper, M. Dworkin, W. Harder, and K. H. Schleifer, eds.). Springer-Verlag, Berlin, pp. 1937–1959.

Razin, S., 1992, Mycoplasma taxonomy and ecology, in: *Mycoplasmas: Molecular Biology and Pathogenesis* (J. Maniloff, R. N. McElhaney, L. R. Finch, and J. B. Baseman, eds.), American Society for Microbiology, Washington, D. C., pp. 3–22.

Razin, S., and Argaman, M., 1962, Susceptibility of Mycoplasma (PPLO) and bacterial protoplasts to lysis by various agents, *Nature* **193:**502–503.

Razin, S., and Argaman, M., 1963, Lysis of Mycoplasma, bacterial protoplasts, spheroplasts and L-forms by various agents, *J. Gen. Microbiol.* **30:**155–172.

Razin, S., and Barash, V., 1969, Solubilization of Mycoplasma membranes by the nonionic detergent Triton X-100, *FEBS Lett.* **3:**217–220.

Razin, S., and Barile, M. F., (eds.), 1985, *The Mycoplasmas*, Volume IV: *Mycoplasma Pathogenicity*, Academic Press, New York.

Razin, S., and Cohen, A., 1963, Nutritional requirements and metabolism of *Mycoplasma laidlawii*, *J. Gen. Microbiol.* **30:**141–154.

Razin, S., and Cosenza, B. J., 1966, Growth phases of Mycoplasma in liquid media observed with the phase-contrast microscope, *J. Bacteriol.* **91:**858–869.

Razin, S., and Jacobs, E., 1992, Mycoplasma adhesion, *J. Gen. Microbiol.* **138:**407–422.

Razin, S., and Kahane, I., 1969, Hybridization of solubilized membrane components from different Mycoplasma species by reaggregation, *Nature* **223:**863–864.

Razin, S., and Knight, B. C. J. G., 1960a, A partially defined medium for the growth of Mycoplasma, *J. Gen. Microbiol.* **22:**492–503.

Razin, S., and Knight, B. C. J. G., 1960b, The effects of ribonucleic acid and deoxyribonucleic acid on the growth of Mycoplasma, *J. Gen. Microbiol.* **22:**504–519.

Razin, S., and Oliver, O., 1961, Morphogenesis of Mycoplasma and bacterial L-form colonies, *J. Gen. Microbiol.* **24:**225–237.

Razin, S., and Rottem, S., 1963, Fatty acid requirements of *Mycoplasma laidlawii*, *J. Gen. Microbiol.* **33:**459–470.

Razin, S., and Rottem, S., 1974, Isolation, solubilization and reconstitution of mycoplasma membranes, *Methods Enzymol.* **32B:**459–468.

Razin, S., and Rottem, S., 1978, Cholesterol in membranes: Studies with mycoplasmas, *Trends Biochem. Sci.* **3:**51–55.

Razin, S., and Yogev, D., 1989, Molecular approaches to characterization of mycoplasmal adhesins, in: *Molecular Mechanisms of Microbial Adhesion* (M. Hook, L. Switalski, R. D. Wells, and E. H. Beachy, eds.), Springer-Verlag, Berlin, pp. 52–76.

Razin, S., Argaman, M., and Avigan, J., 1963, Chemical composition of mycoplasma cells and membranes, *J. Gen. Microbiol.* **33:**477–487.

Razin, S., Morowitz, H. J., and Terry, T. M., 1965, Membrane subunits of *Mycoplasma laidlawii* and their assembly to membrane-like structures, *Proc. Natl. Acad. Sci. USA* **54:**219–225.

Razin, S., Cosenza, B. J., and Tourtellotte, M. E., 1966a, Variations in Mycoplasma morphology induced by long-chain fatty acids, *J. Gen. Microbiol.* **42:**139–145.

Razin, S., Tourtellotte, M. E., McElhaney, R. N., and Pollack, J. D., 1966b, Influence of lipid components of *Mycoplasma laidlawii* membranes on osmotic fragility of cells, *J. Bacteriol.* **91:**609–616.

Razin, S., Ne'eman, Z., and Ohad, I., 1969, Selective reaggregation of solubilized mycoplasma membrane proteins and the kinetics of membrane reformation, *Biochim. Biophys. Acta* **193:**277–293.

Razin, S., Prescott, B., and Chanock, R. M., 1970, Immunogenicity of *Mycoplasma pneumoniae* glycolipids. A novel approach to the production of antisera to membrane lipids, *Proc. Natl. Acad. Sci. USA* **67:**590–597.

Razin, S., Chanock, R. M., Graf, L., and Rapport, M. M., 1971a, Immunogenicity of cytolipin H aggregated with *Acholeplasma laidlawii* membrane proteins, *Proc. Soc. Exp. Biol. Med.* **138:**404–407.

Razin, S., Prescott, B., James, W. D., Caldes, G., Valdesuso, J., and Chanock, R. M., 1971b, Production and properties of antisera to membrane glycolipids of *Mycoplasma pneumoniae, Infect. Immun.* **3:**420–423.

Razin, S., Kahane, I., and Kovartovsky, J., 1972, Immunochemistry of mycoplasma membranes, in: *Ciba Symposium on Pathogenic Mycoplasmas* (J. Birch, ed.), Elsevier, Amsterdam, pp. 93–122.

Razin, S., Kahane, I., Banai, M., and Bredt, W., 1981, Adhesion of mycoplasmas to eukaryotic cells, *Ciba Found. Symp.* **80:**98–113.

Reinert, J. C., and Steim, J. M., 1970, Calorimetric detection of a membrane-lipid phase transition in living cells, *Science* **168:**1580–1582.

Rodwell, A. W., 1963, The steroid growth requirements of *Mycoplasma mycoides, J. Gen. Microbiol.* **32:**91–101.

Rodwell, A. W., and Rodwell, E. S., 1978, Relationship between strains of *Mycoplasma mycoides* subsp. *mycoides* and *capri* studied by two-dimensional gel electrophoresis of cell proteins, *J. Gen. Microbiol.* **109:**259–263.

Rodwell, A. W., Razin, S., Rottem, S., and Argaman, M., 1967, Association of protein and lipid in *Mycoplasma laidlawii* membranes disaggregated by detergents, *Arch. Biochem. Biophys.* **122:**621–628.

Rottem, S., 1980, Membrane lipids of mycoplasmas, *Biochim. Biophys. Acta* **604:**65–90.

Rottem, S., 1982, Transbilayer distribution of lipids in microbial membranes, in: *Microbial Membrane Lipids* (S. Razin and S. Rottem, eds.), Academic Press, New York, pp. 235–262.

Rottem, S., and Razin, S., 1966, Adenosine triphosphatase activity of Mycoplasma membranes, *J. Bacteriol.* **92:**714–722.

Rottem, S., and Razin, S., 1967, Electrophoretic patterns of membrane proteins of Mycoplasma, *J. Bacteriol.* **94:**359–364.

Rottem, S., and Razin, S., 1972, Isolation of mycoplasma membranes by digitonin, *J. Bacteriol.* **110:**699–705.

Rottem, S., Stein, O., and Razin, S., 1968, Reassembly of mycoplasma membranes disaggregated by detergents, *Arch. Biochem. Biophys.* **125:**46–56.

Rottem, S., Hubbell, W. L., Hayflick, L., and McConnell, H. M., 1970, Motion of fatty acid spin labels in the plasma membrane of *Mycoplasma, Biochim. Biophys. Acta* **219:**104–113.

Rottem, S., Pfendt, E. A., and Hayflick, L., 1971, Sterol requirements of T-strain mycoplasmas, *J. Bacteriol.* **105:**323–330.

Rottem, S., Cirillo, V. P., de Kruyff, B., Shinitzky, M., and Razin, S., 1973a, Cholesterol in mycoplasma membranes. Correlation of enzymic and transport activities with physical state of lipids in membranes of *Mycoplasma mycoides* var. *capri* adapted to grow with low cholesterol concentrations, *Biochim. Biophys. Acta* **323:**509–519.

Rottem, S., Yashouv, J., Ne'eman, Z., and Razin, S., 1973b, Cholesterol in Mycoplasma membranes. Composition, ultrastructure and biological properties of membranes from *Mycoplasma mycoides* var. *capri* cells adapted to grow with low cholesterol concentrations, *Biochim. Biophys. Acta* **323:**495–508.

Rottem, S., Slutzky, G., and Bittman, R., 1978, Cholesterol distribution and movement in the *Mycoplasma gallisepticum* cell membrane, *Biochemistry* **17**:2723–2732.

Shirvan, M. H., Gross, Z., Ne'eman, Z., and Rottem, S., 1982, Isolation of *Mycoplasma gallisepticum* membranes by a mild alkaline induced lysis of non-energized cells, *Curr. Microbiol.* **7**:367–370.

Singer, S. J., and Nicolson, G. L., 1972, The fluid mosaic model of the structure of cell membranes, *Science* **175**:720–731.

Slutzky, G. M., Razin, S., Kahane, I., and Eisenberg, S., 1976, Serum lipoproteins as cholesterol donors to mycoplasma membranes, *Biochem. Biophys. Res. Commun.* **68**:529–536.

Slutzky, G. M., Razin, S., Kahane, I., and Eisenberg, S., 1977, Cholesterol transfer from serum lipoproteins to mycoplasma membranes, *Biochemistry* **16**:5158–5163.

Smith, P. F., 1984, Lipoglycans from mycoplasmas, *Crit. Rev. Microbiol.* **11**:157–186.

Smith, P. F., Langworthy, T. A., and Mayberry, W. R., 1976, Distribution and composition of lipopolysaccharides from mycoplasmas, *J. Bacteriol.* **125**:916–922.

Steim, J. M., Tourtellotte, M.E. Reinert, J. C., McElhaney, R. N., and Rader, R. L., 1969, Calorimetric evidence for the liquid-crystalline state of lipids in a biomembrane, *Proc. Natl. Acad. Sci. USA* **63**:104–109.

Su, C. J., Tryon, V. V., and Baseman, J. B., 1987, Cloning and sequence analysis of cytadhesin P1 gene from *Mycoplasma pneumoniae, Infect. Immun.* **55**:3023–3029.

Takayama, K., MacLennan, D. H., Tzagoloff, A., and Stoner, C. D., 1966, Studies on the electron transfer system. LXVII. Polyacrylamide gel electrophoresis of the mitochondrial transfer complexes, *Arch. Biochem. Biophys.* **114**:223–230.

Tarshis, M. A., and Kapitanov, A. B., 1978, Symport carbohydrate transport into *Acholeplasma laidlawii* cells, *FEBS Lett.* **89**:73–77.

Terry, T. M., Engelman, D. M., and Morowitz, H. J., 1967, Characterization of the plasma membrane of *Mycoplasma laidlawii*. II. Modes of disaggregation of solubilized membrane components, *Biochim. Biophys. Acta* **135**:391–405.

Tillack, T. W., Carter, R., and Razin, S., 1970, Native and reformed *Mycoplasma laidlawii* membranes compared by freeze-etching, *Biochim. Biophys. Acta* **219**:123–130.

Tully, J. G., and Whitcomb, R. F., (eds.), 1979, *The Mycoplasmas*, Volume II: *Human and Animal Mycoplasmas,* Academic Press, New York.

van Iterson, W., and Ruys, A. C., 1960, The fine structure of the *Mycoplasmataceae* (microorganisms of the pleuropneumonia group—P.P.L.O.). I. *Mycoplasma hominis, M. fermentans* and *M. salivarium, J. Ultrastruct. Res.* **3**:282–301.

Verkleij, A. J., Ververgaert, P. H. J., Van Deenen, L. L. M., and Elbers, P. F., 1972, Phase transitions of phospholipid bilayers and membranes of *Acholeplasma laidlawii* B visualized by freeze fracturing electron microscopy, *Biochim. Biophys. Acta* **288**:326–332.

Weisburg, W. G., Tully, J. G., Rose, D. C., Petzel, J. P., Oyaizu, H., Yang, D., Mandelco, L., Sechrest, J., Lawrence, T. G., Van Etten, J., Maniloff, J., and Woese, C. R., 1989, A phylogenetic analysis of the mycoplasmas: Basis for their classification, *J. Bacteriol.* **171**:6455–6467.

Whitcomb, R. F., and Tully, J. G., (eds.), 1979, *The Mycoplasmas*, Volume III: *Plant and Insect Mycoplasmas,* Academic Press, New York.

Whitcomb, R. F., and Tully, J. G., (eds.), 1989, *The Mycoplasmas,* Volume V: *Spiroplasmas, Acholeplasmas and Mycoplasmas of Plants and Arthropods,* Academic Press, New York.

Williams, M. H., and Taylor-Robinson, D., 1967, Antigenicity of mycoplasma membranes, *Nature* **215**:973–974.

Woese, C. R., 1987, Bacterial evolution, *Microbiol. Rev.* **51**:221–271.

Chapter 2

Mycoplasma Membrane Lipids

Chemical Composition and Transbilayer Distribution

Robert Bittman

1. INTRODUCTION

The goals of the present review are to summarize the lipid composition of mycoplasmas and to analyze the results obtained about the transbilayer distribution of lipids in mycoplasma and acholeplasma cell membranes. Previous review articles have summarized the lipid composition of mycoplasma membranes (Smith, 1971, 1979; Razin, 1973; Rottem, 1980; McElhaney, 1984). The transbilayer distribution of mycoplasma lipids has also been reviewed (for the most recent reviews, see Rottem and Davis, 1986; Bittman, 1988). The present review gives the structures of the principal lipid components of mycoplasma membranes, including the structures of newly discovered lipids, and discusses data

Abbreviations used in this chapter: PC, phosphatidylcholine; PG, phosphatidylglycerol; DPG, diphosphatidylglycerol; MGDG, monoglucosyl diacylglycerol [1,2-diacyl-3-*O*-(α-D-glucopyranosyl)-*sn*-glycerol]; DGDG, diglucosyl diacylglycerol; GP-DGDG, glycerophosphoryl diglucosyl diacylglycerol.

Robert Bittman   Department of Chemistry and Biochemistry, Queens College of The City University of New York, Flushing, New York 11367.
Subcellular Biochemistry, Volume 20: Mycoplasma Cell Membranes, edited by Shlomo Rottem and Itzhak Kahane. Plenum Press, New York, 1993.

obtained on transbilayer distribution and movement of lipids in mycoplasmas since this topic was last reviewed.

The lipids of mycoplasmas are located almost exclusively in the cell membrane. Since mycoplasmas lack a cell wall, the cell membrane is isolated easily, usually by lysis of the cells in a hypotonic solution. The sterol-requiring mycoplasmas, spiroplasmas, and ureaplasmas depend on an external supply of fatty acids for growth and are unable to alter the fatty acids they take up. This feature has proven to be very useful in elucidating the effect of fatty acyl chains on membrane structure and physical properties. In many cases, controlled changes in the fatty acid composition of a variety of mycoplasmas were achieved by regulating the fatty acid composition of the medium. In some strains it has been possible to produce membranes with essentially a single fatty acyl group (Rodwell and Peterson, 1971; Silvius and McElhaney, 1978). *Acholeplasma* species are capable of saturated fatty acid synthesis from acetate. This ability varies among the various species depending on the level of acyl carrier protein in the cells (Rottem *et al.*, 1973a). *Acholeplasma* species are able to elongate short-chain saturated and unsaturated fatty acids, but are unable to synthesize unsaturated fatty acids; therefore, *A. laidlawii* A depends on an external supply of some unsaturated fatty acids (Razin and Rottem, 1963; Panos and Rottem, 1970).

The influence of lipid interconversions within endogenous lipid pools (such as PG and DPG in *M. capricolum* and MGDG and DGDG in acholeplasmas) or as a result of uptake of exogenous fatty lipids on various aspects of membrane function and physical properties have been investigated intensively using *Acholeplasma* and *Mycoplasma* species (for recent reviews, see McElhaney, 1989, and Chapters 3 and 4 in this volume). Other features that have made these organisms attractive for studies of the roles of lipids in membranes are the inability of mycoplasmas to carry out pinocytosis and the lack of a cell wall, making possible direct contact between the outer leaflet of the cell membrane and the external environment.

2. LIPID COMPOSITION OF MYCOPLASMAS

The main components of the cell membrane of *A. laidlawii* are glycolipids, phospholipids, and neutral lipids. The neutral glycolipids MGDG (or α-GlcDG) and DGDG (for structures, see Figure 1) together usually account for more than 50% by weight of the total membrane lipid, and the phospholipid PG accounts for about 30% by weight of the total lipids in cells supplemented with either palmitate (Bhakoo *et al.*, 1987) or palmitate and oleate (Bevers *et al.*, 1977). Phosphoglycolipids account for most of the remaining membrane material of *Acholeplasma* species (Table I).

Mono- and diglycosyl diacylglycerols, which are commonly found in gram-

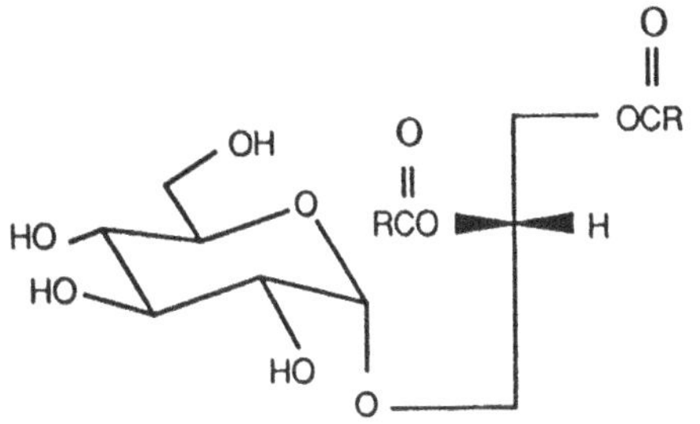

MGDG of A. laidlawii

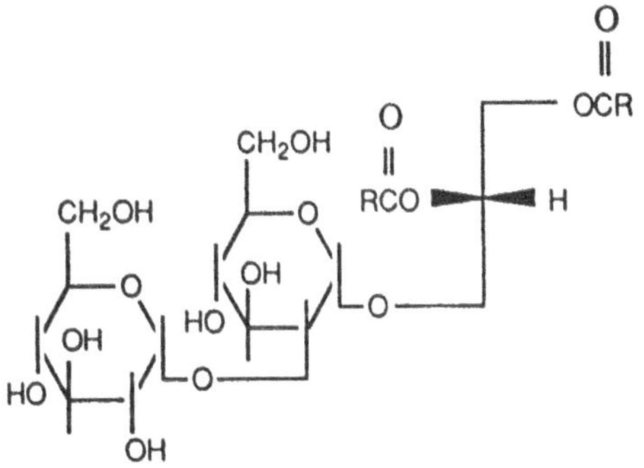

DGDG of A. laidlawii

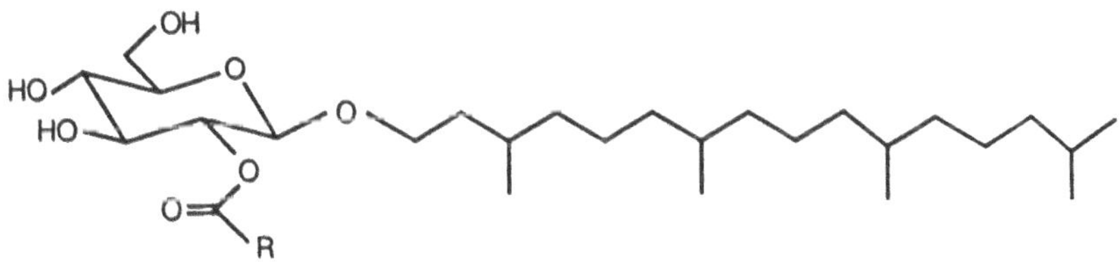

Polyprenylacylglucopyranoside of A. laidlawii

FIGURE 1. Chemical structures of the principal glycolipids (MGDG, DGDG, and 1-polyprenyl-2-acyl-α-D-glucopyranoside) of acholeplasmas. [The structure of the latter glycolipid, which was previously referred to as glycolipid X, was proposed recently (McElhaney, 1989).]

Table I

Representative Compositions of *de Novo*-Synthesized Lipids in Various *Acholeplasma* and *Mycoplasma* Species

Lipid component	A. laidlawii					M. capricolum California kid[f]	M. hominis ATCC 15056[g]	M. gallisepticum strain 5969[h]
	Strain B		Strain A[c]	Strain 992[d]	Oral strain[e]			
	a	b						
					(wt% of total lipids)			
Neutral lipids								
MGDG	38	52	28	17	16			
DGDG	18	12	32	11	13			
Polyprenyl-glc	6			13				
Cholesterol[i]						40	26	32
Cholesteryl esters[i]							11	
Diacylglycerols						2		
Triacylglycerols						14		
Free FAs							6	
Polar lipids								
PG	31	21	21	6	14	32	33	64
DPG				24	26	15		
GP-MGDG				22	8			
GD-DGDG	4		18	6	23			
Unidentified PL				9				
Amino-PL						4		

[a] Bevers *et al.* (1977). Oleate-grown cells were grown to the late log phase.

[b] McDonough *et al.* (1983). Cells were grown with palmitic acid supplementation.

[c] Wieslander and Rilfors (1977). *A. laidlawii* A (EF22) was grown for 24 hr with a 1 : 1 mixture of palmitic and oleic acids.

[d] Efrati *et al.* (1986). Cells were grown to late exponential phase (A_{640} 0.3–0.4) with a mixture of palmitic and elaidic acids.

[e] Gross and Rottem (1979).

[f] Bittman *et al.* (1990). Cells were grown in Edward medium supplemented with 1% BSA, 10 μg cholesterol/ml, no PC, and a 1 : 1 mixture of palmitic and oleic acids.

[g] Rottem and Razin (1973). Cells were grown in Edward medium with horse serum and 20 mM arginine.

[h] Clejan and Bittman (1984a). Cells were grown to midexponential phase in modified Edward medium containing 1% albumin, palmitic and oleic acids (10 μg/ml of each), and cholesterol (10 μg/ml). When grown on serum, the cells take up PC and sphingomyelin, and PG represents about 50% of the total phospholipids (see text).

[i] Cholesterol and cholesteryl esters are not *de novo*-synthesized lipids.

positive bacteria and the *Acholeplasma* species, are not present in *M. galliseptium*, *M. capricolum*, and *M. hominis*. In *M. hominis*, neutral lipids account for 60% by weight of the total membrane lipids, and polar lipids account for the remaining 40% (see Table I). The principal neutral lipids in membranes of *M. hominis* cells grown with 2% (v/v) serum and 20 mM arginine are cholesterol (43% by weight of the neutral lipid fraction), cholesteryl esters (19%), triacylglycerols (23%), and free fatty acids (10%) (Rottem and Razin, 1973). Most of the phospholipid fraction consists of PG, with no DPG present. The content of saturated fatty acyl chains in the polar lipid fraction of *M. hominis* was much higher than the content of unsaturated fatty acyl chains, whereas the reverse was found in the neutral lipid fraction. See Table I for representative membrane lipid compositions of *A. laidlawii* and *Mycoplasma* species.

2.1. Glycolipids

The ratio of MGDG to DGDG in *A. laidlawii* varies with growth conditions such as temperature, type of fatty acid enrichment, and extent of exogenous cholesterol incorporation (see Chapter 5). The decrease in MGDG/DGDG ratio in response to lipid compositions that tend to favor a transition from the lamellar to a hexagonal II phase has been considered to represent a mechanism by which *A. laidlawii* A cells avoid the formation of a nonlamellar phase (Wieslander and Rilfors, 1977; Christiansson and Wieslander, 1978; Wieslander *et al.*, 1980).

When *A. laidlawii* B is grown in the presence of exogenous palmitic acid under conditions of glucose limitation stress, the non-bilayer-forming lipid 1-polyprenyl-2-acyl-α-D-glucopyranoside (for structure, see Figure 1), previously called glycolipid X, is formed at the expense of MGDG (Bhakoo *et al.*, 1987; McElhaney, 1989). It was recently reported that this monoether, monoacylated derivative of MGDG represents about 60 mole% of the total membrane lipid fraction of *A. laidlawii* B grown in palmitate-enriched medium even when normal glucose levels are present (Lewis *et al.*, 1990). Although the polyprenyl-containing glycolipid and MGDG are both non-bilayer-forming lipids, the former has a lower melting gel phase; thus, replacement of MGDG by the polyprenyl-containing lipid makes possible the formation of a lamellar liquid-crystalline phase at 37°C in membranes of *A. laidlawii* B cells containing palmitic acid chains. In different strains of *A. laidlawii* other acylated glucosyl derivatives have been identified (Smith, 1986), but their physical properties have not been investigated.

Figure 1 shows the structures of the major glycolipids of *A. laidlawii*. MGDG of *A. laidlawii* has the α anomeric configuration, whereas in *M. neurolyticum* the β anomer of MGDG is formed (Smith, 1972). In *A. axanthum*, monoglycosyl diglyceride has the α anomeric configuration, and the sugar is galactofuranoside (Mayberry and Smith, 1983). In *M. mycoides*, β-galactofuranosyl diacylglycerol is formed (Plackett, 1967).

Many functions have been proposed for glycoglycerolipids in bacterial, plant, and animal cell membranes. In addition to serving as external surface receptor sites, glycoglycerolipids may participate in the regulation of membrane fluidity (Pask-Hughes *et al.*, 1977) and modulation of protein conformations (Ishizuka and Yamakawa, 1985). The uncharged glycolipids may also serve as spacer molecules, separating the negative charges of acidic phospholipids.

2.2. Phospholipids

In common with most other prokaryotes, mycoplasmas contain the acidic phospholipids PG and DPG (cardiolipin) as the principal phospholipids (for structures, see Figure 2). Table I shows that PG represents about 30–35% by weight of total membrane lipid of *A. laidlawii* B supplemented with palmitate and oleate. GP-DGDG (see Figure 2) accounts for about 5% by weight of the total membrane lipid of *A. laidlawii*. Other phospholipids that are present in smaller amounts in *A. laidlawii* are GP-MGDG and *O*-amino acid esters of PG. Figure 2 shows *O*-alanyl-PG with the alanyl group esterified to the 3 position of glycerol in the head group of PG; glutamyl, glycyl, leucyl, lysyl, and tyrosyl residues are also formed, depending on the growth conditions (Smith, 1979). The position of the *O*-amino acyl residue is not fully established; 2-*O*- and 3-*O*-esters are both possible. The sum of *O*-amino acyl PG, DPG, and GP-MGDG is typically about 10% by weight (McElhaney, 1984). In *A. laidlawii* A, the main lipids are MGDG, DGDG, PG, and GP-DGDG; when cells are grown in medium supplemented with an equal amount of palmitic and oleic acids, the content of PG decreased with aging of the culture (Wieslander and Rilfors, 1977).

Some unusual mycoplasma phospholipids have been reported. Phosphatidylethanolamine, which has not been detected in acholeplasmas and mycoplasmas but is formed in large amounts in gram-negative bacteria and in the genus *Bacillus* and in certain clostridia, was reported to be formed in ureaplasmas (Romano *et al.*, 1972), and plasmalogens have been found in *Anaeroplasma* (Langworthy *et al.*, 1975). Although sphingolipids are rarely synthesized by prokaryotes, ceramide derivatives are formed by *A. axanthum* (Mayberry *et al.*, 1973).

2.3. Cholesterol Content

The content of unesterified cholesterol is much higher in *Mycoplasma* and *Spiroplasma* species than in *Acholeplasma* species (see Table II). *Mycoplasma, Spiroplasma,* and *Ureaplasma* require cholesterol for growth, but are unable to synthesize or modify cholesterol (Rottem, 1980). Although *Acholeplasma* species do not require cholesterol for growth, they nevertheless have a low cholesterol-binding capacity and incorporate a limited amount of cholesterol into

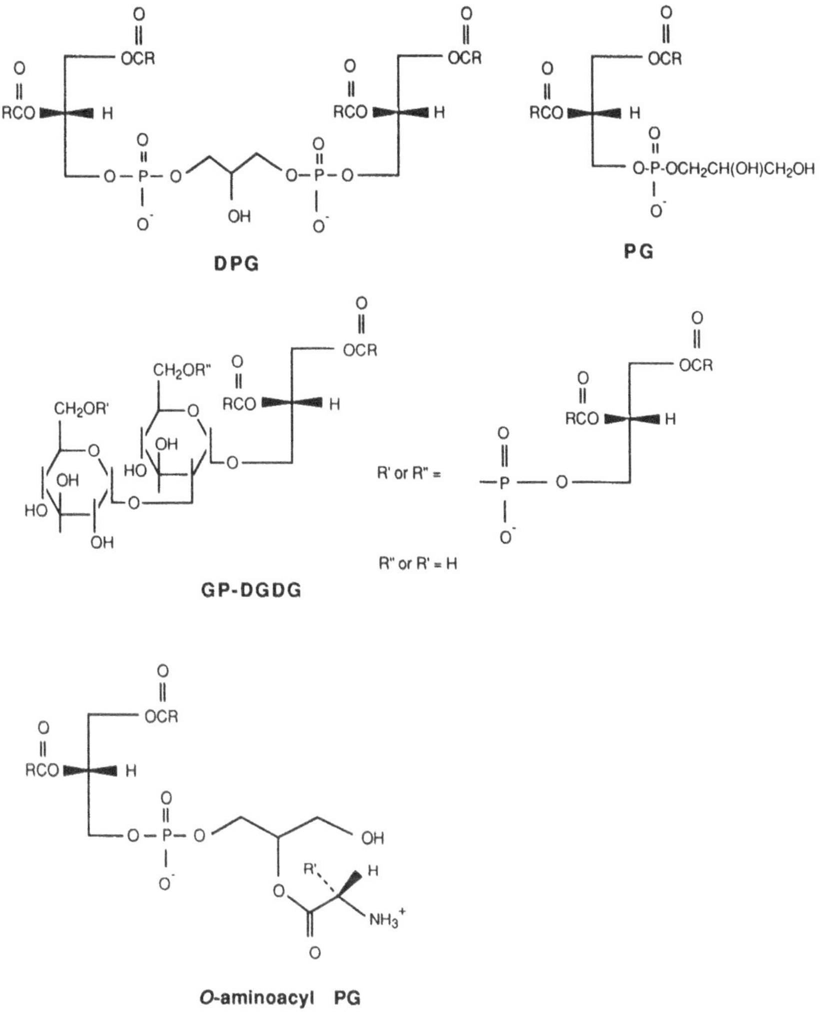

FIGURE 2. Chemical structures of the principal *de novo*-synthesized polar lipids of acholeplasmas (PG, DPG, GP-DGDG, and *O*-aa-PG) and mycoplasmas (PG and DPG).

their cell membrane. Incorporation of exogenous cholesterol increases on aging of *A. laidlawii* cells; aging is accompanied by a decrease in MGDG and 1-polyprenyl-2-acyl-α-D-glucopyranoside content, suggesting that non-lamellar-forming glycolipids may restrict cholesterol incorporation (Efrati *et al.*, 1986). The affinity of glycolipids for cholesterol is assumed to be lower than that of phospholipids for cholesterol. Thus, the low cholesterol content of acholeplasmas is in part a reflection of their membrane lipid composition, since

Table II
Contents of Free Cholesterol and Esterified Cholesterol
of Various *Acholeplasma, Mycoplasma,* and
Spiroplasma Species[a]

	Cholesterol (μmole/mg cell protein)	
Organism	Free cholesterol	Cholesteryl ester
A. granularum	28.5	0
A. axanthum	3.7	0
A. laidlawii	10.2	0
M. capricolum	67.2	67.7
M. arginini	58.2	27.5
M. hominis	76.4	30.6
M. pneumoniae	85.9	58.0
S. floricola	75.0	42.0

[a]Data were obtained for organisms grown in medium containing 5% (v/v) horse serum (Razin *et al.*, 1980), except for *S. floricola* which was grown with 8% (v/v) horse serum (M. Salman, unpublished data). The molar ratios of free cholesterol to total lipid phosphorus in the *Acholeplasma* and *Mycoplasma* species grown on horse serum are < 0.3 and 0.80–0.95, respectively.

glycolipids and phosphoglycolipids constitute the bulk of the *A. laidlawii* membrane, but these lipids are absent in the cholesterol-rich mycoplasmas. Phospholipase A_2-catalyzed hydrolysis of the glycerolipid (PG and DPG) fraction of *A. laidlawii*, which accounts for ~ 30% by weight of the total membrane lipid, resulted in a 55% decrease in the amount of cholesterol incorporated. Differences in membrane protein composition may also contribute to the two- to fourfold lower uptake of exogenous cholesterol by *Acholeplasma* species relative to *Mycoplasma* species. Proteolytic digestion of *M. capricolum* cells resulted in diminished uptake of cholesterol from serum, but the same conditions of trypsin treatment did not affect the limited capacity of *A. laidlawii* to incorporate cholesterol (Efrati *et al.*, 1981). The proposal that mycoplasmas contain membrane proteins that facilitate cholesterol incorporation (Efrati *et al.*, 1981) has not been studied further.

2.4. Cholesteryl Esters

Some mycoplasmas contain surprisingly high contents of cholesteryl esters; Table II shows that cholesteryl esters constitute about one-third of the total cholesterol in *M. hominis, M. mycoides* subsp. *capri, M. capricolum,* and *S. citri* membranes. The uptake of cholesteryl esters into the membrane of growing adapted *M. capricolum* cells occurs rapidly; when a culture of adapted *M. capri-*

colum cells was grown to A_{640} of 0.1 and then transferred to a defined medium rich in free cholesterol and cholesteryl oleate, the contents of free and esterified cholesterol 4 hr after the transfer were 124 and 110 μg/mg membrane protein, respectively (Clejan *et al.*, 1981). The cholesteryl esters may be localized in droplets within mycoplasma membranes (Melchior and Rottem, 1981).

2.5. Adaptation to Low Cholesterol Content

Some mycoplasmas have been adapted to grow on low amounts of exogenous cholesterol, allowing analyses to be made of the effects of cholesterol on various membrane properties. *M. gallisepticum* strain A5969 was adapted to grow in media containing less than 10 μg/ml of cholesterol by progressively decreasing the sterol concentration from 10 to 7.0, 5.5, 3.6, and 2.0 μg/ml after satisfactory growth was achieved at each sterol concentration (Clejan and Bittman, 1984a). The growth medium contained palmitic and oleic acids (10 μg/ml of each) and 1% (w/v) fatty-acid-poor albumin instead of serum, and at each cholesterol concentration at least four passages were made before transfer to the next lower cholesterol-containing medium. The adapted strain grew more slowly than the native strain, and the adapted cells were more fragile; nevertheless, the adapted strain could be stored at $-70°C$ for up to 2 months and new cultures could be started directly without repeating the adaptation process. Table III shows that the cholesterol/phospholipid molar ratio decreases from 0.92 in the native strain to 0.25 in the adapted strain grown on 2 μg/ml of cholesterol, with the total phospholipid content of the cell membrane remaining at ~ 210 μg/mg membrane protein.

Adapted *M. gallisepticum*, *M. capricolum*, and *M. mycoides* subsp. *capri* cells have a markedly lower degree of membrane lipid order at 37°C than the

Table III
Lipid Composition of the Cell Membranes of *M. gallisepticum* and *M. capricolum* Cells Adapted to Grow on Low Exogenous Cholesterol[a]

Free cholesterol in growth medium (μg/ml)	Cholesterol/PL molar ratio	Free cholesterol	Total PL content
		(μg/mg membrane protein)	
10 (native strain)	0.92	100	210
7.0	0.63	71	214
5.5	0.50	54	210
4.0	0.36	38	202
2.0	0.25	27	210

[a]Strain A5969 was grown with 1% albumin, palmitic and oleic acids (10 μg/ml of each), and the indicated cholesterol concentrations. Data from Clejan and Bittman (1984a).

native strains, as measured using fluorescence and electron spin resonance probes and osmotic swelling rates (Le Grimellec *et al.*, 1981; Romano *et al.*, 1986; Rottem *et al.*, 1973b). Adapted *M. capricolum* cells have an increased water volume, which may arise from changes in the surface area of the cell membrane or from increased freedom of motion of membrane lipids (Romano *et al.*, 1986). Freeze-fracture electron micrographs showed that membranes of adapted *M. mycoides* subsp. *capri* cells became aggregated at low temperature whereas membranes of the native strain retained a random distribution of particles on the fracture faces, indicating that cholesterol prevents phospholipid fatty acyl chains from crystallizing (Rottem *et al.*, 1973b). The ability of cholesterol to broaden the phospholipid phase temperature was also supported by studies made with the native strain of *M. arginini*, which tends to incorporate saturated fatty acids preferentially into PG, the principal *de novo*-synthesized phospholipid (Rottem, 1981). It is of interest to note that the small amounts of cholesterol that are required to support the growth of adapted mycoplasma strains would be unable to exert a significant effect on bulk lipid membrane order as a result of its interaction with phospholipids or as a result of its role as a spacer molecule, separating adjacent charged phospholipids; small amounts of cholesterol may have a regulatory role in phospholipid biosynthesis, such as activation of enzymes involved in fatty acid uptake (Dahl *et al.*, 1981; Chapter 5). When lanosterol replaced cholesterol in the growth medium, the growth rate of *M. capricolum* cells was reduced and the membrane lipids were less highly ordered (Huang *et al.*, 1991).

2.6.   Growth with Various Δ^5-Sterols

Cholesterol has been replaced by various other sterols in the membranes of *M. capricolum* and *M. gallisepticum* cells by growing the cells in a defined medium in which the desired sterol is added to the medium together with the fatty acids as an ethanolic solution, followed by albumin (2% w/v). The desired 3β-hydroxysterol can replace cholesterol after multiple passages. Stigmasterol and 4,6-cholestadien-3β-ol were incorporated into the membrane of *M. capricolum* (California kid) cells in amounts similar to those obtained with cholesterol uptake (130–160 μg of sterol/per mg of membrane protein), whereas sitosterol, ergosterol, and cholestanol were incorporated to a somewhat lower extent (85–100 μg/mg membrane protein) (Clejan *et al.*, 1981). The dependence of growth support on sterol structure was not marked, since the maximum slopes of the logarithmic phases of growth were very similar with all of the sterols investigated. Indeed, a very broad range of sterols can support growth of *M. capricolum* (Odriozola *et al.*, 1978; Lala *et al.*, 1979; Efrati *et al.*, 1980). The membrane of growing *M. gallisepticum* cells takes up sitosterol, cholesta-5,22E, 24-trien-3β-ol, and *cis*-22-dehydrocholesterol in high amounts, and these sterols supported cell growth (Clejan and Bittman, 1984b). However, some sterols were poor growth supporters, e.g., a stereoisomer of cholesterol having the *S* configuration at C-20 instead of the usual R configuration, and cholesterol analogues with a very short or very long side chain in place of the isooctyl group.

2.7. Carotenoids

The carotenoid content of the neutral lipid fraction of acholeplasmas is elevated by the presence in the growth medium of pantetheine, which stimulates acyl carrier protein activity (Christiansson and Wieslander, 1980). The content of colored carotenoids in membranes of *A. laidlawii* was much higher when cells were grown with oleic acid supplementation than with elaidic acid supplementation; addition of 0.5% (v/v) sodium propionate to the growth medium inhibited carotenoid formation (Rottem and Markowitz, 1979a). Carotenoids may have a sterol-like role in regulating membrane fluidity of acholeplasmas (Huang and Haug, 1974; Rottem and Markowitz, 1979a).

2.8. Uptake of Exogenous Phospholipids

Exogenous phospholipids are taken up into the cell membrane of various mycoplasmas and spiroplasmas, but not into acholeplasma membranes. PC and sphingomyelin were detected in membranes of *M. pneumoniae* (Beckman and Kenny, 1968; Razin *et al.*, 1970) and *Spiroplasma citri* (Freeman *et al.*, 1976) grown in serum-containing medium. Ceramide-containing phospholipids were reported in membranes of mycoplasma strain S743 (Plackett *et al.*, 1970). Sphingomyelin is taken up into *M. gallisepticum* without modification, but PC in the growth medium is modified by *M. gallisepticum* to give a disaturated PC (Rottem and Markowitz, 1979b).

In the absence of exogenous phospholipids, PG (the only *de novo*-synthesized phospholipid of this organism) comprises about 95% of the phospholipids of *M. gallisepticum;* however, when the native or adapted strain of *M. gallisepticum* is grown with exogenous egg PC (10 µg/ml), PG comprises about 49% of the total phospholipids in the cell membrane (Clejan and Bittman, 1984a). Growth of adapted *M. gallisepticum* cells in the presence of 10 µg/ml of dipalmitoyl-PC gave a membrane phospholipid composition of 49% PG and 47% PC, and growth with 10 µg/ml of *N*-palmitoyl- or bovine brain sphingomyelin produced membranes with 42–48% sphingomyelin (Clejan and Bittman, 1984c).

In membranes of *M. capricolum* grown in a defined medium, PC constituted 20, 30, and 33% of the total phospholipids when the cells were grown in medium containing 5, 10, and 15 µg/ml of PC, respectively; under these conditions, the free cholesterol content of the membrane did not change (47–54 µg/mg cell protein) (Clejan *et al.*, 1981). The polar lipid content of the *M. capricolum* membrane doubled when the concentration of horse serum in the medium was increased from 0.5 to 5% (v/v) (Gross *et al.*, 1982). The proportions of the two major *de novo*-synthesized phospholipids of *M. capricolum* were modified on incorporation of PC and sphingomyelin from serum, with the amount of PG decreasing and that of DPG increasing. The amounts of the minor phospholipids of *M. capricolum* (an amino-phospholipid and an unidentified phospholipid) did not change. A marked decrease in the PG/DPG ratio of *M. capricolum* membranes was also found when cells were grown in medium containing isopalmitic (Bittman and Clejan, 1987). Changes in the relative proportions of non-bilayer-preferring DPG to bilayer-preferring PG in *M. capricolum*, like the changes in the molar ratio of MGDG/DGDG in *A. laidlawii* in response to increased growth temperature, fatty acyl unsaturation, and cholesterol content (Wieslander *et al.*, 1980; Rilfors *et al.*, 1987) or various exogenous molecules (Wieslander *et al.*, 1986), may represent adjustments in the lipid packing that help maintain the bilayer structure required for many membrane-mediated processes (Gross *et al.*, 1982; Bittman *et al.*, 1990).

2.9. Changes in Lipid Composition Associated with Culture Aging

The ratio of PG to other phospholipids has been measured under different growth conditions of *M. hominis* (Rottem and Greenberg, 1975) and *M. capricolum* (Gross and Rottem, 1986). The total phospholipid content decreased on aging of *M. hominis* cultures. A decrease in PG was accompanied by the formation of a lipid with a higher R_f value, which was tentatively identified as phosphatidic acid (Rottem and Greenberg, 1975). In contrast to *M. hominis, M. capricolum* contains both PG and DPG as the principal *de novo*-synthesized phospholipids. *M. capricolum* cells did not show a variation of PG/DPG ratio on progression of growth from the early to late log phase when grown with a high palmitic acid content relative to oleic acid; however, the PG/DPG ratio decreased on culture aging when the oleic acid content of the growth medium exceeded the palmitic acid content (Gross and Rottem, 1986). The molar ratio of PG/DPG of *M. capricolum* membranes increased when the growth medium was supplemented with a low concentration of calcium (0.5 mM calcium chloride), primarily as a result of a decrease in the DPG content from 25% by weight of total phospholipids in the absence of calcium to 17% in the presence of 0.5 mM calcium in the medium (Bittman *et al.*, 1990). The unsaturated fatty acid content of DPG is higher than that of PG, suggesting that an unsaturated-enriched pool of PG serves as the precursor of DPG during biosynthesis (Gross and Rottem, 1986; Bittman *et al.*, 1990).

Changes in the lipid composition of *A. laidlawii* cells (strain IEM-1) were also examined as a function of cell growth (Kapitanov *et al.*, 1990). The ratio of saturated to unsaturated fatty acids in the total membrane lipids increased on culture aging, the cholesterol content increased, and the contents of glycolipids, phospholipids, and carotenoids decreased as cells approached the late log phase.

2.10. Lipoglycans

Lipoglycans of thermoacidophilic bacteria and *Acholeplasma* and *Anaeroplasma* species differ in chemical composition from the classical lipopolysaccharides found in gram-negative bacteria (for a review, see Smith, 1984). For example, the lipoglycan of *T. acidophilum* consists of 24 mannose residues linked to one glucopyranoside, which in turn is linked to a diglycerol tetraether bearing isopranoid chains (Smith, 1980). Thus, the basic lipoglycan structure of *T. acidophilum* is a linear oligosaccharide chain terminating in a novel lipid moiety. [The unusual lipid structures of thermoacidophilic bacteria are not considered in this chapter since thermoplasmas do not belong to the class Mollicutes; for reviews, see Langworthy (1982, 1985).] The lipoglycan of *A. granularum* also contains a terminal lipid moiety, which is a diacylglycerol (Smith, 1981). This lipoglycan is the smallest of the acholeplasma lipoglycans. In addition to glucose and mannose, it contains *N*-acyl-amino sugars. In the lipoglycan of *A. axanthum,* the lipid residues are distributed along the oligosaccharide chain, rather than only being linked to one end of the molecule (Smith, 1983). The ratio of neutral sugars to amino sugars is 10:9. The neutral sugars are glucose (80%) and galactose (20%), whereas the amino sugars are fucosamine (2-deoxy-2-amino-6-deoxygalactosamine) and quinovosamine (2-deoxy-2-amino-6-deoxyglucosamine) in a 2:1 ratio. All of the amino groups are *N*-acylated, with acetyl comprising 75% of the acyl content. The other important acyl groups are myristoyl, (*R*)-3-hydroxytetradecanoyl, and (*R*)-3-hydroxyhexadecanoyl. One glycerol moiety is present per five neutral sugar moieties. Phosphate esters are present in the glycerol and galactose moieties. Table IV presents a summary of

Table IV
Composition of Lipoglycans of Various Acholeplasma Species

Organism	% dry cell wt	Monomer size	Neutral sugars	Amino sugars	Lipids
A. laidlawii	~1	150,000	Glucose, mannose	Fucosamine, quinovosamine	FAs, glycerol
A. axanthum	~1	100,000	Glucose, galactose	Fucosamine, quinovosamine	3-Hydroxy-FAs, myristate, glycerol
A. granularum	~1	20,000	Mannose, glucose	Fucosamine, glucosamine	Diacylglycerol

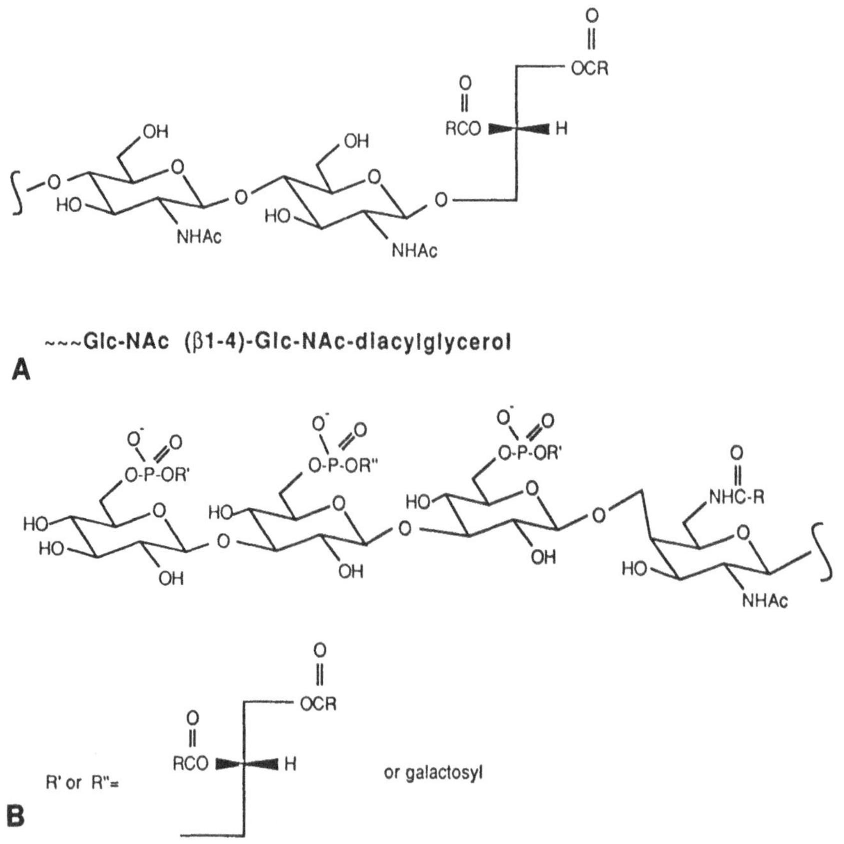

~~~Glc-NAc (β1-4)-Glc-NAc-diacylglycerol

A

B

FIGURE 3. Partial structures of the lipoglycans of (A) *A. granularum* and (B) *A. axanthum*. For *A. granularum*, the β1–3 glycosidic linkage also occurs.

the compositions of the lipoglycans of *Acholeplasma* species, and Figure 3 shows the structures.

3. TRANSBILAYER DISTRIBUTION OF LIPIDS IN *A. LAIDLAWII* MEMBRANES

3.1. Phosphatidylglycerol

The susceptibility of PG to hydrolytic action by purified pig pancreatic phospholipase A_2 has been used to estimate the extent of localization of PG in the outer half of the lipid bilayer of the *A. laidlawii* membrane. PG is the only lipid in this organism that is a substrate for this enzyme, and hydrolysis takes place only when the membrane lipids are in a fluid state (or, in the case of acholeplasma grown with methyl-branched fatty acids, at least in a loosely packed gel phase) (Bouvier *et al.*, 1981). At temperatures above 30°C, all of the PG is hydrolyzed in *A. laidlawii* membranes enriched with palmitic, elaidic, or oleic acids. The possibility that translocation of PG molecules from the inner to the outer leaflet of the bilayer is activated when the outer-leaflet PG is depleted by enzymatic action makes it difficult to reach firm conclusions about the distribution of PG between the two leaflets (Bevers *et al.*, 1977). A pool of 30% of the total PG is resistant to hydrolysis in isolated membranes below 30°C. The less accessible pool may represent protein-bound PG; at high temperature, protein-bound PG is considered to undergo exchange with other lipids. Changes in the state of energization of the cell affected the extent of PG hydrolysis, with complete and rapid hydrolysis taking place in the nonenergized state (Bevers *et al.*, 1978). It may be concluded that the extent of PG localized in the outer monolayer of the membrane of growing *A. laidlawii* B cells is in the range of ~40 or 50% to 80% of the total PG (McElhaney, 1984).

3.2. MGDG, DGDG, PG, and DPG

The transbilayer distribution of phosphoglycolipids and glycolipids has been studied in *A. laidlawii* (oral strain) by using the lactoperoxidase-mediated radioiodination technique (Gross and Rottem, 1979), which was previously applied in mycoplasma membranes to determination of polypeptide transbilayer distribution (Razin, 1978). The radioiodinated labeling intensities of lipids were compared in intact cells and isolated membranes; the labeling agent has access to lipids in both halves of the bilayer in isolated membranes, but only lipids in the outer half of the bilayer of intact cells can be labeled, assuming that the reactive iodinating species does not penetrate into the membrane. It was concluded that all of the glycolipids, MGDG and DGDG, and about 60% of the glycerolipids, PG and DPG, were localized in the outer leaflet of the bilayer. The phosphoglycolipids were distributed approximately equally between the two halves of the membrane.

3.3. Cholesterol

The rates of spontaneous cholesterol transfer from *A. laidlawii* (strain 1012) cells to sonicated lipid vesicles were measured at 37°C (Davis *et al.*, 1984). The cells were grown in medium supplemented with elaidic acid (20 μg/ml), albumin, and cholesterol (0.2 μg/ml) containing a trace of [^{14}C]cholesterol. An excess of vesicles prepared from an equimolar mixture of egg PC and unlabeled cholesterol was used as the acceptors of radiolabeled cholesterol. The time course of cholesterol transfer was found to fit two kinetic curves in both intact cells and unsealed, isolated membranes; the rapidly transferring pool had a half-time of about 30 min. Cells were treated with phospholipase A_2, resulting in a decrease of 86% of the PG content of the membrane. Radiolabeled cholesterol underwent transfer kinetics from the phospholipase-treated cells to vesicles in only a single exponential curve, with a half-time of about 3 hr. It was suggested that the two kinetic pools of cholesterol of untreated intact cells and isolated membranes represent cholesterol molecules associated with glycolipids (which are more available for transfer to acceptor vesicles) and cholesterol molecules associated with phospholipids (which are less available for transfer). The possibility that the fast and slow pools represent cholesterol molecules localized in the outer and inner leaflets, respectively, of the cell membrane, with a slow rate of transbilayer movement, was considered unlikely, since both sides of unsealed membranes would be available for transfer to acceptor vesicles. One cannot exclude the possibility that the biphasic kinetics represent cholesterol molecules segregated into different cholesterol/lipid molar ratios.

4. TRANSBILAYER DISTRIBUTION OF STEROLS IN MEMBRANES OF *MYCOPLASMA* SPECIES

The transbilayer distributions of proteins, carbohydrates, and phospholipids have been studied extensively in many membrane systems. The transbilayer distribution of the remaining major component of mammalian membranes, cholesterol, has not been studied in detail because the small polar group of cholesterol (the hydroxy group) does not lend itself to spectral assays that are possible with other membrane components. Mycoplasmas are ideal for studying cholesterol localization and movement, since cholesterol is an essential membrane component and the organisms cannot synthesize or modify cholesterol. The initial rates of binding of the polyene antibiotic filipin to cholesterol in intact cells of *M. gallisepticum* and *M. capricolum*, which are measured by stopped-flow kinetics, were used to estimate the transbilayer distribution of cholesterol between the inner and outer halves of these membranes (for reviews, see Bittman *et al.*, 1983; Bittman, 1988). Stopped-flow techniques are required in order to minimize membrane perturbations caused by the polyene antibiotic. Filipin reacts rapidly with free cholesterol or other 3β-hydroxysterols, but not with cholesteryl esters, phospholipids, or proteins. The interaction with cholesterol is accompanied by a change in the absorption spectrum of filipin. The relative amounts of cholesterol on the outside and inside halves of the bilayer were estimated from the ratio of second-order rate constants for filipin–cholesterol association in intact cells versus unsealed membranes isolated from the cells. Cholesterol was found to be distributed symmetrically in the bilayer of *M. gallisepticum* cells grown to the early log phase. In *M. capricolum*, about two-thirds of the free cholesterol was estimated to be localized in the outer leaflet.

To examine whether exogenous cholesterol moves rapidly from the outer leaflet to the inner leaflet of growing mycoplasma cells, an early logarithmic culture of adapted *M. capricolum* was transferred to a cholesterol-rich medium in order to insert cholesterol into their cell membrane (Clejan *et al.,* 1978). Despite the dramatic increase in cholesterol content, the distribution of free cholesterol in the two halves of the bilayer did not change after the first hour of incubation with the cholesterol-rich medium at 37°C, as estimated by the filipin-binding technique. Thus, the rate of translocation of free cholesterol is very rapid in growing cells. At low temperature or in the presence of ionophores, cholesterol was localized predominantly in the outer leaflet of the bilayer; therefore, translocation from the outer to the inner leaflet appears to be facilitated in actively growing cells at 37°C where a membrane potential gradient is maintained and/or macromolecular synthesis occurs. On the other hand, intermembrane movement of cholesterol between mycoplasma cells and acceptor lipid vesicles is independent of cell viability (Rottem *et al.,* 1981).

The relative abilities of cholesterol, sitosterol, and stigmasterol to undergo rapid translocation across the two leaflets of the cell membrane of adapted *M. capricolum* cells were compared (Clejan and Bittman, 1984b). Stopped-flow measurements of filipin binding to sterols in intact cells and isolated membranes were made at intervals during the growth stimulation and sterol incorporation phase that occur when adapted cells are transferred to a medium containing 10 μg/ml of sterol, 2% (w/v) albumin, and palmitic and oleic acids (10 μg/ml of each). Sitosterol and stigmasterol remained localized predominantly in the outer leaflet, indicating that an ethyl group at C-24 (as in sitosterol) or an ethyl group at C-24 and a double bond at C-22 (as in stigmasterol) create steric bulk that impedes sterol translocation. It should be noted that the rate of intermembrane movement of sitosterol is much lower than that of cholesterol (Kan and Bittman, 1990, 1991).

The results of transbilayer cholesterol distribution in *M. gallisepticum* and *M. capricolum* membranes obtained by stopped-flow kinetics were confirmed by measurements of radiolabeled cholesterol intermembrane exchange, i.e., from intact cells to an excess of acceptors such as lipid vesicles (for a review, see Bittman, 1988). Since [^{14}C]cholesterol was present in the growth medium in almost identical amounts throughout the growth cycle, the specific activities of the cholesterol pools in the membrane are assumed to be equal. The high osmotic stability of *M. gallisepticum* cells makes them particularly well suited for exchange studies, since prolonged incubations with acceptor membranes are needed for complete exchangeability. During this time, no significant lysis takes place, and little fusion with acceptor vesicles is found, as measured with a nonexchangeable marker incorporated into either the donor or acceptor species. Two kinetic pools of exchangeable cholesterol, each representing one-half of the total cholesterol, were found in resting *M. gallisepticum* cells, but more than 90% of the cholesterol in isolated membranes was exchanged in one pool (Rottem *et al.*, 1978, 1981). The fraction of total free cholesterol in each pool did not change when *M. gallisepticum* cells were grown in serum (giving membranes that contain 50% PG, 37% PC, and 10% sphingomyelin) or in serum-free medium (giving membranes that contain PG as 95% of the total membrane phospholipids) (Clejan and Bittman, 1984c).

The molecular basis for the observation of two cholesterol pools in exchange studies with mycoplasmas is not fully clear. A rapid equilibration between the two pools is inhibited because of interactions with membrane proteins or phospholipids. The two pools may represent cholesterol molecules localized in the outer and inner halves of the bilayer, with the molecules in the inner half representing the slowly exchanging pool since desorption must take place after flip-flop to the external surface. In support of this hypothesis, [^{14}C]cholesterol exchange data with open membranes of both *M. gallisepticum* and *M. capricolum* are fit by a single exponential curve (Rottem *et al.*, 1978, 1981; Clejan and Bittman, 1984a; Bittman *et al.*, 1990). Preferential phospholipid–cholesterol interactions have been postulated to be the source of the two pools in *M. gallisepticum* membranes (Rottem and Davis, 1986). However, this conclusion remains to be resolved in view of the finding that the sizes of the two pools did not change on incorporation of PC and sphingomyelin into mycoplasma membranes (Clejan and Bittman, 1984a) and the lack of detailed information about phospholipid transbilayer asymmetry in mycoplasma membranes. Biphasic exchange kinetics have been detected for cholesterol in other biological membranes, such as brush border membrane vesicles (Bloj and Zilversmit, 1982) and *Torpedo* electroplax membranes (Leibel *et al.*, 1987). Although the nature of the two pools could not be defined rigorously, the slowly exchanging pool was taken to represent cholesterol bound tightly to membrane proteins. Indeed, the possibility has been considered that proteins may be involved in preventing rapid equilibration of the two pools of cholesterol in mycoplasma membranes (Bittman, 1988). Treatment of *M. gallisepticum* cells with phenylenediamine caused extensive cross-linking of membrane proteins and impeded cholesterol exchange (Clejan and Bittman, 1984a). Depletion of membrane protein affected the rate of the slowly exchanging pool more than that of the rapidly exchanging pool.

Diamide-induced oxidative cross-linking of sulfhydryl groups of cysteine residues in *M. gallisepticum* membranes stimulated the rate of phospholipid movement from mycoplasma membranes to lipid vesicles, but did not affect the rate of cholesterol exchange (Bittman *et al.*, 1985). This observation suggests that relatively discrete regions in the membrane are involved in the desorption of cholesterol and phospholipid molecules. In addition, the rates of spontaneous cholesterol and phospholipid exchange between *M. capricolum* membranes and lipid vesicles were enhanced to different extents when cells were grown with calcium supplementation to induce structural defects in the membrane bilayer (Bittman *et al.*, 1990).

The effects of aging of mycoplasma cultures on transbilayer cholesterol distribution have also been studied by both filipin-binding and radiolabeled cholesterol exchange measurements. In *M. gallisepticum*, cell aging was accompanied by an enrichment of cholesterol in the outer leaflet of the bilayer (Rottem *et al.*, 1978; Bittman *et al.*, 1981), assuming that the two cholesterol pools represent cholesterol molecules localized in the outer and inner leaflets of the bilayer. Increases in the rigidity of the lipid domain of *M. gallisepticum* (Rottem and Greenberg, 1975; Rottem *et al.*, 1981) and in the membrane protein-to-phospholipid ratio were found on aging (Amar *et al.*, 1976), without significant changes in fatty acid composition. The increased membrane rigidity may impede the translocation of cholesterol from the outer to the inner half of the bilayer. Partial proteolysis of membrane proteins resulted in the disappearance of the change in pool size noted on aging of *M. gallisepticum* (Bittman *et al.*, 1981). In *M. capricolum* membranes the content of unsaturated fatty acyl chains in phospholipids increases on aging, which may compensate for the increase in protein-to-lipid ratio to produce little variation in membrane fluidity (Gross and Rottem, 1986). Filipin-binding studies indicated that the transbilayer distribution of cholesterol did not change on aging of *M. capricolum* cells (Bittman *et al.*, 1981).

Interactions between cholesterol and nearest neighbor molecules in the bilayer affect the rate of cholesterol exchange. For example, the rate of cholesterol exchange between *M. gallisepticum* cell membranes and lipid vesicles decreased with an increase in cholesterol/phospholipid molar ratio (Clejan and Bittman, 1984a), on incorporation of synthetic cross-linked PE derivatives (Bittman *et al.*, 1985), and on incorporation of sphingomyelin (Clejan and Bittman, 1984c). The ability of sphingomyelin to lower the rate of cholesterol exchange between mycoplasma membranes and acceptor species is explained by the greater lateral packing density in the lipid–water interface when sphingomyelin is present (Kan *et al.*, 1991; Grönberg *et al.*, 1991). The rates of both cholesterol and phospholipid movement between mycoplasma membranes and vesicles are enhanced in *M. capricolum* membranes that have the potential to undergo a transition from the lamellar to a nonlamellar phase (Bittman *et al.*, 1990). Microdefects in lipid packing caused by changes in membrane lipid composition brought about by varying the conditions of growth supplementation (such as addition of PC to alter the PG/DPG ratio or of calcium ions) facilitate desorption of cholesterol and PC from mycoplasma membranes. In other biological membranes, functional modifications have been observed when bilayer instability is induced by a pretransitional state leading to the onset of a nonlamellar phase.

The transbilayer distribution of various sterols has been measured in *M. capricolum* and *M. gallisepticum* membranes by using the filipin–sterol binding technique and by the kinetics of radiolabeled sterol exchange between mycoplasma cells and an excess of lipid vesicles. Cholestanol and 4,6-cholestadien-3β-ol, which have the same side chain as cholesterol but differ in the structure of the steroid nucleus, have the same transbilayer distribution in the membrane of *M. capricolum* as cholesterol (two-thirds in the outer leaflet and one-third in the inner leaflet) (Clejan *et al.*, 1981). About 85–90% of the ergosterol, sitosterol, and stigmasterol molecules are present in the outer leaflet, indicating that alkylation at C-24 and unsaturation at Δ^{22} of the sterol side chain interfere with the movement of the sterol molecule across the lipid bilayer. The same characteristic of accumulating sterols with bulky alkyl substituents or unsaturation in the side chain in the outer half of the bilayer was found in *M. gallisepticum* membranes (Clejan and Bittman, 1984b). Although the translocation process is inhibited by the insertion of double bonds or alkyl groups into the sterol side chain, changes in the length (as opposed to the bulkiness) of the sterol side chain did not alter the transbilayer sterol distribution relative to cholesterol.

The physiological role of extensive sterol localization in the two halves of the mycoplasma bilayer remains to be established. Growth-supporting sterols do not need to be translocated extensively into the inner half of the bilayer, since sitosterol, cholesta-5,22*E*,24-trien-3β-ol, and *cis*-22-dehydrocholesterol supported growth of *M. gallisepticum* but had a different transbilayer distribution than cholesterol (Clejan and Bittman, 1984b). In *M. capricolum*, sitosterol, ergosterol, and stigmasterol supported growth nearly as well as cholesterol but were localized in the two leaflets of the bilayer differently than cholesterol (Clejan *et al.*, 1981). Sterols that differed from cholesterol in the structure of the steroid nucleus were distributed in the *M. capricolum* bilayer identically as cholesterol but were poorer growth supporters (Clejan *et al.*, 1981); in *M. gallisepticum*, 20-isocholesterol and a cholesterol analogue with a side chain containing eight methylene groups had the same transbilayer distribution as cholesterol but were poor growth supporters and were incorporated at lower levels than cholesterol (Clejan and Bittman, 1984b). No information is available about whether a certain minimal amount of sterol is needed in the inner leaflet of mycoplasmas for optimal membrane function.

ACKNOWLEDGMENT. Work from the author's laboratory cited herein was supported in part by NIH Grant HL 16660.

5. REFERENCES

Amar, A., Rottem, S., Kahane, I., and Razin, S., 1976, Characterization of the mycoplasma membrane proteins. Composition and disposition of proteins in membranes from aging *Mycoplasma hominis* cultures, *Biochim. Biophys. Acta* **426**:258–270.

Beckman, B. L., and Kenny, G. E., 1968, Immunochemical analysis of serologically active lipids of *Mycoplasma pneumoniae*, *J. Bacteriol.* **96**:1171–1180.

Bevers, E. M., Singal, S. A., Op den Kamp, J. A. F., and van Deenen, L. L. M., 1977, Recognition of different pools of phosphatidylglycerol in intact cells and isolated membranes of *Acholeplasma laidlawii* by phospholipase A_2, *Biochemistry* **16**:1290–1295.

Bevers, E. M., Leblanc, G., Le Grimellec, C., Op den Kamp, J. A. F., and van Deenen, L. L. M., 1978, Disposition of phosphatidylglycerol in metabolizing cells of *Acholeplasma laidlawii*, *FEBS Lett.* **87**:49–51.

Bhakoo, M., Lewis, R. N. A. H., and McElhaney, R. N., 1987, Isolation and characterization of a novel monoacylated glucopyranosyl neutral lipid from the plasma membrane of *Acholeplasma laidlawii*, B, *Biochim. Biophys. Acta* **922**:34–45.

Bittman, R., 1988, Sterol exchange between mycoplasma membranes and vesicles, in: *Biology of Cholesterol* (P. L. Yeagle, ed.), CRC Press, Boca Raton, Fla., pp. 173–195.

Bittman, R., and Clejan, S., 1987, Kinetics of cholesterol and phospholipid exchange between mycoplasma membranes and lipid vesicles, *Isr. J. Med. Sci.* **23**:398–402.

Bittman, R. Blau, L., Clejan, S., and Rottem, S., 1981, Determination of cholesterol asymmetry by rapid kinetics of filipin–cholesterol association: Effect of modification in lipids and proteins, *Biochemistry* **20**:2425–2432.

Bittman, R., Clejan, S., and Rottem, S., 1983, Transbilayer distribution of sterols in mycoplasma membranes: A review, *Yale J. Biol. Med.* **56**:397–403.

Bittman, R., Clejan, S., Robinson, B. P., and Witzke, N. M., 1985, Kinetics of cholesterol and phospholipid exchange from membranes containing cross-linked proteins or cross-linked phosphatidylethanolamines, *Biochemistry* **24**:1403-1409.

Bittman, R., Clejan, S., and Hui, S. W., 1990, Increased rates of lipid exchange between *Mycoplasma capricolum* membranes and vesicles in relation to the propensity of forming non-bilayer lipid structures, *J. Biol. Chem.* **265**:15110–15117.

Bloj, B., and Zilversmit, D. B., 1982, Heterogeneity of rabbit intestine brush border plasma membrane cholesterol, *J. Biol. Chem.* **257**:7608–7614.

Bouvier, P., Op den Kamp, J. A. F., and van Deenen, L.L.M., 1981, Studies on *Acholeplasma laidlawii* grown on branched-chain fatty acids, *Arch. Biochem. Biophys.* **208**:242–247.

Christiansson, A., and Wieslander, Å., 1978, Membrane lipid metabolism in *Acholeplasma laidlawii* A EF 22. Influence of cholesterol and temperature shift-down on the incorporation of fatty acids and synthesis of membrane lipid species, *Eur. J. Biochem.* **85**:65–76.

Christiansson, A., and Wieslander, Å., 1980, Control of membrane polar lipid composition in *Acholeplasma laidlawii* A by the extent of saturated fatty acid synthesis, *Biochim. Biophys. Acta* **595**:189–199.

Clejan, S., and Bittman, R., 1984a, Kinetics of cholesterol and phospholipid exchange between *Mycoplasma gallisepticum* cells and lipid vesicles. Alterations in membrane cholesterol and protein content, *J. Biol. Chem.* **259**:441–448.

Clejan, S., and Bittman, R., 1984b, Distribution and movement of sterols with different side chain structures between the two leaflets of the membrane bilayer of mycoplasma cells, *J. Biol. Chem.* **259**:449–455.

Clejan, S., and Bittman, R., 1984c, Decreases in rates of lipid exchange between *Mycoplasma gallisepticum* cells and unilamellar vesicles by incorporation of sphingomyelin, *J. Biol. Chem.* **259**:10823–10826.

Clejan, S., Bittman, R., and Rottem, S., 1978, Uptake, transbilayer distribution, and movement of cholesterol in growing *Mycoplasma capricolum* cells, *Biochemistry* **17**:4579–4583.

Clejan, S., Bittman, R., and Rottem, S., 1981, Effects of sterol structure and exogenous lipids on the transbilayer distribution of sterols in the membrane of *Mycoplasma capricolum*, *Biochemistry* **20**:2200–2204.

Dahl, J. S., Dahl, C. E., and Bloch, K., 1981, Effect of cholesterol on macromolecular synthesis and fatty acid uptake by *Mycoplasma capricolum*, *J. Biol. Chem.* **256**:87–91.

Davis, P. J., Efrati, H., Razin, S., and Rottem, S., 1984, Two pools of cholesterol in *Acholeplasma laidlawii* membranes, *FEBS Lett.* **175**:51–54.

Efrati, H., Shinitzky, M., and Razin, S., 1980, Effects of charged cholesteryl esters on mycoplasma growth, *FEBS Lett.* **122**:59–63.

Efrati, H., Rottem, S., and Razin, S., 1981, Lipid and protein membrane components associated with cholesterol uptake by mycoplasmas, *Biochim. Biophys. Acta* **641**:386–394.

Efrati, H., Wax, Y., and Rottem, S., 1986, Cholesterol uptake capacity of *Acholeplasma laidlawii* is affected by the composition and content of membrane glycolipids, *Arch. Biochem. Biophys.* **248**:282–288.

Freeman, B. A., Sissenstein, R., McManus, T. T., Woodward, J. E., Lee, I. M., and Mudd, J. B., 1976, Lipid composition and lipid metabolism of *Spiroplasma citri*, *J. Bacteriol.* **125**:946–954.

Grönberg, L., Ruan, Z.-S., Bittman, R., and Slotte, J. P., 1991, Interaction of cholesterol with synthetic sphingomyelin derivatives in mixed monolayers, *Biochemistry* **30**:10746–10754.

Gross, Z., and Rottem, S., 1979, Lipid distribution in *Acholeplasma laidlawii* membrane. A study using the lactoperoxidase-mediated iodination, *Biochim. Biophys. Acta* **555**:547–552.

Gross, Z., and Rottem, S., 1986, Lipid interconversions in aging *Mycoplasma capricolum* cultures, *J. Bacteriol.* **167**:986–991.

Gross, Z., Rottem, S., and Bittman, R., 1982, Phospholipid interconversions in *Mycoplasma capricolum*, *Eur. J. Biochem.* **122**:169–174.

Huang, L., and Haug, A., 1974, Regulation of membrane lipid fluidity in *Acholeplasma laidlawii*: Effect of carotenoid pigment content, *Biochim. Biophys. Acta* **352**:361–370.

Huang, T.-H., DeSiervo, A. J., and Yang, Q.-X., 1991, Effect of cholesterol and lanosterol on the structure and dynamics of the cell membrane of *Mycoplasma capricolum*. Deuterium nuclear magnetic resonance study, *Biophys. J.* **59**:691–702.

Ishizuka, I., and Yamakawa, T., 1985, in *Glycoglycerolipids* (H. Wiegandt, ed.), Elsevier and Academic Press, New York, pp. 101–197.

Kan, C.-C., and Bittman, R., 1990, Constraint of the spontaneous intermembrane movement of sitosterol by its 24α-ethyl group, *J. Am. Chem. Soc.* **112**:884–886.

Kan, C.-C., and Bittman, R., 1991, Spontaneous rates of sitosterol and cholesterol exchange between phospholipid vesicles and between lysophospholipid dispersions: Evidence that desorption rate is impeded by the 24α-ethyl group of sitosterol, *J. Am. Chem. Soc.* **113**:6650–6656.

Kan, C.-C., Ruan, Z.-S., and Bittman, R., 1991, Interaction of cholesterol with sphingomyelin in bilayer membranes: Evidence that the hydroxy group of sphingomyelin does not modulate the rate of cholesterol exchange between vesicles, *Biochemistry* **30**:7759–7766.

Kapitanov, A. B., Ivanova, V. F., and Ladygina, V. G., 1990, Cholesterol accumulation in plasma membrane and changes of membrane enzyme activity of *Acholeplasma laidlawii* cells during culture ageing, *Mech. Ageing Dev.* **55**:161–169.

Lala, A. K., Buttke, T. M., and Bloch, K., 1979, On the role of the sterol hydroxyl group in membranes, *J. Biol. Chem.* **254**:10582–10585.

Langworthy, T. A., 1982, Lipids of bacteria living in extreme environments, *Curr. Top. Membr. Transp.* **17**:45–77.

Langworthy, T. A., 1985, Lipids of archaebacteria, in: *The Bacteria*, Volume 8 (J. R. Spkatch and L. N. Ornston, eds.), Academic Press, New York, pp. 459–497.

Langworthy, T. A., Mayberry, W. R., Smith, P. F., and Robinson, I. M., 1975, Plasmalogen composition of *Anaeroplasma*, *J. Bacteriol.* **122**:785–787.

Le Grimellec, C., Cardinal, J., Giocondi, M.-C., and Carriére, S., 1981, Control of membrane lipids in *Mycoplasma gallisepticum*: Effect on lipid order, *J. Bacteriol.* **146**:155–162.

Leibel, W. S., Firestone, L. L., Legler, D. C., Braswell, L. M., and Miller, K. W., 1987, Two pools of cholesterol in acetylcholine receptor-rich membranes from *Torpedo*, *Biochim. Biophys. Acta* **897**:249–260.

Lewis, R. N. A. H., Yue, A. W. B., McElhaney, R. N., Turner, D. C., and Gruner, S. M., 1990, Thermotropic characterization of the 2-*O*-acyl, polyprenyl α-D-glucopyranoside isolated from palmitate-enriched *Acholeplasma laidlawii* B membranes, *Biochim. Biophys. Acta* **1026**:21–28.

McDonough, B., Macdonald, P. M., Sykes, B. D., and McElhaney, R. N., 1983, F-19 NMR studies of lipid fatty acyl chain order and dynamics in *Acholeplasma laidlawii* B membranes. A physical, biochemical, and biological evaluation of monofluoropalmitic acids as membrane probes, *Biochemistry* **22**:5097–5103.

McElhaney, R. N., 1984, The structure and function of the *Acholeplasma laidlawii* plasma membrane, *Biochim. Biophys. Acta* **779**:1–42.

McElhaney, R. N., 1989, The influence of membrane lipid composition and physical properties of membrane structure and function in *Acholeplasma laidlawii*, *CRC Crit. Rev. Microbiol.* **17**:1–32.

Mayberry, W. R., and Smith, P. F., 1983, Structures and properties of acyl diglucosylcholesterol and galactofuranosyl diacylglycerol from *Acholeplasma axanthum*, *Biochim. Biophys. Acta* **752**:434–443.

Mayberry, W. R., Smith, P. F., Langworthy, T. A., and Plackett, P., 1973, Identification of the amide-linked fatty acids of *Acholeplasma axanthum* S473 as D-(−)-3-hydroxyhexadecanoate and its homologues, *J. Bacteriol.* **116**:1091–1095.

Melchior, D., and Rottem, S., 1981, The organization of cholesterol esters in *Mycoplasma capricolum* membranes, *Eur. J. Biochem.* **117**:147–153.

Odriozola, J. M., Waitzkin, E., Smith, T. L., and Bloch, K., 1978, Sterol requirement of *Mycoplasma capricolum*, *Proc. Natl. Acad. Sci. USA* **75**:4107–4109.

Panos, C., and Rottem, S., 1970, Incorporation and elongation of fatty acids isomers by *Mycoplasma laidlawii* A, *Biochemistry* **9**:407–412.

Pask-Hughes, R. A., Mozaffary, H., and Shaw, N., 1977, Glycolipids in prokaryotic cells, *Biochem. Soc. Trans.* **5**:1675–1677.

Plackett, P., 1967, The glycerolipids of *Mycoplasma mycoides*, *Biochemistry* **6**:2746–2754.

Plackett, P., Smith, P. F., and Mayberry, W. R., 1970, Lipids of a sterol-nonrequiring Mycoplasma, *J. Bacteriol.* **104**:798–807.

Razin, S., 1973, Physiology of mycoplasmas, *Adv. Microb. Physiol.* **10**:1–80.

Razin, S., 1978, The mycoplasmas, *Microbiol. Rev.* **42**:414–470.

Razin, S., and Rottem, S., 1963, Fatty acid requirements of *Mycoplasma laidlawii*, *J. Gen. Microbiol.* **33**:459–470.

Razin, S., Prescott, B., Caldes, G., James, W. D., and Chanock, R. M., 1970, Role of glycolipids and phosphatidylglycerol in the serological activity of *Mycoplasma pneumoniae*, *Infect. Immun.* **1**:408–418.

Razin, S., Kutner, S., Efrati, H., and Rottem, S., 1980, Phospholipid and cholesterol uptake by mycoplasma cells and membranes, *Biochim. Biophys. Acta* **598**:628–640.

Rilfors, L., Wikander, G., and Wieslander, Å., 1987, Lipid acyl-chain dependent effects of sterols in *Acholeplasma laidlawii* membranes, *J. Bacteriol.* **169**:830–838.

Rodwell, A. W., and Peterson, J. E., 1971, The effect of straight-chain saturated, monoenoic and branched-chain fatty acids on growth and fatty acid composition of mycoplasma strain, *J. Gen. Microbiol.* **68**:173–186.

Romano, N., Smith, P. F., and Mayberry, W. R., 1972, Lipids of a T strain of *Mycoplasma*, *J. Bacteriol.* **109**:565–569.

Romano, N., Shirvan, M. H., and Rottem, S., 1986, Changes in membrane lipid composition of *Mycoplasma capricolum* affect the cell volume, *J. Bacteriol.* **167**:1089–1091.

Rottem, S., 1980, Membrane lipids of mycoplasmas, *Biochim. Biophys. Acta* **604**:65–90.

Rottem, S., 1981, Cholesterol is required to prevent crystallization of *Mycoplasma arginini* phospholipids at physiological temperature, *FEBS Lett.* **133**:161–164.

Rottem, S., and Davis, P. J., 1986, Cholesterol pools in mycoplasma membranes. Modifications in phospholipid composition affect the kinetics of cholesterol exchange with lipid vesicles, in: *Enzymes of Lipid Metabolism, Part II* (L. Freysz, H. Dreyfus, R. Massarelli, and S. Gatt, eds.), Plenum Press, New York, pp. 421–428.

Rottem, S., and Greenberg, A. S., 1975, Changes in composition, biosynthesis, and physical state of membrane lipids occurring upon aging of *Mycoplasma hominis* cultures, *J. Bacteriol.* **121**:631–639.

Rottem, S., and Markowitz, O., 1979a, Carotenoids act as reinforcers of the *Acholeplasma laidlawii* lipid bilayer, *J. Bacteriol.* **140**:944–948.

Rottem, S., and Markowitz, O., 1979b, Membrane lipids of *Mycoplasma gallisepticum:* A disaturated phosphatidylcholine and a phosphatidylglycerol with an unusual positional distribution of fatty acids, *Biochemistry* **18**:2930–2935.

Rottem, S., and Razin, S., 1973, Membrane lipids of *Mycoplasma hominis*, *J. Bacteriol.* **113**:565–571.

Rottem, S., Muhsam-Peled, O., and Razin, S., 1973a, Acyl carrier protein in mycoplasmas, *J. Bacteriol.* **113**:586–591.

Rottem, S., Yashouv, J., Ne'eman, Z., and Razin, S., 1973b, Cholesterol in mycoplasma membranes. Composition, ultrastructure, and biological properties of membranes from *Mycoplasma mycoides* var. *capri* cells adapted to grow with low cholesterol concentrations, *Biochim. Biophys. Acta* **323**:495–508.

Rottem, S., Slutzky, G., and Bittman, R., 1978, Cholesterol distribution and movement in the *Mycoplasma gallisepticum* cell membrane, *Biochemistry* **17**:2723–2726.

Rottem, S., Shinar, D., and Bittman, R., 1981, Symmetrical distribution and rapid transbilayer movement of cholesterol in *Mycoplasma gallisepticum* membranes, *Biochim. Biophys. Acta* **649**:572–580.

Silvius, J. R., and McElhaney, R. N., 1978, Lipid compositional manipulation in *Acholeplasma laidlawii* B. Effect of exogenous fatty acid composition and cell growth when endogenous fatty acid production is inhibited, *Can. J. Biochem.* **56**:462–469.

Smith, P. F., 1971, *The Biology of Mycoplasmas*, Academic Press, New York.

Smith, P. F., 1972, Lipid composition of *Mycoplasma neurolyticum*, *J. Bacteriol.* **112**:554–558.

Smith, P. F., 1979, The composition of membrane lipids and lipopolysaccharides, in: *The Mycoplasmas*, Volume 1 (M. F. Barile and S. Razin, eds.), Academic Press, New York, pp. 231–257.

Smith, P. F., 1980, Sequence and glycosidic bond arrangement of sugars in lipopolysaccharide from *Thermoplasma acidophilum*, *Biochim. Biophys. Acta* **619**:367–373.

Smith, P. F., 1981, Structure of the oligosaccharide chain of lipoglycan from *Acholeplasma granularum*, *Biochim. Biophys. Acta* **665**:92–99.

Smith, P. F., 1983, Structural characteristics of the lipoglycan from *Acholeplasma axanthum*, *Biochim. Biophys. Acta* **752**:271–276.

Smith, P. F., 1984, Lipoglycans from mycoplasmas, *Crit. Rev. Microbiol.* **11**:157–186.

Smith, P. F., 1986, Structures of unidentified lipids in *A. laidlawii*, strain A-EF 22, *Biochim. Biophys. Acta* **879**:107–112.

Wieslander, Å., and Rilfors, L., 1977, Qualitative and quantitative variations of membrane lipid species in *Acholeplasma laidlawii* A, *Biochim. Biophys. Acta* **466**:336–346.

Wieslander, Å., Christiansson, A., Rilfors, L., and Lindblom, G., 1980, Lipid bilayer stability in membranes. Regulation of lipid composition in *Acholeplasma laidlawii* is governed by molecular shape, *Biochemistry* **19**:3650–3655.

Wieslander, Å., Rilfors, L., and Lindblom, G., 1986, Metabolic changes of membrane lipid composition in *Acholeplasma laidlawii* by hydrocarbons, alcohols, and detergents: Arguments for effects on lipid packing, *Biochemistry* **25**:7511–7517.

Chapter 3

Physical Studies of Lipid Organization and Dynamics in Mycoplasma Membranes

Ronald N. McElhaney

1. INTRODUCTION

The mycoplasmas are a diverse group of prokaryotic microorganisms that lack a cell wall. Since the mycoplasmas are genetically and morphologically the simplest organisms capable of autonomous replication, they provide useful models for the study of a number of problems in molecular and cellular biology. Mycoplasmas are particularly valuable for studies of the structure and function of cell membranes. Being nonphotosynthetic prokaryotes as well as lacking a cell wall or "outer membrane," mycoplasma cells possess only a single membrane, the limiting or plasma membrane. This membrane contains essentially all the cellular lipid and, because these cells are small, a substantial fraction of the total cellular protein as well. Due to the absence of a cell wall, substantial quantities of highly purified membranes can usually be easily prepared by gentle osmotic lysis followed by differential centrifugation, a practical advantage not offered by

Ronald N. McElhaney   Department of Biochemistry, University of Alberta, Edmonton, Alberta, Canada T6G 2H7.

Subcellular Biochemistry, Volume 20: Mycoplasma Cell Membranes, edited by Shlomo Rottem and Itzhak Kahane. Plenum Press, New York, 1993.

other prokaryotic microorganisms. For a thorough discussion of the isolation and characterization of mycoplasma membranes, the reader is referred to a previous review by Razin (1979b).

Another useful property of mycoplasmas is the ability to induce dramatic yet controlled variations in the fatty acid composition of their membrane lipids. Thus, relatively large quantities of a number of exogenous saturated, unsaturated, branched chain, or alicyclic fatty acids can be biosynthetically incorporated into the membrane phospho- and glycolipids of these organisms. In cases where *de novo* fatty acid biosynthesis is either inhibited or absent, fatty acid-homogeneous membranes (membranes whose glycerolipids contain only a single species of fatty acyl chain) can sometimes be produced. Moreover, by growing mycoplasmas in the presence or absence of various quantities of cholesterol or other sterols, the amount of these compounds present in the membrane can be dramatically altered. The ability to manipulate membrane lipid fatty acid composition and cholesterol content, and thus to alter the phase state and fluidity of the membrane lipid bilayer, makes these organisms ideal for studying the roles of lipids in biological membranes.

The unique advantages of mycoplasmas for membrane studies, especially for studies of membrane lipid organization and dynamics, have induced a large number of investigators to study these microorganisms using a wide variety of physical techniques. For this reason, we probably know more about the roles of lipids in mycoplasma membranes in general, and in the *Acholeplasma laidlawii* membrane in particular, than in any other biological membrane. The aim of the present review is to provide a comprehensive and up-to-date summary of these studies and to offer a critical analysis of their validity and significance.

This review will focus generally on the molecular aspects of mycoplasma membrane structure and particularly on physical studies of membrane lipid organization and dynamics. More detailed treatments of various mycoplasma membrane functions are presented elsewhere in this volume. Since, to the best of the author's knowledge, no detailed molecular studies of the structure of membranes from *Anaeroplasma*, *Spiroplasma*, or *Ureaplasma* species have been published, I deal here only with the membrane of *Acholeplasma laidlawii* and with the membranes of several *Mycoplasma* species. Although preliminary studies of membrane structure in several *Thermoplasma* species have been reported (see Langworthy, 1979), these members of the Archaebacteria are beyond the scope of this review. For a more comprehensive and general coverage of both *Acholeplasma* and *Mycoplasma* membrane structure and function, the reader is referred to earlier reviews by Rottem (1979) and Razin (1979a, 1982); for more detailed summaries of investigations of the *A. laidlawii* membrane, the reader is referred to earlier reviews by McElhaney (1984b, 1989).

2. PHYSICAL STUDIES OF LIPID ORIENTATION AND DYNAMICS

A wide variety of physical techniques have been applied to the *A. laidlawii* plasma membrane, particularly to investigate the organization and dynamics of the lipid bilayer found therein. The results obtained by these various techniques are summarized and critically evaluated in the following section. The more circumscribed physical studies of the membranes of various *Mycoplasma* species are also reviewed here.

2.1. Differential Scanning Calorimetry and Differential Thermal Analysis

The unique properties of the *A. laidlawii* membrane were utilized by Steim *et al.* (1969) to show for the first time that biological membranes can undergo a gel-to-liquid-crystalline lipid phase transition similar to that previously reported for lamellar phospholipid–water systems. These workers demonstrated that when whole cells or isolated membranes were analyzed by differential scanning calorimetry (DSC), two relatively broad endothermic transitions are observed on the initial heating scan (see Figure 1). The lower-temperature transition is fully reversible, varies markedly in position with changes in the chain length and degree of unsaturation of the membrane lipid fatty acyl chains, is broadened and eventually abolished by cholesterol incorporation, and exhibits a transition enthalpy characteristic of the mixed-acid synthetic phospholipids. Moreover, an

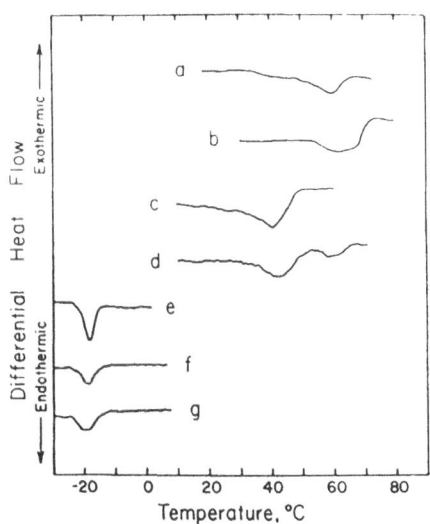

FIGURE 1. DSC scans of *A. laidlawii* lipids, membranes, and whole cells. (a) Total membrane lipids from cells grown in tryptose with added stearate; (b) membranes from stearate-supplemented tryptose; (c) total membrane lipids from cells grown in unsupplemented tryptose; (d) membranes from unsupplemented tryptose; (e) total membrane lipids from cells grown in tryptose with added oleate; (f) membranes from oleate-supplemented tryptose; (g) whole cells from oleate-supplemented tryptose. The first four preparations were suspended in water; for the latter three scans, the solvent was 50% ethylene glycol containing 0.15 M NaCl. (From Steim *et al.*, 1969.)

endothermic transition having essentially identical properties is observed for the protein-free total membrane lipid extract dispersed in excess water or aqueous buffer, indicating that the presence of membrane proteins has little effect on the thermotropic phase behavior of most of the membrane lipids. The higher-temperature transition, in contrast, is irreversible, is independent of membrane lipid fatty acid composition or cholesterol content, and is absent in total membrane lipid extracts, indicating that it is due to an irreversible thermal denaturation of the membrane proteins. A comparison of the enthalpies of transition of the lipids in the membrane and in water dispersions indicates that at least 75% of the total membrane lipids participate in this transition. Evidence was also presented that the lipids must be predominantly in the fluid state to support normal growth. These results were later confirmed and extended by Reinert and Steim (1970) and Melchior *et al.* (1970), who showed that the gel-to-liquid-crystalline lipid phase transition is a property of living cells. The former authors also demonstrated that only the enthalpy of the higher-temperature transition is reduced by protease treatment of isolated membranes and that changes in the circular dichroism spectra in the 200–300 nm range accompany this transition, confirming that it is indeed due to protein denaturation. They also presented more extensive data on the relative transition enthalpies of the lipid in the membrane and in aqueous dispersion, concluding that about 90% of the lipids participate in the gel-to-liquid-crystalline phase transition. Although this latter interpretation was challenged by Chapman and Urbina (1971), its correctness has subsequently been confirmed by a number of other studies using a variety of other techniques. These studies provided perhaps the first strong, direct experimental evidence for the hypothesis that lipids are organized as a liquid-crystalline bilayer in biological membranes, a basic feature of the currently well-accepted fluid-mosaic model of membrane structure (Singer and Nicolson, 1972).

The thermotropic phase behavior of the lipids of the *A. laidlawii* membrane was subsequently investigated by McElhaney (1974a,b) using differential thermal analysis (DTA), who demonstrated that the phase transition midpoint temperature of the gel-to-liquid-crystalline lipid phase transition could be varied from −20°C to greater than +37°C in membranes enriched (50–80%) in a variety of exogenous fatty acids (see Figure 2). Under such conditions, the lipid phase transitions observed are relatively broad, typically 20–30°C in width. Subsequent studies by Silvius and McElhaney (1978) and Silvius *et al.* (1980) have shown that about one-half of this broadness can be attributed to fatty acid heterogeneity, since fatty acid-homogeneous membranes exhibit transition ranges of only 10–15°C. The remaining broadness has been shown to be due to polar head-group heterogeneity. These workers also demonstrated that the thermotropic behavior of the individual membrane lipids varies considerably. Comparing the phospho- and glycolipids of fatty acid-homogeneous membranes, it was found that isolated phosphatidylglycerol (PG) exhibits comparatively simple

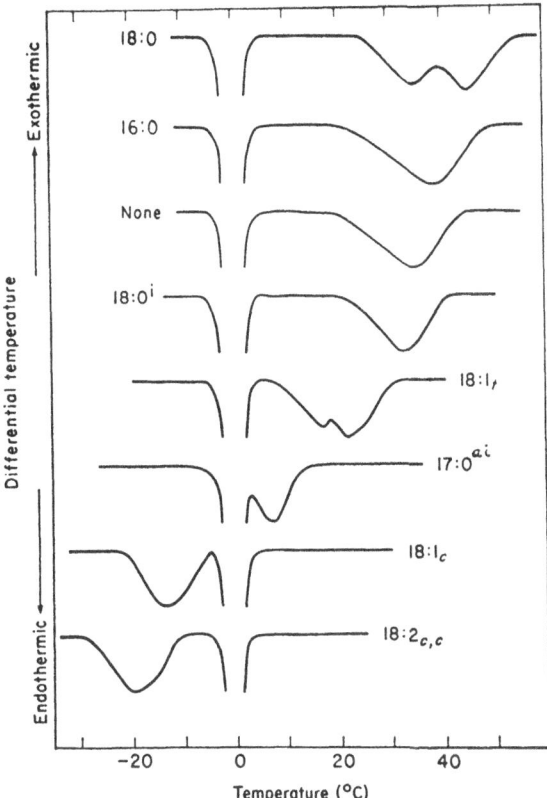

FIGURE 2. Temperature-base DTA thermograms of isolated *A. laidlawii* plasma membranes grown without fatty acid supplementation or in the presence of various exogenous fatty acids. The temperature differentials between the samples and inert reference material are plotted as a function of the temperature of the reference, using a heating rate of 5°C/min. Cooling the sample at a rate of 5°C/min results in essentially identical curves, except that the entire thermogram is shifted by 2 to 3°C to lower temperatures. The large endotherm centered around 0°C is due to melting of the ice from excess water associated with the membrane preparations. (From McElhaney, 1974b.)

In this and the following figures, fatty acids are designated by the number of carbon atoms followed by the number of double bonds, if any, present in the molecule; the subscripts *c* and *t* denote the *cis* or *trans* configuration, respectively, of these double bonds. The superscripts *i* and *ai* indicate a methyl group attached to the penultimate carbon atom (an isobranched fatty acid) and the antepenultimate carbon atom (an anteisobranched acid), respectively.

phase behavior that is nearly identical to that of the total membrane lipids dispersed in water, while the total phosphate-containing lipids [containing small amounts of *O*-amino acid esters of PG and a glycerylphosphoryl diglucosyl diacylglyceride (GPDGDG) as well as PG] exhibit a similar single endotherm but centered at a slightly lower temperature. In contrast, the isolated neutral glycolipids, monoglucosyl diacylglycerol (MGDG) and diglucosyl diacylglycerol (DGDG), show complex thermotropic behavior, with multiple endotherms all centered at higher temperatures than observed for the total membrane lipids. Interestingly, however, *mixtures* of MGDG and DGDG exhibit a much simpler behavior than do each component separately and in fact the total neutral gly-colipid fraction gives a single major endotherm (with a low-temperature shoulder) which is centered at a temperature similar to that observed for the total membrane lipids and for isolated PG. Thus, although the individual polar lipids of the *A. laidlawii* membrane can exhibit somewhat different thermotropic phase behavior, mixtures of these lipid components nevertheless appear to exhibit an appreciable amount of mutual miscibility in both the gel and liquid-crystalline states. Another interesting finding made in this study was the lack of a measur-able effect of the presence of carotenoids or Mg^{2+} on the thermotropic phase behavior of aqueous dispersions of the total membrane lipids.

These early DSC studies of the lipid thermotropic phase behavior in *A. laidlawii* membranes utilized cells whose membrane lipids were only moderately enriched in various exogenous fatty acids and low-sensitivity calorimeters were employed. The resultant broad lipid phase transitions (due primarily to fatty acid compositional heterogeneity) and the relatively poor quality of the DSC traces obtained (due to baseline instability and noise) could have obscured subtle differ-ences in lipid thermotropic phase behavior in intact cells, isolated membranes, and total membrane lipid dispersions. Indeed, Mantsch and co-workers, using Fourier transform infrared (FTIR) spectroscopy, recently reported that the gel-to-liquid-crystalline phase transition in intact cells highly enriched in saturated fatty acids occurs some 5 to 10°C below that of isolated membranes derived from them, suggesting that the organization of the lipids in the membranes of living cells differs from that of the isolated membranes (see Section 2.6). This result is in contrast to the finding of the earlier DSC studies, which showed that the lipid chain-melting transition in living cells, isolated membranes, and lipid disper-sions is essentially identical, except that in the former systems about 10% of the lipid is prevented from participating in this cooperative phase transition by their interaction with membrane proteins.

In order to resolve this apparent discrepancy in results and to confirm or refute the original DSC findings, Seguin *et al.* (1987) recently repeated the experiments of Stein *et al.* (1969) using fatty acid-homogeneous *A. laidlawii* B cells (to remove fatty acid compositional heterogeneity) and a modern, high-sensitivity calorimeter (to improve the quality of the DSC traces obtained). These workers find that the fully reversible gel-to-liquid-crystalline lipid phase transi-tions observed in elaidic acid-homogeneous cells and membranes have essen-tially identical phase transition temperatures, enthalpies, and degrees of coopera-tivity, suggesting that membrane lipid organization in these two samples is very similar or identical (see Figure 3). In contrast, the midpoint of the chain-melting

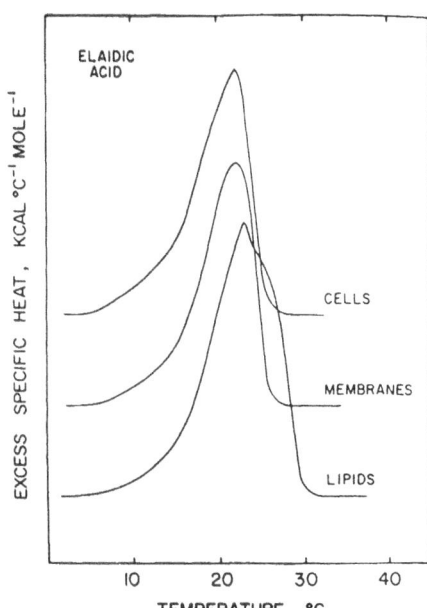

FIGURE 3. Initial DSC heating scans of *A. laidlawii* B elaidic acid-homogeneous intact cells, isolated membranes, and extracted total membrane lipids in water. The heating scan rate is 30°C/hr, at which the sample and reference are in thermal equilibrium throughout the DSC run. (From Seguin *et al.*, 1987.)

transition of the membrane lipid dispersion is shifted to a higher temperature, exhibits a greater enthalpy, and is considerably less cooperative than in cells or membranes, suggesting that native membrane lipid organization has been perturbed during extraction and resuspension of the membrane lipids in water. They also find that the thermal denaturation of the proteins in the cells and membranes has absolutely no effect on the peak temperature or cooperativity of the lipid phase transition but does increase the transition enthalpy, suggesting some decrease in the number of lipid molecules interacting with the membrane proteins. However, about 15% of the lipids do not participate in the cooperative gel-to-liquid-crystalline phase transition in both the cells and native membranes, presumably because their cooperative phase behavior is abolished by interaction with the transmembrane regions of integral membrane proteins. Alternatively, a larger proportion of the membrane lipids may interact with the membrane proteins but have their cooperative melting behavior only partially perturbed, thereby leading to the 15% reduction in the transition enthalpy observed. The fact that the gel-to-liquid-crystalline lipid phase transition in cells and membranes exhibits a similar temperature maximum and a *higher* cooperativity than does the membrane lipid dispersion favors the former interpretation. In general, the results obtained with intact cells and membranes support the earlier DSC studies which reported a nearly identical lipid thermotropic phase behavior in both systems, and not the IR spectroscopic results of Mantsch and coworkers, who reported significantly different phase behavior in these systems. We feel that this difference in results may be due at least in part to the existence of a thermal history-dependent gel-state lipid polymorphism (see below).

When similar calorimetric experiments were performed with isopalmitic acid-homogeneous *A. laidlawii* B cells, membranes, and lipids, two well-resolved endotherms are observed in all three systems (Seguin *et al.*, 1987) (see Figure 4). The properties of the lower-enthalpy lipid transition centered at 8 to 9°C are dependent on the heating scan rate and the thermal history of the sample. In particular, the apparent transition temperature increases with increasing scan rate and annealing the sample at 0°C for 24 hr before beginning the DSC run results in two to threefold increase in the observed calorimetric enthalpy. Since similar hysteresis is typically observed in the formation and interconversions of highly ordered gel phases in bilayers of synthetic phospholipids, the lower-temperature endotherm was tentatively identified as a phase transition between a more highly ordered and a less highly ordered gel state. In contrast, the properties of the higher-enthalpy transition centered at 21 to 22°C exhibit no dependence on heating scan rate or on thermal history, indicating that this is the typical gel-to-liquid-crystalline or chain-melting transition previously observed in this organism by a variety of techniques.

The structural changes associated with each of the two lipid phase transitions detected by DSC were investigated by FTIR and ^{31}P-nuclear magnetic

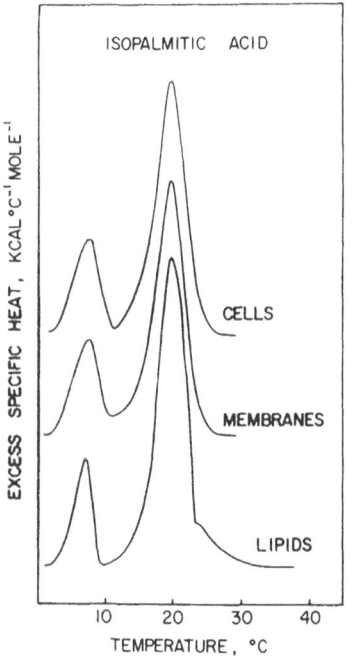

FIGURE 4. Initial DSC heating scans of *A. laidlawii* B isopalmitic acid-homogeneous intact cells, isolated membranes, and extracted total membrane lipids in water. Heating scan rate is 30°C/hr. (From Seguin *et al.*, 1987.)

resonance (NMR) spectroscopy. These spectroscopic techniques confirm that the lower-temperature endotherm is due to a transition from a highly ordered gel phase (in which the all-*trans* lipid hydrocarbon chains are very closely packed, the bilayer interfacial region is partly dehydrated, and the phospholipid polar head groups are undergoing "slow" axially asymmetric motion) to a disordered gel phase (in which the lipid hydrocarbon chains, while still largely extended, are more loosely packed, the interfacial region is fully hydrated, and the phospholipid polar head groups are undergoing fast, axially symmetric motion). These spectroscopic techniques also confirm that the higher-temperature transition corresponds to a conversion from a loosely packed gel state to the liquid-crystalline state, in which the lipid hydrocarbon chains are conformationally disordered and contain a number of *gauche* conformers. All three physical techniques indicate that at least 80% of the total membrane lipid participate in both the gel/gel and gel/liquid-crystalline phase transitions (Seguin *et al.*, 1987).

The finding that gel-phase polymorphism can exist in *A. laidlawii* B membranes is quite surprising in view of the fact that most binary mixtures of synthetic phospholipids do not exhibit multiple gel states, even when they contain identical fatty acyl chains. Thus, the ability of the *A. laidlawii* membrane lipids to form a highly ordered gel phase seems all the more remarkable, since this organism contains three major and two minor lipid classes including both phospho- and glycoglycerolipids. These results thus imply that the *A. laidlawii* B membrane lipid classes are highly miscible in all three lipid phase states detected, a result compatible with the earlier DTA study of mixtures of the individual membrane lipid classes. Moreover, gel-state polymorphism in this organism is not restricted to membranes containing a single methyl isobranched fatty acid, as *A. laidlawii* B membranes made homogeneous with members of most fatty acid classes tested, or containing two different classes of fatty acids, may also exhibit multiple gel-state phase transitions. It thus seems clear the gel-state polymorphism is not restricted to single-component lipid model membranes, but can occur in lipid bilayers and in biological membranes containing appreciable polar head-group and fatty acyl chain compositional heterogeneity as well.

The effect of the incorporation of cholesterol into the *A. laidlawii* membrane on lipid thermotropic phase behavior was studied by de Kruyff *et al.* (1972, 1973) using low-sensitivity DSC. In membranes enriched with several different fatty acids, the incorporation of cholesterol was reported to reduce the phase transition temperature slightly, to reduce the transition enthalpy markedly, and to have little effect on the cooperativity or on the temperature range of the gel-to-liquid-crystalline phase transition. Very similar results were reported when cholesterol was incorporated into vesicles made from the synthetic phospholipid 1-stearoyl-2-oleoyl-phosphatidylcholine. A puzzling aspect of these studies was the rather high transition enthalpy values reported, ranging from 9.8 ± 06 to 11.3 ± 2.4 cal/g, depending on fatty acid supplementation. Previous DSC studies of aqueous dispersions of *A. laidlawii* total membranes lipids have all yielded values of 3.8–4.0 ± 0.2 cal/g lipid.

Macdonald and Cossins (1983) examined the effect of hydrostatic pressure and of alcohols on the lipid phase transition temperature of isolated *A. laidlawii* membranes using turbidimetric, fluorescence, and DTA methods. These workers found that increases in pressure increase the phase transition temperature by 16–17°C per 1000 atm, values in broad agreement with both model lipid bilayers (17–24°C/1000 atm) and other biological membranes (18–27°C/1000 atm) (see Wong *et al.*, 1988). In contrast, the presence of 100 mM pentanol or benzyl alcohol decreases the phase transition temperature by 8.4 and 9.2°C, respectively, again in agreement with studies on pure phospholipid vesicles. These authors conclude that the lipid bilayer of the *A. laidlawii* membranes responds to pressure and alcohol incorporation in agreement with thermodynamic theory and that the presence of lipid fatty acid or polar head-group heterogeneity and membrane protein has little if any effect on the response of the weakly cooperative gel-to-liquid-crystalline phase transition to these variables.

In contrast to *A. laidlawii* membranes, there have been only a few preliminary thermal analytical studies of the membranes of *Mycoplasma* species. Rottem *et al.* (1973) studied lipid thermotropic phase behavior in isolated membranes from *M. mycoides* var. *capri* cells grown at 37°C with high and low levels of exogenous cholesterol. In membranes containing 20–25 wt% cholesterol, no cooperative lipid phase transition could be detected by low-sensitivity DSC, presumably because the high levels of cholesterol present either abolished the cooperative gel-to-liquid-crystalline phase transition entirely or broadened it beyond detectability, as occurs in model phospholipid bilayers containing large amounts of cholesterol. In membranes containing 3–4 wt% cholesterol, a reversible endothermic phase transition occurring over the range 22–29°C and centered at about 25°C was reported. The unexpectedly small range over which the lipid phase transition appeared to occur is probably due to the poor quality of the DSC trace, which makes accurate detection of the onset and completion temperatures difficult. Although no enthalpy values are reported, these authors state that the energy content of the lipid transition in cholesterol-poor *M. mycoides* var. *capri* is very low compared with that of *A. laidlawii* membranes. Rottem (1981) has also studied the thermotropic phase behavior of an aqueous dispersion of the total membrane lipid extract from *Mycoplasma arginini,* grown in the presence of high levels of exogenous cholesterol, with the same lipid extract from which the cholesterol component has been removed chromatographically. The total membrane lipid dispersion, which contains 25–30 wt% cholesterol, exhibits no detectable cooperative thermotropic phase transition by conventional DSC. In contrast, the cholesterol-free lipid extract exhibits a broad, ramplike phase transition occurring over the range 35–54°C and centered at about 45°C. This means that in the absence of high levels of incorporated cholesterol, the phospho- and gly-colipids in *M. arginini* membranes would exist almost exclusively in the gel state at the optimal growth temperature of 37°C. Since mycoplasma and bacterial cells cannot grow properly when most of their membrane lipid exists in the gel state (see McElhaney, 1984a), cholesterol is presumably required by this organism in order to fluidize the otherwise solidlike membrane glycerolipids, thus permitting a reasonable level of membrane function to occur at normal growth temperatures (see McElhaney, 1985).

2.2. Differential Scanning Dilatometry

Melchior *et al.* (1977) utilized differential scanning dilatometry to measure the change in volume of isolated *A. laidlawii* membranes as a function of temperature. These workers found a differential increase in the specific volume of aqueous membrane dispersions of about 2.1% over the temperature range 20–45°C, a temperature range corresponding to the gel-to-liquid-crystalline phase transition monitored by DSC. Since synthetic lipid bilayers typically exhibit volume increases of 3.5–4.0% at their chain-melting phase transitions, and since membrane lipids account for only about one-third by weight of the *A. laidlawii* membranes, a volume increase of this magnitude for the membrane as a whole seems reasonable, although perhaps a bit larger than expected. It is unfortunate that an aqueous dispersion of the total membrane lipids was not also examined by this technique.

2.3. X-ray Diffraction

The x-ray diffraction studies of Engelman (1970, 1971) provided direct structural confirmation of the DSC studies just discussed. In these studies the x-ray diffraction patterns from intact, isolated *A. laidlawii* membranes enriched in several different fatty acids were collected over a range of temperatures. At temperatures below the lower boundary of the calorimetrically determined gel-to-liquid-crystalline phase transition, membranes exhibit a sharp, wide-angle diffraction at 4.15 Å. This diffraction line, which arises from parallel hydrocarbon chains packed in a close hexagonal array with an axis-to-axis spacing of 4.80 Å, had previously been observed in lamellar gel phases of phospholipid–water mixtures. At temperatures within the phase transition range, the sharp diffraction at 4.15 Å gradually gives way to a diffuse reflection at 4.6 Å, which is characteristic of the more widely spaced, more disordered hydrocarbon chain packing occurring in liquid-crystalline lamellar phospholipid–water phases. At temperatures above the phase transition upper boundary, only the diffuse 4.5-Å reflection is observed. Similar changes in hydrocarbon chain packing were noted in intact

A. laidlawii cells and in the isolated total membrane lipids dispersed in water. In all cases at least 80% of the lipids participate in the transition. Furthermore, the diffracting regions giving rise to the 4.15-Å reflection are comparatively large, about 400 Å in extent, indicating that relatively large domains of well-ordered lipid exist below the transition temperature, as shown also by freeze-fracture electron microscopy. Above the phase transition temperature, the diffracting regions are much smaller, indicating a loss of long-range order.

Useful information was also obtained from the low-angle x-ray scattering patterns from randomly dispersed and partially oriented membranes. In both cases the electron density profile obtained is that characteristic of a lipid bilayer, with only a minor contribution from the membrane protein, indicating that the majority of the membrane protein is not present in a regular, repeating structural arrangement. The thickness of the lipid bilayer varies in the expected way with the chain length and structure of the lipid fatty acyl chains, in all cases decreasing substantially in the liquid-crystalline state. In palmitate-enriched membranes below the phase transition temperature, the hydrocarbon chains appear to be oriented perpendicular to the plane of the membrane, while in oleate-enriched membranes the hydrocarbon chains seem to be tilted about 20°C from perpendicular, presumably due to the presence of a "kink" produced by the *cis* double bond. The area per lipid molecule in the membrane is 40–45 Å^2 in the gel state and 60–70 Å^2 in the liquid-crystalline state, just as in the lipid–water dispersions. These results indicate that the bilayer is the predominant or exclusive lipid structure in the *A. laidlawii* membrane and that lipid–protein interactions have little effect on the average spacing and orientation of the lipid molecules.

It has been reported that *A. laidlawii* membranes, which have been solubilized by detergent and separated into their lipid and protein components, can reaggregate under appropriate conditions to form structures which appear similar to the native membranes (see Razin, 1982). J. C. Metcalfe *et al.* (1971) and S. M. Metcalfe *et al.* (1971) have studied native and reaggregated membranes by x-ray diffraction and by several other physical techniques (see later sections). These workers report that a lipid bilayer structure of similar thickness is present in both systems, and that this bilayer undergoes a gel-to-liquid-crystalline phase transition over the same temperature range in both native and reaggregated membranes. However, in the reaggregated membranes the size of the coherently reflecting regions of the lipid bilayer is smaller than in the native membranes and the broad diffraction band due to the presence of membrane protein is altered in position and intensity. These and other results led Metcalfe and co-workers to conclude that although native and reaggregated membranes are indistinguishable in terms of composition, density, electron microscopic appearance, and basic structure, the lipid bilayer is less extensive and the organization and disposition of the membrane proteins is considerably different in the reaggregated as compared with the native membranes.

2.4. NMR Spectroscopy

The technique of NMR spectroscopy has proven of great value in studies of lipid orientation and dynamics in lipid bilayers and natural membranes (for reviews, see Davis, 1983; Macdonald *et al.*, 1984a; Seelig and Macdonald, 1987; Seelig and Seelig, 1980; Smith and Jarrell, 1983). Unlike the nitroxide-containing ESR probes and those probes usually utilized in fluorescence polarization studies, most NMR probes do not significantly perturb their microenvironments. Moreover, the theoretical analyses of NMR spectra are generally more rigorous and based on sounder theoretical frameworks than is usually the case for ESR and fluorescence polarization spectroscopy. Finally, NMR spectroscopy is responsive to relatively slow molecular motions, generally in the range 10^3–10^5/sec. Motions in this domain are most likely to have direct biological relevance.

2.4.1. ^{2}H-NMR Spectroscopy of Membrane Lipids

The technique of deuterium NMR (^{2}H-NMR) has been widely used to study lipid hydrocarbon chain orientational order in model and biological membranes, particularly the membrane of *A. laidlawii*. In addition to being relatively nonperturbing, the deuterium nucleus, having a low natural abundance, can be selectively placed at various positions in the lipid molecule or in the fatty acyl chain. Furthermore, the electric quadruple moment of deuterium allows a direct measurement of the molecular order parameter, a measure of the time-averaged orientation relative to the bilayer normal, from the observed quadruple splittings. Direct measurements of the rates of motion (relaxation times) of the hydrocarbon chains of the *A. laidlawii* membrane lipids, the other component of fluidity, have not yet been made by ^{2}H-NMR. The only significant disadvantage of deuterium is its low sensitivity, which until recently required the presence of relatively high probe levels (typically 50 mole% or more) in the membrane of interest (see Davis, 1983; Seelig and Macdonald, 1987; Seelig and Seelig, 1980; Smith and Jarrell, 1983).

Oldfield *et al.* (1972) were the first to apply ^{2}H-NMR to the *A. laidlawii* membrane. These workers selectively labeled the entire membrane lipid hydrocarbon chains by growing this organism in the presence of exogenous, fully deuterated lauric or palmitic acids. The ^{2}H-NMR spectrum obtained from isolated membranes, recorded at the growth temperature of 37°C, consists of a broad, unstructured envelope of numerous overlapping resonances. Nevertheless, the spectral shape observed is qualitatively that expected from the simultaneous presence of both gel and liquid-crystalline lipid phases, in agreement with previous DSC, DTA, and x-ray diffraction studies for *A. laidlawii* membranes enriched in saturated fatty acids.

Stockton *et al.* (1975) pioneered the use of specifically deuterated fatty acids in NMR studies of biological membranes. *A. laidlawii* membranes were highly enriched by the biosynthetic incorporation of exogenous palmitic acid labeled only at the terminal methyl group with deuterium and ^{2}H-NMR spectra were recorded at a variety of temperatures. Although instrumental limitations precluded direct observation of the broad gel-phase signal present at lower temperatures, the intensity of the liquid-crystalline spectrum, which first appears at 20°C, increases with increasing temperature from 20 to 44°C, leveling off at temperatures above 44°C. At 37°C, about half the lipid appears to exist in the fluid state. Above 44°C, the orientational order decreases fairly rapidly with increasing temperature. Interestingly, the ESR order parameter, derived from the spectra of the intercalated 5-doxyl stearic acids in these same membranes, exhibits a sharp jump discontinuity at 37°C, the midpoint of the broad lipid phase transition detected by ^{2}H-NMR and other physical techniques. The ESR probe, however, is insensitive to the boundaries of the gel-to-liquid-crystalline transition.

Stockton *et al.* (1977) later extended the above study to include palmitic acid probes labeled at a variety of positions from C-2 through C-16 of the hydrocarbon chain. In this study, both the gel and liquid-crystalline spectra could be observed directly and the phase boundaries assigned in the previous study could thus be confirmed. In addition, the orientational order parameter profile for lipids existing just above the upper boundary of the lipid phase transition could be determined. It was found that a plot of order parameter versus the position of the deuteron in the chain reveals a "plateau" region of roughly constant order extending from C-2 through C-10, after which the order parameter declines progressively more rapidly toward the methyl terminus (see Figure 5), just as

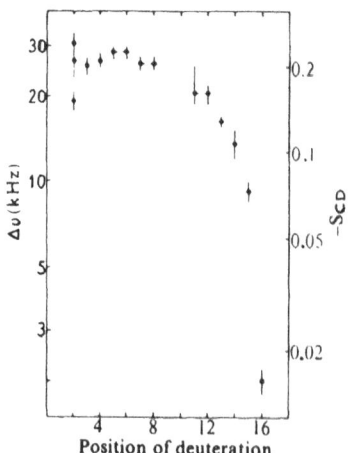

FIGURE 5. The distance between the peaks in the powder pattern, $\Delta\nu$, versus position of deuteration of the palmitate chains at 42°C. Points connected by dotted lines were obtained from samples labeled in two adjacent positions for which the two powder patterns are not resolved. Error bars are estimated from the spectra and do not include possible effects resulting from variation in membrane composition. (From Stockton *et al.*, 1977.)

observed with model membranes composed of dipalmitoylphosphatidylcholine above its phase transition temperature (see Seelig and Seelig, 1980). Using the fully deuterated palmitic acid probe, it was also shown that the effect of cholesterol on the liquid-crystalline membrane lipids is to increase orientational order, particularly in the plateau region.

The spectrum of membranes enriched in 2,2-dideuteropalmitate differed from all other probes tested in revealing an unusual line shape which appears to consist of three overlapping powder doublets, whereas all other positions produce single doublet signals (see Figure 5). These authors suggested that these multiple signals, which are also observed in C-2-labeled synthetic phospholipid liquid-crystalline bilayer systems, could be due to differences in the initial conformation of the two acyl chains, to differences in the polar lipid head groups, or to the presence of membrane protein. This first suggestion has since been confirmed by ^{2}H-NMR and by x-ray and neutron diffraction studies of model membranes, which have revealed that the fatty acyl chain at position 1 of the glycerol backbone projects directly downward toward the bilayer core, while the chain esterified at position 2 begins nearly parallel to the bilayer plane before binding to become perpendicular to the bilayer plane at the C-2 position of the hydrocarbon chain (for a review, see Davis, 1983). Three signals are observed because the C-2 chain may exist in one of two forms within the generally preferred conformation, the orientational orders of these two forms being significantly different. Subsequent studies by Rance *et al.* (1983) demonstrated that the conformations of the membrane lipids in the region of the C-2 position are qualitatively similar for all the various lipid classes and that the presence of membrane protein has little if any effect on these conformations.

The properties of the gel and liquid-crystalline lipid domains of the *A. laidlawii* membrane were subsequently studied in more detail by Smith *et al.* (1979). Membranes enriched in 13,13-dideuteropalmitic acid at 45°C, just above the calorimetrically determined upper phase transition boundary, exhibit an almost perfect powder pattern characteristic of fluid lipid with only a single quadruple splitting, just as in the case of the total membrane lipids dispersed in water. This means that the presence of the membrane protein does not perturb the average orientational order of the lipid hydrocarbon chains and that lipid molecules must be exchanging rapidly on the NMR time scale between the bulk and protein boundary lipid domains. These results are in contrast to those obtained by ESR using nitroxide fatty acid probes, where the presence of membrane protein appears to immobilize and disorder the lipid hydrocarbon chains and where the exchange between bulk and boundary lipid domains is slow on the ESR time scale (see Jost *et al.*, 1973; Marsh, 1985). Within the calorimetrically determined phase transition boundaries, separate gel and liquid-crystalline spectral components coexist, indicating that the lipids of these domains are in slow exchange ($< 10,000$ times/sec). At the lower boundary of the phase transition, the gel-state spectrum indicates a distribution of order parameters, the average being about one-half of the theoretical maximum for totally immobilized chains in the all-*trans* extended conformation. As the temperature is lowered still further, the observed quadruple splitting eventually approaches its theoretical maximum, but only at temperatures well below the phase transition lower boundary. These authors suggested that the membrane lipid hydrocarbon chains exist in the all-*trans* extended conformation as they enter the gel state but continue to undergo rapid rotational motion about their long axes, this rotational motion being slowly decreased with further decreases in temperature. This interpretation has been challenged by Pink and Zuckerman (1980), who have presented theoretical and reviewed experimental data indicating that transitions between different chain conformational states, through the formation of *gauche* bonds, do occur in gel-state lipids, although on a time scale that is rapid compared with the ^{2}H-NMR time scale. However, the early Raman spectroscopic studies of model membranes on which this argument is based, which indicate substantial numbers of *gauche* conformers in the gel state, are not correct, and recent FTIR spectroscopic studies indicate that few if any *gauche* conformers are present below the phase transition temperature (Casal and McElhaney, 1990; Mendelsohn *et al.*, 1989). Thus, the original ^{2}H-NMR interpretations appear to be correct. Both groups agree, however, that rapid rotational motion does occur in the gel state until quite low temperatures are reached.

Generally similar results were reported for a ^{2}H-NMR study of *A. laidlawii* membranes highly enriched ($\geq$ 90 mole%) in myristic acid by Jarrell *et al.* (1982), except that the length of the plateau region of the order parameter profile appeared to be shortened somewhat compared with membranes enriched in palmitic acid. In addition, although the average order parameters of liquid-crystalline hydrocarbon chains were the same in intact membranes and in an aqueous dispersion of membrane lipids, an increase in *linewidth* of about 20% was detected in the membrane, indicating that the presence of membrane proteins increases the heterogeneity of order distribution without affecting its average value. Furthermore, in the gel state a small fraction of lipid hydrocarbon chains in the membrane (but not in the lipid–water dispersion) remain disordered near their methyl terminal region, even at very low temperatures, indicating a small membrane protein disordering effect. Finally, the gel state of membranes highly enriched in myristic acid was found to be more highly ordered than previously observed for membranes less highly enriched in myristic acid, suggesting that fatty acid heterogeneity may affect lipid gel-state organization more than the presence of protein.

Rance *et al.* (1980) have also investigated the orientational order of *A. laidlawii* membranes enriched in unsaturated instead of saturated fatty acids, utilizing biosynthetically incorporated, specifically deuterated oleic acid. The orientational order of the oleoyl chain was determined at a variety of temperatures from $-50°C$ to $+41°C$. Above $10-15°C$, a single sharp quadrupolar powder pattern was observed for all C^2H_2 segments except the C-9 and C-10 positions (and of course the C-2 segment). The C-9–C-10 spectra appear to consist of two overlapping powder patterns of equal integrated intensity, indicating motional inequivalence of the C-9 and C-10 deuterons. This could only occur if the double bond is not parallel to the bilayer normal, but instead is slightly tilted. Below $10-15°C$, a second, broader spectral component, due to gel-phase lipid, begins to appear and grow at the expense of the liquid-crystalline component as the temperature is reduced. These spectra indicate that the center of the solid–fluid phase transition is about $12°C$, in good agreement with DSC studies. As noted in previous ^2H-NMR studies, some relatively rapid hydrocarbon chain rotation, at least in the terminal half of the chain, persists down to -30 to $-35°C$, well below the lower boundary of the calorimetrically detectable phase transition, and this was again ascribed to a disordering effect of the membrane protein.

A plot of orientational order parameter versus oleoyl chain position revealed a profile parameter somewhat different from that previously observed with palmitate-enriched *A. laidlawii* membranes (see Figure 6). Instead of a plateau region extending from C-2 to C-10, the oleate-enriched membranes exhibit a shortened plateau region extending only to about C-7, followed by a "dip" in the profile with a local minimum at C-10, after which the order parameters again increase before falling off toward the methyl terminus of the oleoyl chain. Although this profile would appear to indicate a markedly decreased order in the center of the oleoyl chain as compared with palmitate, in fact this behavior is due

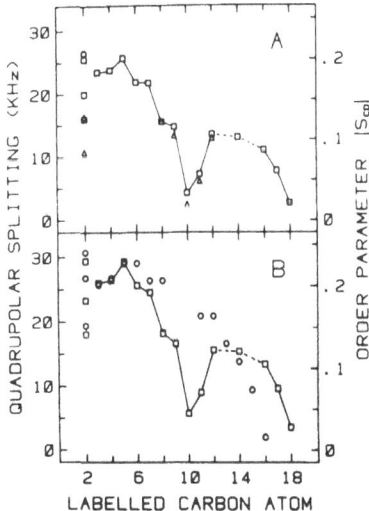

FIGURE 6. Variation of the quadrupolar splitting and deuterium order parameter $[S_{C^2H}]$ with labeled carbon atom. (A) □, oleate-labeled *A. laidlawii* membranes at 25°C; △, deuterium label attached to oleate chain of 1-palmitoyl-2-oleoyl-phosphatidylcholine (POPC) at 27°C; ○, deuterium label attached to palmitate chain of POPC at 27°C; (B) □, oleate-labeled *A. laidlawii* membranes at 0°C; ○, palmitate-labeled *A. laidlawii* membranes at 42°C. (From Rance *et al.*, 1980.)

largely to the geometry of the double bond and its titled alignment with respect to the bilayer normal. Actually, the fluctuations about the average orientation of this segment of the oleoyl chain appear to be rather similar to the corresponding region of the palmitoyl chain if compared at the same reduced temperature (i.e., at the same temperature relative to the phase transition temperature). The major effect of the presence of the double bond then seems to be to cause a local organizational perturbation in its immediate vicinity, which, although not profound, is sufficiently strong to cause the formation of predominantly fluid rather than predominantly gel-state lipid at physiological temperatures.

The temperature dependence of the order parameters of the various C^2H_2 segments of the oleoyl chains was also determined in the study summarized above. The temperature dependence in the liquid-crystalline state is approximately linear, with order decreasing with increasing temperature. In absolute terms, the change in order parameter with temperature was greatest in the plateau region (here C-2–C-6 or C-7) and smallest at the methyl end of the hydrocarbon chain. The opposite is true, however, if the change in order parameter with temperature is expressed as a percentage change in the quadrupolar splittings observed. Similar findings were also made for *A. laidlawii* membranes highly enriched in fully perdeuterated palmitic acid (Davis *et al.*, 1980).

Jarrell *et al.* (1983) studied the orientational order and dynamics of *A. laidlawii* membranes enriched with a cyclopropyl-containing fatty acid by ^2H-NMR. Specifically, dihydrosterculic acid (*cis*-9,10-methyleneoctadecanoic acid, $19:cp,c\Delta^9$), specifically deuterated at several positions along the chain, was biosynthetically incorporated into the membrane lipids of this organism. The transition from the gel to the liquid-crystalline phase was determined to occur from -15 to $0°C$, a range somewhat narrower than, but with a midpoint similar to, that found for membranes enriched in oleic acid. The acyl chains of $19:cp,c\Delta^9$-containing membranes are less mobile in the gel and in the liquid-crystalline state than those of oleic acid-containing membranes. Above $0°C$, the lipids are in the liquid-crystalline phase and give rise to powder spectra characteristic of axially symmetric motion. The overall ordering is greater everywhere than that in the case of oleoyl chains and features a maximum at the cyclopropyl moiety, in sharp contrast to the plateau found with saturated chains. Detailed analysis of the data for the cyclopropane ring indicates that the C-9–C-10 bond is inclined at 89° relative to the direction of motional averaging, in sharp contrast to the 3° estimated for oleic acid in the same membranes. These authors suggest that the replacement of a *cis*-double bond by a *cis*-cyclopropane ring in the lipid fatty acyl chains of eubacterial membranes gives rise to a less fluid lipid bilayer with generally similar but not identical physical properties. It should be noted that in all of the ^2H-NMR studies reviewed thus far, the *A. laidlawii* membranes highly enriched in palmitic acid residues behaved in all respects quite similarly to dipalmitoylphosphatidylcholine model membranes, while the *A. laidlawii* membranes enriched in palmitate and oleate were very similar in behavior to bilayers of 1-palmitoyl-2-oleoyl-phosphatidylcholine (for reviews, see Davis, 1983; Seelig and Seelig, 1980). This finding supports the suitability of simple phospholipid bilayer membranes as reasonable models for more complex biological membranes, at least as far as hydrocarbon chain orientation and dynamics are concerned. Moreover, Kang *et al.* (1981) showed that the ^{2}H-NMR spectra of freshly isolated or lyophilized membranes (enriched with specifically deuterated myristic or palmitic acids) and aqueous dispersions of the total membrane lipids are identical. Together, these findings emphasize that the presence of membrane proteins has only a small effect on the average organization of membrane lipid fatty acyl chains, at least on the NMR time scale.

The effect of the presence of cholesterol on the orientational order of palmitate-enriched, oleate-enriched, and dihydrosterculate-enriched *A. laidlawii* membranes has been studied in some detail by Davis *et al.* (1980), Rance *et al.* (1982), and Jarrell *et al.* (1983), respectively. In fully perdeuterated palmitate-containing membranes, the incorporation of relatively large amounts of cholesterol (reported as about 39 mole%) essentially abolishes a discrete gel-to-liquid-crystalline phase transition, as detected both by ^{2}H-NMR and by DSC. Between 20 and 45°C, the normal boundaries of the lipid phase transition in palmitate-enriched membranes not containing cholesterol, the cholesterol-enriched membranes exhibit an order parameter profile qualitatively similar to that normally observed for the liquid-crystalline phase in the absence of cholesterol. However, the cholesterol-containing membranes have a higher average order and an extended plateau region. In absolute terms, the increase in order is greatest in the plateau region and smallest at the methyl end of the chains, although the reverse is true if the increase in order is expressed in percentage terms. Below 20°C, the ^{2}H-NMR spectra of cholesterol-enriched membranes are multicomponent, suggestive of complex motional and/or phase behavior. In oleate-enriched and dihydrosterculate-enriched *A. laidlawii* membranes, the order parameters of the various segments of the oleoyl or dihydrosterculoyl chain increase more or less linearly with cholesterol concentration from 0 to 27 mole%. The effect of cholesterol on the liquid-crystalline order parameter profile is quite similar to that just described for palmitate-enriched membranes. Interestingly, the temperature dependence of the ^{2}H-NMR spectra, and the NMR-determined phase transition position and width characteristic of oleic acid-enriched and dihydrosterculic acid-enriched membranes lacking cholesterol, is very little affected by the presence of up to 27 mole% cholesterol. This is in contrast to the ^{2}H-NMR and calorimetrically reported behavior of palmitate-enriched membranes, and to the behavior of cholesterol in dipalmitoylphosphatidylcholine model membranes as determined by DSC and other techniques, where similar amounts of cholesterol significantly broaden the gel-to-liquid-crystalline phase transition and may also alter the phase transition midpoint temperature (for a review, see McElhaney, 1982).

Eriksson *et al.* (1991) recently studied the hydrocarbon chain orientational order and dynamics of the major glucoglycerolipids of the *A. laidlawii* A membrane using ^{2}H-NMR spectroscopy and biosynthetically incorporated perdeuterated palmitic acid. The hydrocarbon chains of the MGDG in lipid–water dispersions were found to exhibit a higher degree of orientational order than those of the DGDG but the MGDG chains appeared to undergo a larger amplitude of slow reorientational motion than do the DGDG chains, due perhaps to an increased rate of bilayer fluctuation or lateral diffusion over a curved bilayer surface. Interestingly, similar results have been reported for aqueous dispersions of synthetic phosphatidylethanolamines and phosphatidylcholines, with the former exhibiting the larger degree of hydrocarbon chain order and slow-motion fluctuation. Since both MGDG and phosphatidylethanolamines are membrane lipids with relatively small, poorly hydrated but strongly interacting polar head groups which tend to form nonlamellar phases at higher temperatures, these authors suggest the relatively high orientational order of the MGDG hydrocarbon chains and their greater amplitude of collective motions may be related to the tendency of this lipid to induce curvature and instability into the lipid-crystalline bilayer phase, due to its inverted cone shape (see Chapter 4 for further discussion). Wieslander *et al.* (1981) had previously used ^{2}H-NMR (and ^{1}H-NMR) spectroscopy to show that MGDG (or mixtures of MGDG and DGDG), isolated from *A. laidlawii* membranes enriched in oleic acid, can form reversed cubic as well as reversed hexagonal phases when dispersed in water.

Huang *et al.* (1991) recently employed ^{2}H-NMR spectroscopy to study the effect of cholesterol and lanosterol on the orientational order and dynamics of the lipid hydrocarbon chains of the *Mycoplasma capricolum* membrane. As reported previously for *A. laidlawii*, the incorporation of increasing quantities of cholesterol increases the order of the lipid hydrocarbon chain at the optimal growth temperature of 37°C and abolishes the formation of gel-state lipid at lower temperatures. As well, the presence of high levels of cholesterol abolishes the temperature dependence of the spin lattice (T_1) and the transverse (T_{2e}) relaxation times. In contrast, the incorporation of comparable amounts of lanosterol is much less effective in this regard, as reported earlier by Dahl *et al.* (1980a,b) using fluorescence polarization spectroscopy. Also as reported earlier for *A. laidlawii*, the average order and T_1 values of the hydrocarbon chains in isolated membranes and total membrane lipid dispersions are comparable at a given temperature and cholesterol level, indicating that the presence of membrane protein has little effect on the average orientation or on the rates of the fast motions of the lipid hydrocarbon chains. In contrast, the T_{2e} values of the isolated membranes are much smaller than those of the total membrane lipid dispersions, indicating that the presence of membrane proteins introduces a slow motion of the phospholipid molecules not present in their absence. Huang and co-workers also report that the growth rates of *M. capricolum* cells are positively correlated with the relatively more ordered and less dynamic state of the membrane lipid hydrocarbon chains induced by the incorporation of increasing quantities of cholesterol, results again in agreement with the earlier studies of C. E. Dahl *et al.* (1980) and J. S. Dahl *et al.* (1980).

2.4.2. ^{19}F-NMR Spectroscopy of Membrane Lipids

The technique of ^{19}F-NMR spectroscopy has recently been applied to the *A. laidlawii* membranes by Macdonald *et al.* (1983) and McDonough *et al.* (1983). In these studies small amounts of palmitic acid probes containing a single fluorine atom at various positions along the hydrocarbon chain were biosynthetically incorporated in the membrane lipids. The much greater sensitivity of the ^{19}F as compared with the ^{2}H nucleus in the NMR experiment allows usable information to be collected with only small incorporations of probe (5–10 mole%), thus allowing the study of membranes which are very highly enriched in a variety of other exogenous fatty acids. This technique also avoids the laborious and expensive synthesis of a complete series of specifically deuterated fatty acids for each exogenous fatty acid to be studied, which has been necessary in ^{2}H-NMR studies. Other potential advantages include the ability to determine order parameters in the gel state and to study the rates of motion of various segments of the hydrocarbon chains via relaxation measurements. Physical, biochemical, and biological evidence was presented that the biosynthetic incorporation of these monofluoropalmitic acid probes does not perturb the structure or function of the *A. laidlawii* membrane.

The orientational order parameter of *A. laidlawii* membranes highly enriched in pentadecanoic, methyl isopalmitic, methyl anteisopalmitic, and palmitelaidic acids were studied by ^{19}F-NMR as a function of temperature in the liquid-crystalline state (Macdonald *et al.*, 1983). In this series of fatty acids, which have nearly the same effective chain lengths, the effect of the presence of a methyl group substitution or of a *trans*-double bond on chain order could be determined. At all temperatures the *n*-saturated and methyl branched fatty acid-enriched membranes exhibit the typical plateau profile already described for palmitic acid-containing membranes, whereas the palmitelaidic acid-enriched membranes show a progressive decrease in order from the carbonyl function toward the methyl terminus without a clear-cut plateau region (see Figure 7). At the growth temperature of 37°C, the chain average order parameter value decreases in the order pentadecanoic > isopalmitic > anteisopalmitic > palmitelaidic acid-enriched membranes, which is also the order of decreasing phase transition temperatures as determined by DTA. Thus, at a constant absolute

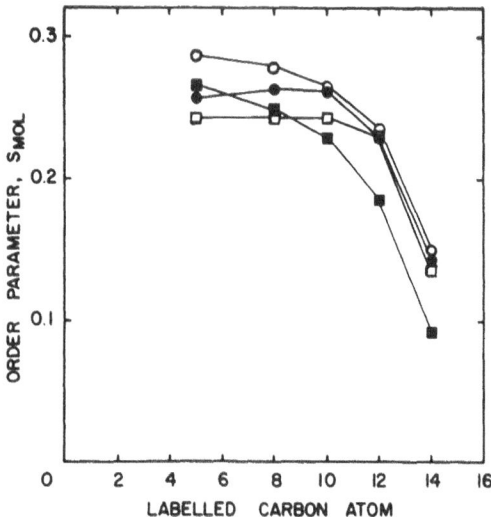

FIGURE 7. ^{19}F-NMR hydrocarbon chain order parameter profiles at 37°C of membranes of *A. laidlawii B* enriched with 15:0 (○), 16:0[i] (●), 16:0[ai] (□), and 16:1*t*Δ9 (■). (From Macdonald *et al.*, 1983.)

temperature (37°C), the introduction of a methyl isobranch, a methyl ante-isobranch, or of a *trans*-double bond into an *n*-saturated fatty acyl chain results in progressively more disorder of the liquid-crystalline bilayer hydrocarbon core. Interestingly, however, if the chain average order parameters are compared at comparable reduced temperatures (i.e., at similar temperatures relative to their phase transition temperatures), then the results are exactly opposite to those obtained at 37°C.

Experimental and theoretical results indicate that the rotation of the membrane lipid hydrocarbon chains in the gel state remains sufficiently rapid to allow for a complete motional averaging and thus axially symmetric spectra on the ^{19}F-NMR but not the ^{2}H-NMR time scale (Macdonald *et al.*, 1984a,b). This permits a detailed determination of the orientational order of the C–F bond in fluoropalmitic acid-labeled gel-state lipid in model and biological membranes, which is not possible with ^{2}H-NMR. Macdonald *et al.* (1984c, 1985a–d) thus determined the orientational order parameter profiles of *A. laidlawii* membranes highly enriched in a number of linear saturated, methyl iso- and anteisobranched, and *cis*- and *trans*-cyclopropyl fatty acids. The effect of the position of the *cis*- or *trans*-double bond within the membrane lipid hydrocarbon chain on gel-state order was determined as well (Macdonald *et al.*, 1985a,b). These workers found that at temperatures below the phase transition temperature, all types of hydrocarbon chains are relatively highly ordered but that there are much larger differences between the chain average order of different classes of fatty acids in the gel than in the liquid-crystalline state. At comparable reduced temperatures in the gel state (see Figure 8), order parameter values decrease in the order pentadecanoic > isopalmitic > palmitelaidic > anteisopalmitic > palmitoleic acid. This decreasing sequence of orientational order correlates well with the decreasing phase transition temperatures of the membrane lipids determined by DSC. Thus, as expected, membrane lipids whose hydrocarbon chains can pack in the most highly ordered array exhibit the most stable, highest melting gel states. What was not expected, however, was the finding that membrane lipids with the highest phase transition temperature from the *least* ordered liquid-crystalline states once melting occurs. This is probably because at the higher temperatures necessary to induce melting of the more stable gel phases, the increased thermal energy possessed by the fluid hydrocarbon chains produces higher rates of motion and lower orientational order than is the case for membranes having lower phase transition temperatures. Although the detailed chemical structure of the fatty acyl

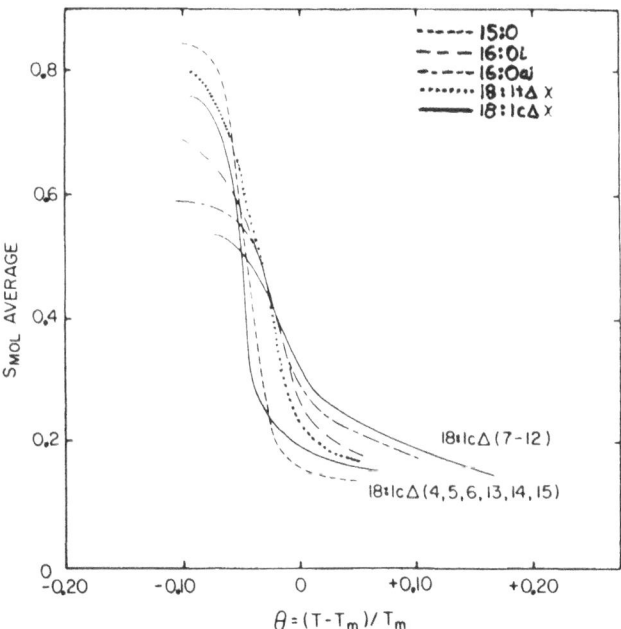

FIGURE 8. A comparison of the effect of the lipid phase transition on overall orientational hydrocarbon chain order in different classes of fatty acid structures as determined by ^{19}F-NMR. For the purpose of this comparison, isomeric *cis*-octadecenoic acids have been divided into two classes: (1) those with the site of unsaturation near the center of the chain and (2) those with the site of unsaturation near either the carbonyl head group or the methyl terminus of the acyl chain. (From Macdonald *et al.*, 1984a.)

group also has an effect on chain order, the lipid phase transition is the prime determinant of this parameter. For a more detailed summary of ^{19}F-NMR studies of the relationship between fatty acyl chain structure, orientational order, and lipid phase state in *A. laidlawii* membranes, the reader is referred to reviews by Macdonald *et al.* (1984a,c).

Macdonald *et al.* (1983) also compared the orientational order parameter profiles and the chain average order values for intact *A. laidlawii* cells, isolated membranes, and total membrane lipid dispersions having the same lipid polar head group and fatty acid compositions. They found that all three systems behave identically within the experimental error of their determinations. Thus, the presence of cytoplasmic or membrane protein does not appear to significantly perturb membrane lipid hydrocarbon chain organization in either the gel or liquid-crystalline states, a result in agreement with the ^{2}H-NMR studies reviewed earlier.

2.4.3. ^{13}C-NMR Spectroscopy of Membrane Lipids

Metcalfe *et al.* (1972) have conducted a preliminary ^{13}C-NMR spectroscopic study of *A. laidlawii* membrane biosynthetically enriched in [1-^{13}C]palmitic acid. The carboxyl carbon resonance consists of a doublet of equal intensity, presumably arising from the two conformationally nonequivalent fatty acyl chains, since a similar doublet spectrum was observed for aqueous dispersions both of the total membrane lipids and of synthetic dipalmitoylphosphatidylcholine. At temperatures below about 30–35°C, where the membrane lipid exists predominantly in the gel state, no carboxyl spectrum is observed for either the intact membranes or the membrane lipid dispersion. The intensity of this resonance progressively increases with temperature until 45–50°C, where the lipid exists entirely in the liquid-crystalline state. No change in either the position or intensity of the carboxyl resonance occurs upon exposure of the membranes to a temperature of 65°C, which thermally denatured the membrane proteins. These results thus agree with those from subsequent ^{2}H-NMR studies of the lipid hydrocarbon chains indicating that the membrane proteins have only very minor effects on lipid organization in the *A. laidlawii* membrane.

2.4.4. ^{31}P-NMR Spectroscopy of Membrane Lipids

de Kruyff *et al.* (1976) have utilized ^{31}P-NMR spectroscopy to study the behavior of the phosphate head groups in the PG and GPDGDG of the *A. laidlawii* membrane and in aqueous dispersions of the total membrane lipids. The ^{31}P-NMR spectra of both the intact membranes and aqueous membrane lipid dispersions are essentially identical. Both systems exhibit a typical "solid state" spectrum, characteristic of a phospholipid bilayer, in which the major contribution to the linewidth is made by the chemical shift anisotropy. Complete degradation of the PG by phospholipases does not change the nature but only reduces the intensity of the ^{31}P-NMR spectrum, indicating that the phosphate groups in PG and GPDGDG in native membranes have similar orientations and motions. This observation is somewhat surprising in view of the quite different chemical environments and probable locations in the bilayer of these two phosphate groups, and in fact Lindblom *et al.* (1986) report different ^{31}P-NMR chemical shift anisotropies for these two phospholipids. In both intact membranes and lipid dispersions, a broadening of the ^{31}P resonance is observed below the lipid phase transition temperature, indicating that some restriction in the motion of the phosphate groups occurs in the gel state. Interestingly, pronase digestion of up to 60% of the protein of the intact membrane, and the subsequent binding of cytochrome c to these deproteinated membranes, does not affect the ^{31}P-NMR spectrum, suggesting that no strong lipid polar head group–protein interactions occur, or at least that these lipid–protein complexes undergo fast rotation about an axis perpendicular to the membrane plane. However, these workers also report that Ca^{2+} binding either to isolated membranes or to membrane lipid dispersions also does not affect the ^{31}P-NMR spectrum. This is a surprising result, since Ca^{2+} has been reported to bind tightly to negatively charged phospholipids, even inducing the formation of solidlike phospholipid–calcium clusters in model membranes by a process of isothermal lateral phase separation (see McElhaney, 1982). However, as discussed earlier, Silvius *et al.* (1980) also report that the addition of Mg^{2+} or Ca^{2+} to *A. laidlawii* membrane lipid dispersions does not increase the phase transition temperature as measured by DTA. Perhaps these anionic lipids, as isolated, already contain bound cations. Alternatively, the neutral glycolipids present may inhibit the binding of Mg^{2+} and Ca^{2+} by the anionic lipid components.

The only other ^{31}P-NMR study of *A. laidlawii* membranes is that of Seguin *et al.* (1987) discussed earlier. In this study, the lipids of isopalmitic acid-homogenous membranes were shown to undergo two phase transitions, a lower-temperature transition from a more highly ordered to a less highly ordered gel state and a higher-temperature transition from a less highly ordered gel state to the liquid-crystalline state, by DSC and FTIR spectroscopy. Below the temperature of the first phase transition, proton-decoupled ^{31}P-NMR spectra have a basal linewidth of about 120 ppm and are nearly symmetrical (see Figure 9). Broad, powder-pattern spectra of this type are indicative of a relatively slow, axially asymmetric motion of the phosphate head group and have been observed in the so-called subgel (L_c) phases of some synthetic phospholipid bilayers. Upon warming to a temperature above the gel/gel phase transition temperature, the basal linewidth narrows to about 90 ppm and the spectra become slightly asymmetric. Spectra of

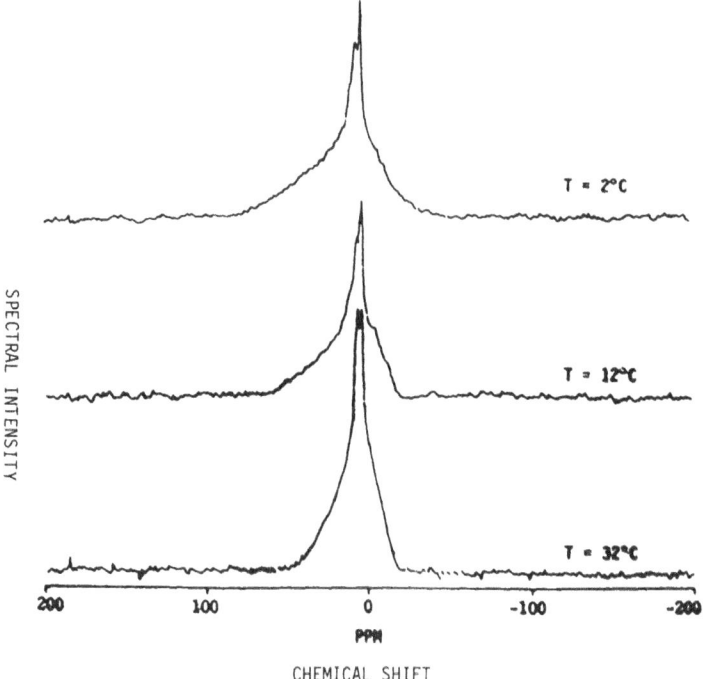

FIGURE 9. Proton-decoupled ^{31}P-NMR spectra of isolated membranes from isopalmitic acid-homogeneous *A. laidlawii* B cells collected at various temperatures. The chemical shifts (in parts per million) of the phospholipid phosphorus atoms are expressed relative to the ^{31}P-NMR signal from the phosphorus atom in an 85% solution of phosphoric acid. (From Seguin *et al.*, 1987.)

this sort are characteristic of the gel (L_β or P_β) phases of synthetic lipid bilayers. Finally, at temperatures above the gel-to-liquid-crystalline phase transition temperature, the ^{31}P-NMR spectra narrow further to about 70 ppm and become markedly asymmetric, exhibiting an upfield peak and a downfield shoulder. Spectra of this type indicate that the phosphate head group is undergoing fast, axially symmetric motion on the NMR time scale and is characteristic of synthetic phospholipid bilayers in the liquid-crystalline (L_α) phase. These workers also report that the ^{31}P-NMR spectra of isolated membranes and of vesicles generated from the total membrane lipid are almost identical. However, in this study some evidence for a separate spectral component due to the phosphate resonance from GPDGDG was presented. These observations indicate that ^{31}P-NMR spectroscopy is sensitive to both the gel/gel and gel/liquid-crystalline phase transition of the membrane lipid bilayer and that the presence of proteins has little effect on the motion and orientations of the phosphate head group in the membrane.

2.4.5. ^{1}H-NMR Spectroscopy of Membrane Lipids

J. C. Metcalfe *et al.* (1971) utilized ^{1}H-NMR spectroscopy of the phenyl protons of a benzyl alcohol probe to study native and reaggregated *A. laidlawii* membranes. These workers report that there are extensive lipid–protein interactions in both membrane systems which exclude some of the bindings sites for the probe molecules that are exposed on the detergent-solubilized and separated lipid and protein components. The mobility of the benzyl alcohol molecules in both systems increases with increasing alcohol concentration as well, implying that both membrane components become more fluid. However, the lipid-associated probe exhibits small differences and the protein-associated probe larger differences in mobility between the reaggregated and native membranes, indicating that these structures, although generally similar, are not structurally identical, supporting the x-ray diffraction results discussed earlier and the fluorescence spectroscopic results to be discussed below.

2.4.6. ^{2}H-NMR Spectroscopy of Membrane Proteins

Kinsey *et al.* (1981) have reported the first observations of amino acid side-chain dynamics in membrane *proteins* using high-field ^{2}H-NMR. Although this study concentrated on the spectrum of an individual membrane protein, bacteriorhodopsin, in the purple membrane of the bacteria *Halobacterium halobium*, some preliminary data on [γ-^{2}H$_{6}$] valine-labeled *A. laidlawii* B membranes were also presented. Similar ^{19}F-NMR studies of the environment, exposure, and mobility of membrane proteins, which have been biosynthetically labeled with specifically fluorinated amino acids, also appear to be feasible in this organism. Thus, in the future, it should be possible to observe in some detail the locations and motions of amino acid side chains in various portions of membrane proteins and to study the effects of lipid composition on protein structure and dynamics generally. Needless to say, this would be extremely valuable, since at present we have almost no data on protein structure and dynamics in the *A. laidlawii* membrane, despite a relative wealth of information on the lipids of this membrane.

In contrast to the many NMR spectroscopic studies of *A. laidlawii* membranes, the membranes of *Mycoplasma* species have not been investigated with this technique.

2.5. Electron Spin Resonance Spectroscopy

A number of groups have applied ESR spectroscopy to study the fluidity and phase state of the lipids of the *A. laidlawii* membrane. The most common ESR probes utilized in these studies were free fatty acids containing a nitroxide radical (usually the doxyl group) substituted at a particular position on the hydro-carbon chain. The fatty acid probes can be physically incorporated (intercalated) as such into isolated membranes by simply allowing them to partition into the lipid bilayer, or they can be added to the growth medium where they are bio-synthetically incorporated into the membrane polar lipids by growing cells. The latter technique is, of course, preferable, as the ESR probes are present as fatty acyl groups rather than as free acids and are presumably distributed relatively evenly among the various membrane glyco- and phospholipids. The "fluidity" of the membrane lipids is usually expressed as a motional parameter, related to the rotational correlation time of the nitroxide group, or as an apparent order parame-ter, which is related to the time-averaged orientation. However, there is some question as to whether or not this technique can effectively resolve the static orientational and dynamic motional contributions which determine membrane lipid fluidity. The ESR technique is sensitive to relatively fast motions on a time scale of 10^7–10^9/sec. Arrhenius plots of the motional or order parameters often show "breaks" (changes in slope) at a particular temperature, and these are assumed to be related to the gel-to-liquid-crystalline phase transition temperature of the membrane lipids. Another group of ESR probes, which can be used only to monitor membrane lipid phase transitions, are small, relatively nonpolar nitrox-ide molecules such as 2,2,6,6-tetramethylpiperidine-N-oxyl (TEMPO). These ESR probes will partition into liquid-crystalline lipid domains in preference to water, but are largely excluded from the gel-state regions of lipid bilayers. Since probes like TEMPO have quite different ESR spectra in lipid and in water, the relative partitioning between the membrane and its aqueous environment as a function of temperature can be determined spectroscopically, thus allowing one to monitor the progress of the entire gel-to-liquid-crystalline phase transition. For a good review of the scope and limitations of ESR spectroscopy as applied to membranes, the reader is referred to Schreier et al. (1978).

The TEMPO partitioning technique has been applied to A. laidlawii mem-branes by Metcalfe et al. (1972). Palmitic acid-enriched membranes undergo a temperature-dependent increase in TEMPO partitioning over the temperature range from 20 to about 50°C, the latter temperature being an estimate since the nitroxide group is chemically reduced at higher temperatures, leading to a loss of resonance intensity (see Figure 10). Aqueous dispersions of the total membrane lipids exhibit essentially identical behavior and moreover bind the same amount of TEMPO at all temperatures above 30°C, although isolated membranes bind about twice as much TEMPO as do the lipid dispersions at temperatures below 20°C. These workers concluded that palmitate-enriched A. laidlawii membranes undergo a gel-to-liquid-crystalline phase transition over the temperature range from 20 to about 50°C in which essentially all of the lipid participates and which is little influenced by the presence of membrane proteins. Similar results were of course obtained with the DSC, x-ray diffraction, and NMR studies already re-

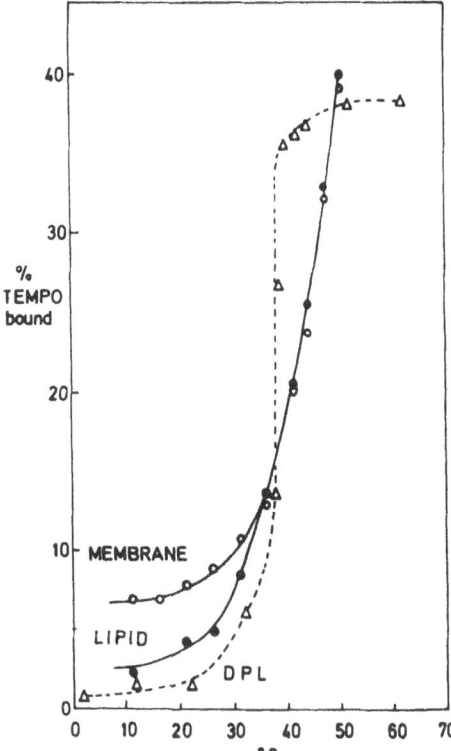

FIGURE 10. Calculated % TEMPO bound by (○) *A. laidlawii* membranes (8.9%, w/w) and (●) extracted lipid vesicles (3.1%, w/w). These preparations contain equivalent lipid concentrations (3.1%, w/w). Total TEMPO concentration 5 × 10⁻⁴ M. (△) % TEMPO bound by 3.1% (w/w) dipalmitoyl lecithin (DPL). (From Metcalfe *et al.*, 1972.)

viewed. Although very little TEMPO was found to bind to *n*-butanol-extracted membrane proteins, it could not be definitely determined whether the excess TEMPO bound to isolated membranes at low temperature was due to the presence of membrane proteins in their native state or to a somewhat more disordered gel state than exists in membrane lipid dispersions. Alternatively, some TEMPO might have preferentially partitioned into the somewhat disordered boundary lipid domain thought to surround integral membrane proteins (see Jost *et al.*, 1973; Marsch, 1985). The TEMPO partitioning technique has also been applied to *A. laidlawii* membranes from cells grown without fatty acid supplementation by Grant and McConnell (1973). Although these investigators claim that no phase transition could be detected by this technique, a careful inspection of their data indicates the presence of a broad phase transition commencing at about 20°C. Since data were collected only until about 45°C, which is the upper boundary of the calorimetrically determined gel-to-liquid-crystalline phase transition, one cannot ascertain whether or not this ESR technique would have accurately detected the completion of the gel-to-liquid-crystalline lipid phase transition in the membranes of this organism. Lindblom *et al.* (1986) successfully used the TEMPO partitioning technique to monitor the gel-to-liquid-crystalline phase transition of aqueous dispersions of the *A. laidlawii* total membrane lipids containing various proportions of palmitic and oleic acids.

The first ESR study of the *A. laidlawii* membrane in which a fatty acid spin probe was utilized was that of Tourtellotte *et al.* (1970), who used biosynthetically incorporated 12-doxyl stearic acid to probe the fluidity and phase state of membranes enriched in either stearic or oleic acids. These workers found that the doxyl group of the fatty acid probe is present in a hydrophobic, semiviscous environment in the isolated membranes, just as in aqueous dispersions of the extracted membrane lipids. However, the spin label in the isolated membranes is slightly but significantly less mobile than that in membranes generated from the protein-free lipid extract (see Figure 11). However, heat denaturation of the membrane proteins does not affect the mobility of the spin label in the membrane. These workers concluded that the mobility of at least a portion of the hydrocarbon chains in the native *A. laidlawii* membranes are slightly reduced by a *weak* association with integral membrane proteins. Although the effect of the proteins on the orientational order of the membrane lipids was not determined, studies in model systems using fatty acid spin probes have shown that the boundary lipids surrounding at least some integral membrane proteins are more orientationally disordered and motionally restricted than are the liquid-crystalline lipids not in contact with these proteins (see Jost *et al.*, 1973; Marsh, 1985).

The rotational correlation time (τ_c) of the nitroxide radical of the fatty acid

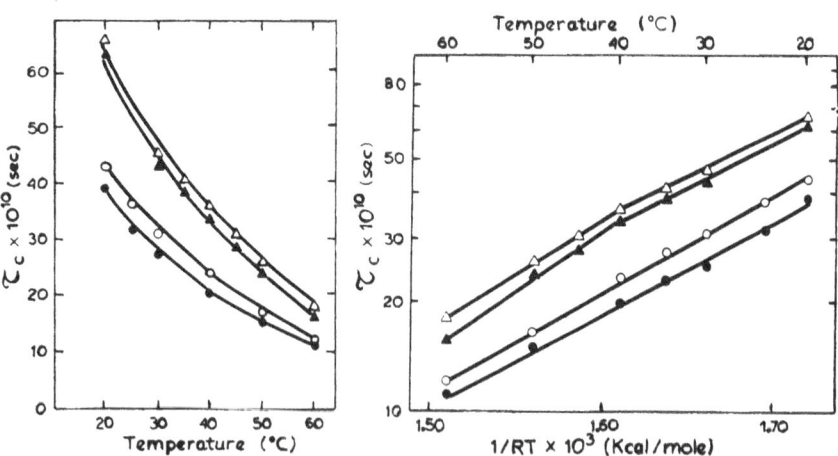

FIGURE 11. (Left) The effects of temperature on 12NS mobility in *A. laidlawii* membranes. △, stearate-enriched membranes; ▲, extracted lipids from stearate-enriched membranes; ○, oleate-enriched membranes; ●, extracted lipids from oleate-enriched membranes. (Right) Arrhenius plots of the data presented in the left panel. (From Tourtellotte *et al.*, 1970.)

spin probe was found to be a decreasing exponential function of temperature over the range 20–60°C. However, the values of the stearate-enriched membranes are nearly twice those of the oleate-enriched membranes at all temperatures, indicating that membranes containing stearic acid are less "fluid" than those containing oleate. The constant ratio of rotational correlation times with temperature is curious, since at 55°C the lipids of both membranes exist exclusively in the liquid-crystalline state, whereas at 20°C the lipids of the stearate-containing membranes are entirely solid, while those of the oleate-containing membranes remain completely fluid. These results suggest that the lipids containing the fatty acid spin probe are excluded from gel-state domains, since the mobility of solid-state lipids should be much lower than that observed. An Arrhenius plot gave a straight line for oleate-enriched membranes but yielded a slight break at 40°C with stearate-enriched membranes. The apparent activation energy of probe motion for oleate-enriched membranes was 4.1–4.2 kcal/mole, while that for stearate-enriched membranes was 3.9 kcal/mole below and 5.0 kcal/mole above the break temperature of 40°C (Tourtellotte *et al.*, 1970).

Rottem *et al.* (1970) studied the motion of a variety of intercalated spin labels in isolated *A. laidlawii* membranes enriched in several different fatty acids. In this study, spin label mobility was expressed as the hyperfine splitting $(2T_\parallel)$, a spectral motion parameter whose units are in gauss. In the isolated membranes from unsupplemented cells, $2T_\parallel$ exhibits an inverse temperature dependence, indicating that molecular motion increases with increasing temperature. In addition, at any given temperature the $2T_\parallel$ value progressively decreased as the nitroxide radical was moved away from the carbonyl function, indicating that molecular motion increases toward the methyl terminal end of the fatty acyl chain. Membranes enriched with elaidic acid exhibit a greater $2T_\parallel$ than those enriched in oleic acid at low temperatures, as expected, but the values for both membranes nearly converge at temperatures above 30°C. Interestingly, membranes from cells grown without fatty acid supplementation were found to be intermediate in mobility between elaidate- and oleate-enriched membranes at temperatures below 30°C, despite the greater proportion of gel-state lipid present in the unsupplemented membranes. However, the membranes derived from unsupplemented cells did exhibit a lower mobility than elaidate- or oleate-enriched membranes at higher temperatures. Although the variation in $2T_\parallel$ with temperature was described as "steep" by these authors, the temperature dependence of this parameter was actually much less than that reported for the rotational correlation time in the study by Tourtellotte *et al.* (1970).

Rottem and Samuni (1973) subsequently investigated the effect of proteins on the motion of several fatty acid spin probes intercalated into the *A. laidlawii* membrane. The digestion of up to 75–80% of the protein of isolated membranes results in a small but significant increase in probe mobility, as indicated by a progressive decrease in the $2T_\parallel$ value. The addition of the soluble, positively charged proteins lysozyme and cytochrome c to the pronase-treated isolated membranes partially reverses the observed increase in the mobility of the spin-labeled fatty acids, and the subsequent removal of these extrinsic proteins by extraction with 1 M NaCl restores the increased probe mobility characteristic of protein-depleted membranes. These workers concluded that membrane proteins, including those bound electrostatically to the membrane surface, restrict the mobility of the lipid fatty acyl chains in native *A. laidlawii* membranes.

Askarova *et al.* (1987) have studied the effect of the aging of *A. laidlawii* cultures on the mobility of both the lipids and proteins in isolated membranes by ESR spectroscopy. Membrane lipid fluidity was probed with the intercalated spin probe 5-nitroxyl stearic acid and membrane protein mobility by a nitroxyl group bound covalently to the exposed sulfhydryl groups of membrane proteins. These workers found that the rotational mobility of both the lipid and protein probes decreases throughout the logarithmic phase of growth but remains constant in the stationary phase. Since the presence of membrane proteins is known to decrease the rate of motion of adjacent lipid and protein molecules in model systems (see Marsh, 1985), the decrease in the lipid/protein ratio known to occur upon the aging of *A. laidlawii* cultures can probably explain both of these results. However, the saturated/unsaturated fatty acid ratio of the membrane lipids of this organism can also increase during the logarithmic phase of growth and this could have contributed to the observed decrease in membrane lipid and protein mobility with culture age.

Huang *et al.* (1974a,b) investigated the effect of changes in fatty acid composition on the mobility of intercalated 5- and 12-nitroxide stearic acid in *A. laidlawii* membranes enriched in arachidic, lauric, or oleic acids. When the relative mobilities of these membranes were compared on the basis of the $T_\parallel$ values of the 5-nitroxide probe, it was found that at the growth temperature (28 or 37°C) the relative mobilities of all membranes were nearly identical. However, at lower temperatures relative mobilities decreased in the order oleic > arachidic > lauric acid-enriched membranes, while at higher temperatures the order was either arachidic > lauric > oleic acid-enriched membranes (28°C growth temperature) or arachidic > oleic > lauric acid-enriched membranes (37°C growth temperature). These results are at variance with other experiments carried out by these workers. For example, at 22°C the relative mobilities, as determined by the values of the 12-nitroxide stearic acid probe, decrease in the order oleic > lauric > arachidic acid-enriched membranes, in contrast to the low-temperature 5-nitroxide stearate results. Moreover, osmotic fragility and glycerol permeability experiments indicate that the relative fluidities should be oleic > lauric > arachidic acid-enriched membranes *at all temperatures,* which is the same order which would be predicted on the basis of the relative phase transition temperatures as determined by DSC or DTA, for example. The ESR results of Huang and co-workers are thus somewhat suspect, as is their conclusion that *A. laidlawii* maintains a constant membrane lipid fluidity at its growth temperature in the face of marked changes in the chain length and degree of unsaturation of the hydrocarbon chains in its membrane lipids.

Butler *et al.* (1978) studied the effect of alterations in fatty acid composition and cholesterol content on the molecular order of *A. laidlawii* membranes using intercalated 5- and 12-nitroxide fatty acid probes. Plots of order parameter versus temperature for the 5-nitroxide probe reveal that palmitate- and stearate-enriched membranes have similar degrees of order within the physiological temperature range, and that the temperature dependence of the order parameter is relatively steep in both systems. In contrast to the results of Huang *et al.* (1974a,b), the order parameter of this probe in oleate-enriched membranes is less temperature-dependent and the degree of orientational order is significantly lower than is the case for membranes enriched in *n*-saturated fatty acids, particularly at lower temperatures. Plots of order parameter versus temperature are linear in both oleate- and palmitate-containing membranes, thus revealing no evidence that the 5-nitroxide probe is sensitive to the gel-to-liquid-crystalline phase transition which is known to take place in the membranes enriched with palmitate. Evidence suggestive of a phase transition was, however, obtained in stearate-enriched membranes. Both probes also report a slightly but significantly *lower* average order in aqueous dispersions of the total membrane lipids than in the isolated membranes, indicating that the presence of membrane protein produces a small *increase* in the order of these unoriented systems, in contrast to the results observed in lipid–protein model systems (see Marsh, 1985). Both probes report that the incorporation of cholesterol produces an increase in the degree of order of palmitate-enriched membranes at all temperatures, although the effect is greatest at higher temperatures. It is interesting to note that the presence of cholesterol reduces the temperature dependence of the order parameter in palmitate-containing membranes, making it quite similar to oleate-containing membranes. Since the incorporation of cholesterol is known to broaden and eventually abolish a cooperative gel-to-liquid-crystalline phase transition, it is tempting to speculate that the presence of a phase transition may sometimes manifest itself as an increased temperature dependence of ESR probe motion or order rather than as an abrupt change in that dependence at some discrete temperature.

Two groups have studied the effect of the presence of carotenoid pigment on lipid fluidity of *A. laidlawii* membranes using ESR spectroscopy. Huang and Haug (1974) utilized intercalated 12-nitroxide stearic acid to investigate the fluidity of arachidic acid-supplemented cells whose carotenoid pigment contents had been varied over 50-fold by adding either acetate or propionate to the growth medium. Acetate has been shown to stimulate carotenoid biosynthesis in this organism while propionate inhibits carotenoid production. When cells were cultured in the presence of acetate, a small decrease in membrane lipid mobility is observed by ESR spectroscopy. Membranes isolated from cells grown in acetate are also characterized by a slightly higher buoyant density, a slightly higher fragility, and a slightly lower glycerol permeability. From these results, Huang and Haug concluded that carotenoid pigments rigidify the lipids of the *A. laidlawii* membrane. A similar conclusion was later reached by Rottem and Markowitz (1979), who utilized intercalated 5- and 12-nitroxide stearic acids to study membrane lipid mobility in oleate-supplemented cells cultured with or without exogenous propionate. Both probes report much less motional restriction, especially at higher temperatures, in membranes derived from cells grown in the presence of oleate and propionate than in those cultured in oleic acid alone. Moreover, as had been observed earlier for cholesterol, the presence of higher levels of carotenoids also decreases the temperature dependence of the motional parameters monitored. In addition, the selective removal of carotenoids from the *A. laidlawii* membranes by incubation with egg phosphatidylcholine vesicles induces a restricted probe mobility in that artificial membrane system.

Rottem (1975) has used ESR spectroscopy to suggest that the fluidity of the exterior and interior regions of *A. laidlawii* (and *M. hominis*) membranes may be different. When intact cells were spin-labeled by incubation with 12-nitroxide stearic acid, the τ_c at 37°C measured from the ESR spectrum was calculated to be 4.25 nsec, whereas when isolated membranes were labeled by the same technique, the τ_c value was 4.55 nsec. When intact cells were labeled but the membranes subsequently isolated in the absence of probe, the τ_c was determined to be 4.30 nsec. If it is assumed that the 12-nitroxide stearic acid does not "flip-flop" across the lipid bilayer of the *A. laidlawii* membrane, then when intact cells are labeled the probe should be localized in the outer half of the bilayer, while when isolated membranes are labeled the probe should be incorporated into both the inner and outer halves of the bilayer. The results presented above thus suggest that a higher degree of fluidity exists in the outer as compared with the inner half of the *A. laidlawii* membranes and that the differences in the observed behavior of intact cells and isolated membranes are not due to alterations in membrane structure produced by the isolation procedure. It was further suggested that presence of a larger amount of protein in the inner region of the membrane of this organism may be responsible for the lower mobility of the probe in the inner half of the bilayer.

The validity of Rottem's study is difficult to assess. The differences between the measured rotational correlation times are quite small and may not be statistically significant. Furthermore, no evidence for the assumption that free fatty acid spin probes do not flip-flop in these membranes was presented. In fact, these probes do flip-flop rapidly in phsopholipid bilayer model membranes. Finally, it should be noted that the results obtained in this study are the opposite of what might be predicted from the lipid transbilayer distribution study of Gross and Rottem (1979), which suggested that the neutral glycolipids are located essentially exclusively in the outer monolayer of the *A. laidlawii* membrane, whereas the phospholipids are located predominantly in the inner monolayer. Since these glycolipids have higher phase transition temperatures than do the corresponding phospholipids (Silvius *et al.*, 1980) and since membranes formed from synthetic glycolipids have both higher phase transition temperatures and higher viscosities than those formed from synthetic phospholipids of similar fatty acid composition (see Iwamoto *et al.*, 1982; Mannock *et al.*, 1988, 1990; Sen *et al.*, 1990), one might expect that the outer region of the lipid bilayer of the membrane of this organism would have a lower rather than a higher "fluidity" relative to the inner region.

Burke *et al.* (1985) studied the interaction of amine local anesthetics with the *A. laidlawii* membrane by ESR spectroscopy. Spin-labeled anesthetics and intercalated 5-doxyl stearic acid were used to probe both the effect on the anesthetic molecule of binding to the membrane and the effect on the membrane lipid bilayer of the bound anesthetic. These workers found that although both cationic and neutral forms of the anesthetic partition strongly into the membrane, the positively charged, membrane-associated form is considerably more motionally restrained than is the neutral form. However, only the neutral form of the local anesthetic appears to fluidize the membrane lipid bilayer at the concentration tested. Since both cationic and neutral amine local anesthetics are thought to intercalate into lipid bilayers with the polar amine head group at the bilayer surface and the hydrophobic tail penetrating into the hydrocarbon region of the bilayer, it is not clear why only the uncharged form appears to fluidize the lipid hydrocarbon chains in the *A. laidlawii* membrane.

As stated earlier, a number of groups have utilized ESR spectroscopy and nitroxide fatty acid probes to determine apparent "phase transition temperatures" through the identification of "breaks" in the Arrhenius plots of some motional or orientational spectral parameter. The agreement between the transition temperature as determined by ESR and by other nonperturbing physical techniques such as DSC, or NMR or FTIR spectroscopy ranges from good to poor (see Table I). In particular, the ESR-determined transition temperatures, while always falling within the phase transition boundaries as determined by these other techniques, can range from well below to somewhat above the actual midpoint of the gel-to-liquid-crystalline phase transition as determined by calorimetry and by other physical techniques. Also, the ESR-determined phase transition always appears to be rather sharp, whereas in fact the actual transitions are quite broad, occurring over a range of 25–30°C. Finally, in several ESR studies *no breaks* in the Arrhenius plots of spectral motional or orientational parameters were observed in *A. laidlawii* membranes where lipid phase transitions have been detected by DTA and by other techniques. It is thus clear that ESR spectroscopic techniques utilizing nitroxide fatty acid probes are not the methods of choice for accurately characterizing lipid phase transitions in biological membranes. On the other hand, TEMPO partitioning *may* be a more suitable ESR technique for monitoring

Table I
Some Gel-to-Liquid-Crystalline Phase Transition
Temperatures Observed in *A. laidlawii* Membranes
by ESR Spectroscopy and DTA

Fatty acid supplementation	ESR-determined T_m (°C)	DTA-determined T_m (and range) (°C)
12:0	25[a]	31 (15–42)[f]
14:0	34[b]	34 (18–45)
16:0	35,[c] N.D.[d]	38 (20–50)
18:0	40[e]	41 (25–55)
20:0	25[a]	40 (20–56)
18:1$_{trans}$	22,[b] 26[c]	21 (5–32)
18:1$_{cis}$	N.D. (<0°C?)[c–e]	−13 (−22 to −4)

[a] Huang *et al.* (1974b).
[b] James and Branton (1973).
[c] Rottem *et al.* (1970).
[d] Butler *et al.* (1978). N.D., not detected above 0°C.
[e] Tourtellote *et al.* (1970).
[f] McElhaney (1974a,b) and unpublished data.

the gel-to-liquid-crystalline phase transitions of biological membranes, provided that technical limitations can be overcome. Generally speaking, however, the simplest, quickest, least expensive, and most reliable method for accurately determining the course of the chain melting transition in model or biological membranes is DSC (see McElhaney, 1982).

A few ESR studies of the membranes of several *Mycoplasma* species have also been carried out. Rottem and Verkleij (1982) used ESR spectroscopy to study the orientational order of 5- and 12-doxyl stearic acid probes intercalated into *M. gallisepticum* membranes. According to these workers, only single-component, bulk bilayer-type ESR spectra are observed despite the fact that these membranes are unique among the mycoplasmas in having a very high protein/lipid ratio (about 4:1 by weight). The apparent absence of a measurable boundary lipid component was ascribed to the fact that the intramembranous protein particles in these membranes are largely clustered even at 37°C as revealed by freeze-fracture electron microscopy. This clustering, suggested to be due to the formation of patches of gel-state disaturated PC, was in turn postulated to minimize lipid–protein interactions. Indeed, as predicted, the order parameter values and their temperature dependencies are essentially identical in 5-doxyl stearate-labeled membranes and membrane lipid dispersions. However, the order parameter values of isolated membranes are considerably lower, and their temperature dependence greater, than in membrane lipid dispersions when the 12-doxyl stearic acid is utilized as a probe. Moreover, a similar result is seen by fluorescence polarization spectroscopy using 1,6-diphenyl-1,3,5-hexatriene (DPH) as a probe. These latter results clearly indicate that in fact the membrane protein does interact with the membrane lipids, disordering the hydrophobic core of the lipid bilayer of this organism. Clearly additional experimental work is required to clarify these apparently contradictory results. Moreover, it seems unlikely that patches of gel-state lipid could form in a membrane which is reported to contain 0.9 mole of cholesterol per mole of phospholipid, since cholesterol has been shown to fluidize lipid below its characteristic phase transition temperature (see Demel and de Kruyff, 1976).

Leon and Panos (1981), in their investigation of the effect of exogenous fatty acid supplementation on the growth, osmotic fragility, and membrane lipid composition and physical properties of *M. pneumoniae,* carried out ESR studies on isolated membranes using both 5- and 12-doxyl stearic acid probes. Not surprisingly, the orientational order reported by both molecules when intercalated into the membrane decreases with increasing temperature and increases with the saturated/unsaturated ratio of the membrane lipid fatty acids. Also, at all temperatures order was higher at the 5- than at the 12-position, as expected, since a gradient of decreasing orientational order and increasing motional rates has been observed in many previous ESR studies of model and biological membranes (see Jost *et al.,* 1973; Marsh, 1985).

Rottem *et al.* (1973), using an intercalated 5-doxyl stearic probe, studied the relative fluidities of isolated membranes of *M. mycoides* var. *capri* containing high and low levels of exogenous cholesterol. In both membranes the motional rates of the nitroxide fatty acid probe increase with temperature, as expected. However, in the high-cholesterol membranes this increase follows a smooth if slightly curvilinear form, whereas in the low-cholesterol membranes a definite change in slope is noted at about 20°C, with the slope being greater at the lower temperatures. Since, as discussed earlier, a cooperative gel to liquid-crystalline lipid centered at about 25°C is detected by DSC only in the case of the cholesterol-poor membranes, the change in the temperature dependence of the ESR motional rate presumably occurs at the lower boundary of that phase transition. In the liquid-crystalline state motional rates decrease slightly even though the saturated/unsaturated fatty acid ratio of the membrane lipids is elevated, indicating that cholesterol decreases the motional freedom of the hydrocarbon chains in fluid bilayers. However, below the calorimetrically determined phase transition temperature cholesterol increases motional rates, indicating that it fluidizes gel-state bilayers. These results are in accord with those of the MNR and ESR studies of the effect of cholesterol on *A. laidlawii* membranes reviewed earlier.

2.6. Fourier Transform Infrared Spectroscopy

Casal *et al.* (1979, 1980, 1982) have utilized FTIR spectroscopy to investigate lipid organization and dynamics in *A. laidlawii* membranes. This atomic vibrational spectroscopic technique yields information on the conformation and on the rates and amplitudes of translational and rotational motion of the lipid hydrocarbon chains on a very short time scale (see Casal and Mantsch, 1984). Some information about membrane protein conformation can also be obtained using this technique (see Surewicz and Mantsch, 1988).

Casal *et al.* (1979, 1980) initially studied isolated membranes from *A. laidlawii* cells biosynthetically enriched in fully perdeuterated palmitic acid and compared their behavior with that of aqueous dispersions of the total membrane lipids. The temperature dependence of the frequency shift of the symmetric and antisymmetric CD_2 stretching modes, which provide information about the distribution of *trans* and *gauche* conformers in the hydrocarbon chain, and the half-bandwidth of the CD_2 stretching mode, which is related primarily to the rates and amplitudes of motion of the chain, were measured in both systems. The former parameter is primarily a measure of orientational order and the second parameter is primarily a measure of motion. Together they should provide a reasonable description of membrane lipid "fluidity." In both the isolated membranes and the aqueous lipid dispersions, a broad, ramplike transition is detected over the temperature range 15–40°C, as expected from earlier physical studies of palmitate-enriched membranes (see Figure 12). However, by FTIR spectroscopy this transition is shown to occur in two overlapping stages. In the lower temperature range the principal change is an increase in the rates and amplitudes of hydrocarbon chain motion without a concomitant increase in the number of *gauche* conformers, which are essentially absent in the gel state. In the upper portion of the phase transition only a small additional increase in chain motion is observed, but a large increase in the *gauche/trans* conformer ratio occurs. Thus, the broad lipid phase transition in the membranes and in the isolated lipids appeared to proceed in two stages, the first consisting primarily of a reduction in the packing density and rigidity of the lipid matrix and the second consisting primarily of the actual melting of the hydrocarbon chains. This broad, two-stage phase transition may in fact actually be due to partially overlapping gel/gel and gel/liquid-crystalline lipid phase transitions, as discussed earlier (see Seguin *et al.*, 1987).

A comparison of the isolated membrane with the membrane lipid dispersion reveals that the presence of membrane protein had only a minor effect on the lipid phase transition. The membrane proteins produce a slight decrease in the rate of acyl chain motion in the liquid-crystalline state, which could be interpreted as a decrease in membrane lipid "fluidity." On the other hand, the membrane proteins also increase the population of *gauche* conformers in both the gel and liquid-crystalline lipid phases, which could be interpreted as a decrease in orientational order and thus as an increase in "fluidity." It thus appears that lipid hydrocarbon chain mobility and orientational order do not always vary in parallel, and one must carefully define the term "membrane lipid fluidity" if confusion is to be avoided. Incidentally, the protein amide band patterns reveal that the membrane proteins are predominantly in the α-helical conformation with some random coil structure also being present, but little or no β-structure is observed.

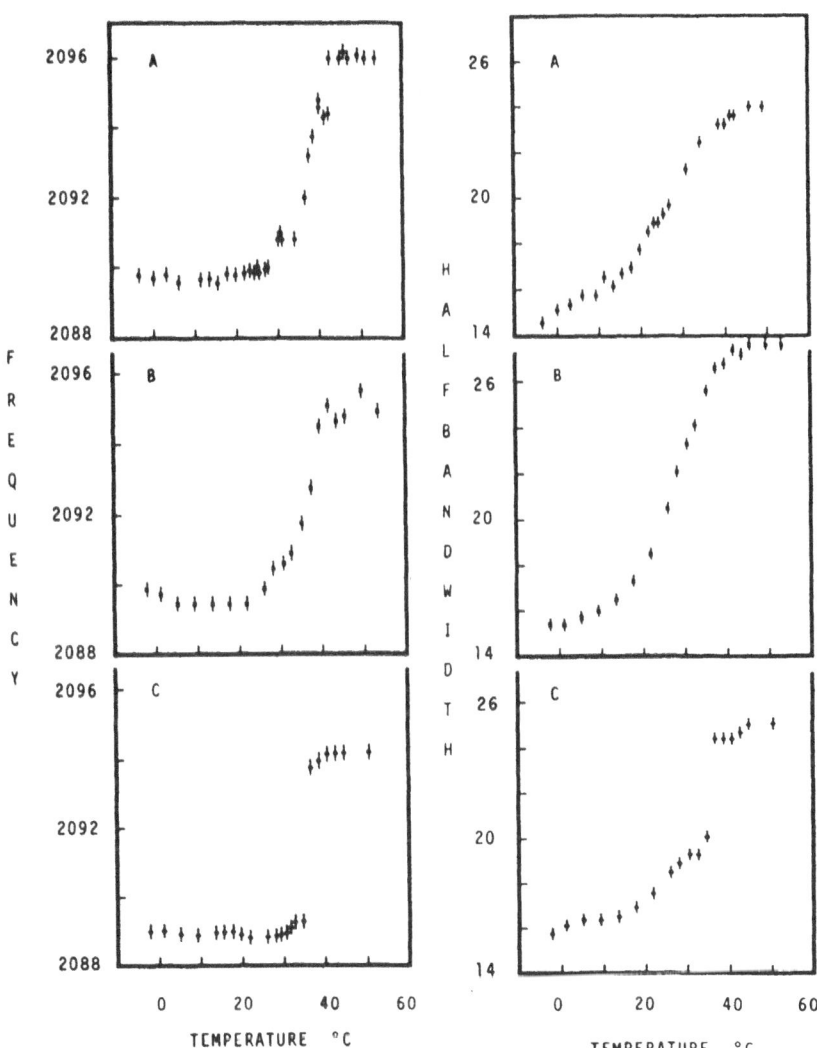

FIGURE 12. (Left) Temperature dependence of the frequency (in cm^{-1}) of the maximum of the CD$_2$ symmetric stretching vibration of the perdeuteropalmitoyl chains in (A) intact *A. laidlawii* plasma membranes, (B) deproteinated plasma membranes, and (C) DPPC-d_{62} model membranes. (Right) Temperature dependence of the half-bandwidths (in cm^{-1}) of the symmetric CD$_2$ stretching vibration of the perdeuteropalmitoyl chains in (A) intact plasma membranes, (B) deproteinated plasma membranes, and (C) DPPC-d_{62} model membranes. (From Cameron *et al.*, 1983.)

The results of a later FTIR spectroscopic study by Cameron *et al.* (1985) of live *A. laidlawii* cells do not appear to be completely in accord with the studies of Casal and co-workers just described. In particular, this later study found that the onset of the formation of *gauche* conformers in the membrane lipid hydrocarbon chains with heating preceded the increase in the motional rates and amplitudes of these chains, a result opposite to that reported by Casal *et al.* (1980). Although in principle this difference could be due to intrinsic differences between the isolated membranes used in the former study and the live cells used in the later study, it is difficult to rationalize this qualitative difference on these grounds alone. In my view the results reported by Cameron *et al.* appear to be less reasonable, since it is difficult to see how even a partial chain melting could be accomplished without an increase in the motion of the lipid hydrocarbon chains.

Casal *et al.* (1982) also studied the phase transition behavior in *A. laidlawii* membranes made nearly homogeneous in fully perdeuterated pentadecanoic acid. Except for the expected increase in the sharpness of the gel-to-liquid-crystalline lipid phase transition, the results obtained were generally similar to those already reported for the palmitate-enriched membrane and membrane lipid dispersion. However, in contrast to the earlier study, the mobility as well as the orientational disorder were both reported to be slightly greater in the isolated membranes than in the membrane lipid dispersions above the phase transition temperature, whereas the presence of membrane proteins appeared to decrease the rate and amplitude of hydrocarbon chain motion in the liquid-crystalline phase in their earlier study. The reason for this difference in results is not clear.

Cameron *et al.* (1983, 1985) have investigated both "live" *A. laidlawii* cells and isolated membranes by FTIR spectroscopy. Although in general the lipid phase transition behavior of these two systems is similar, the gel-to-liquid-crystalline phase transition, and in particular the rise in the *gauche/trans* conformer ratio, occurs at a lower temperature in the live cells than in the isolated membranes, so that at a comparable temperature within the phase transition boundaries the proportion of fluid lipid is always higher in the live cells than in isolated membranes (see Figure 13). Thus, it appears that the process of membrane isolation may alter somewhat the properties of the lipid phase transition. However, the comparative DSC studies of the thermotropic phase behavior of viable *A. laidlawii* cells and of isolated membranes discussed earlier did not reveal any detectable differences between the two systems. Whether this discrepancy in results is due to the different physical techniques utilized in these studies or to different methods of membrane isolation remains to be determined.

Mantsch *et al.* (1988) recently reinvestigated the differences in the gel-to-liquid-crystalline lipid phase transition in intact cells and isolated membranes of *A. laidlawii* by FTIR spectroscopy, using a wider variety of exogenous fatty acids and growth conditions than in the previous studies from this laboratory. Although these workers confirm that the midpoint temperature of the lipid chain-

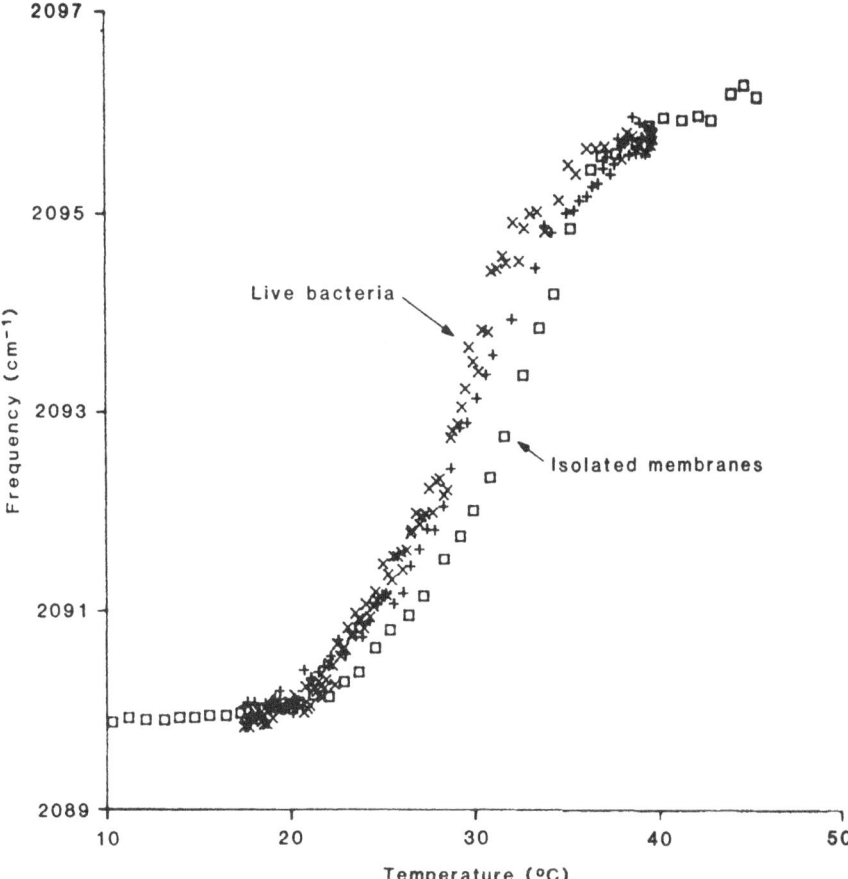

FIGURE 13. Temperature dependence of the frequency of the CD_2 symmetric stretching band of the lipids of *A. laidlawii* B grown at 30°C on perdeuteromyristic acid in the presence of avidin. Shown are frequencies from spectra of live cells with the temperature ascending from 20 to 39°C (+) and descending from 39 to 16°C (X) and frequencies from spectra of isolated membranes with the temperature ascending from 5 to 45°C (□). (From Cameron *et al.*, 1983.)

melting phase transition as monitored by this technique is indeed always lower in cells than in membranes, the difference in midpoint temperature was found to decrease as the chain length of the incorporated fatty acid decreases or as the degree of unsaturation increases. The difference in phase transition temperature midpoints between the two systems also seems to depend on the method of isolation and treatment of the membranes. Membranes prepared by the mechanical fracture of cells in a French press show a smaller elevation of the midpoint phase transition temperature than those prepared by osmotic lysis. Moreover, aqueous dispersions of the total membrane lipids exhibit an even more pronounced increase in transition midpoint temperature than do either of the isolated membrane preparations.

Some insight into the molecular basis of this phenomenon may be provided by the data presented by this group on the effect of temperature on the lipid gel-to-liquid-crystalline phase transition in (initially at least) intact cells. Simply storing cells at −20°C for 1 week was itself found to increase the phase transition midpoint temperature slightly. Also, incubation of cells at 60°C for 1 or 18 hr results in progressively larger elevations in transition temperature. The former result is compatible with the suggestion of Seguin *et al.* (1987) based on their DSC studies, that low-temperature incubation may elevate the phase transition temperature by inducing the formation of higher-melting, subgel-like lipid domains in *A. laidlawii* cells but especially in the membranes and isolated lipid dispersions, thus producing differences in their phase transition temperatures. The latter result, and the comparative studies of the lipid thermotropic phase behavior of the total membrane lipids to membranes and cells, suggest that the presence of membrane protein, especially in its undenatured form, tends to lower the phase transition midpoint temperature. This interpretation is partially compatible with the DSC results of Seguin *et al.* (1987) in that the phase transition temperature of aqueous dispersions of the total membrane lipids was found to be significantly higher than in cells or membranes. However, the results for intact cells and isolated membranes were identical in the DSC study.

Mantsch *et al.* (1988) also studied membrane protein conformation and thermal denaturation in intact cells and isolated membranes by FTIR spectroscopy, utilizing the conformationally sensitive amide-I and amide-II bands of the polypeptide chain (see Surewicz and Mantsch, 1988). At the optimal growth temperature of 37°C, both membrane and total cellular proteins were found to be predominantly in the α-helical conformation but with a substantial amount of β-sheet and random coil structures also being present. The finding of a relatively high α-helical content and some random structure in the *A. laidlawii* membrane proteins was also made in an earlier study from this laboratory (Casal *et al.*, 1980); however, that earlier study indicated little or no β-structure to be present. Only small changes in protein conformation were noted over the physiological temperature range, including that over which the membrane lipid gel-to-liquid-crystalline phase transition occurred. However, for both cells and membranes major changes in protein conformation occur over the temperature range 40 to > 75°C, presumably corresponding to the thermal denaturation of these proteins detected in the earlier DSC studies already discussed.

The accessibility of membrane proteins to water was also probed in this study by replacing H_2O with D_2O and monitoring the exchange of NH to ND by the change in frequency of the amide-I and amide-II bands. At room temperature no more than 25% of the amide protons could be exchanged, indicating that most peptide bonds are buried in hydrophobic domains, either in the globular portions of proteins on the bilayer surface or in transmembrane α-helices. At temperatures between 40 and 50°C, NH to ND exchange begins to increase markedly but only above 90°C does the NH band become negligible. These results confirm that the thermal denaturation of the membrane proteins does indeed involve an unfolding of the polypeptide chain and an exposure of hydrophobic regions of that chain to water (see Surewicz and Mantsch, 1988).

2.7. Fluorescence Spectroscopy

Several different fluorescence spectroscopic techniques have been applied to *A. laidlawii* cells or membranes. S. M. Metcalfe *et al.* (1971) carried out a study of native and detergent-reaggregated membranes using the probe 1-anilinonaphthalene-8-sulfonic acid (ANS), whose fluorescence intensity varies with the polarity of its environment. In both membrane systems the fluorescence of the lipid-associated ANS molecules is similar, whereas the fluorescence of the protein-associated ANS molecules differs significantly. These workers concluded that the lipid bilayer is a major structural feature in both native and detergent-reaggregated membranes, that the organization of the membrane proteins differs in each system, and that lipid bilayer organization is relatively insensitive to the presence or to the conformational state of the membrane proteins. However, as the exact locations of the probe molecules were not determined and because ANS fluorescence can vary with factors other than polarity, little quantitative molecular information can be derived from this study.

Haberer *et al.* (1982) studied the effect of MVL3 virus adsorption and capping on membrane structure in intact *A. laidlawii* cells using the excimer fluorescence technique. In this method small amounts of the probe pyrene lecithin are intercalated into the cell membrane and the fluorescence intensities of monomers and dimers, which fluoresce at different wavelengths, are measured. From this ratio the association constant of probe molecules can be determined, and, since the association constant is dependent on the lateral mobility of the pyrene lecithin molecules in the plane of the membrane, it is also a measure of the fluidity of the membrane lipid bilayer. These workers observe a decrease in dimer formation upon virus association, which could result either from a rigidification of the hydrophobic core of the lipid bilayer or from an increase in the area of the membrane lipid in which the probe is free to diffuse. Because DPH fluorescence polarization spectroscopic measurements indicate no change in membrane lipid bilayer viscosity upon virus binding, they favor the latter interpretation. Their view is that the capping or aggregation of some of the membrane proteins induced by virus binding, which was observed by freeze-fracture electron microscopy, decreases the amount of lipid interacting with the membrane protein, and that this in turn permits additional numbers of lipid molecules to interact with the fluorescent probe, in effect decreasing dimer formation by an effective decrease in probe concentration in the bulk lipid phase. This result implies that the MVL3 virus binds to proteins on the surface of the cell membrane rather than inserting into the lipid bilayer.

Finally, as discussed earlier, Macdonald and Cossins (1983) employed DPH fluorescence polarization spectroscopy, in conjunction with optical transmission and DSC, to demonstrate that pressure increases and alcohol decreases the gel-to-liquid-crystalline phase transition temperature of the lipids of isolated *A. laidlawii* membranes.

The techniques of DPH fluorescence polarization spectroscopy (see Shinitzky and Barenholz, 1978) has been applied in a number of studies of the role of cholesterol or other sterols in the membranes of several *Mycoplasma* species. The first application of this technique was that by Rottem *et al.* (1973), who investigated lipid bilayer organization in isolated membranes of *M. mycoides* var. *capri* containing high or low levels of cholesterol. In cholesterol-enriched membranes, Arrhenius plots of lipid microviscosity (actually primarily lipid orientational order rather than motional rates) are linear over the temperature range 10–45°C with an apparent activation energy of 4.5 kcal/mole (see Figure 14). In contrast, in cholesterol-poor membranes the Arrhenius plots are biphasic linear with a break at about 24°C, very near the calorimetrically determined lipid phase transition midpoint. Above 24°C, the apparent activation energy for probe movement is 3.7 kcal/mole and below 24°C, it is 10.5 kcal/mole. Lipid orientational order in the cholesterol-poor membranes is higher at lower temperatures and lower at higher temperatures as compared with cholesterol-rich membranes as well. These authors conclude that the incorporation of high levels of cholesterol into the *M. mycoides* var. *capri* membranes abolishes a cooperative gel-to-liquid-crystalline phase transition by increasing order in the liquid-crystalline phase and decreasing it in the gel phase of the lipid bilayer, thus producing a state of intermediate fluidity. This interpretation is in accord with that of a number of other studies of the role of cholesterol in model and biological membranes (see Demel and de Kruyff, 1976).

C. E. Dahl *et al.* (1980) utilized DPH fluorescence polarization spectroscopy to study the effect of alkyl-substituted precursors of cholesterol on the organization of model membranes and on the growth and membrane organization of *M. capircolum*. In both the model and *M. capricolum* membranes, the effectiveness of various sterols in ordering the lipid bilayer above its phase transition temperatures increases in the order lanosterol < 4,4-dimethyl cholestanol ≤ 4β-methyl cholestanol < 4α-methyl cholestanol < cholestanol < cholesterol. Significantly, this is also the order of the effectiveness of these sterols in promoting the growth of this sterol auxotropic organism and the order in which these compounds are formed in the cholesterol biosynthetic pathway. These authors conclude that the sequential biosynthetic removal of the three methyl groups from the lanosterol

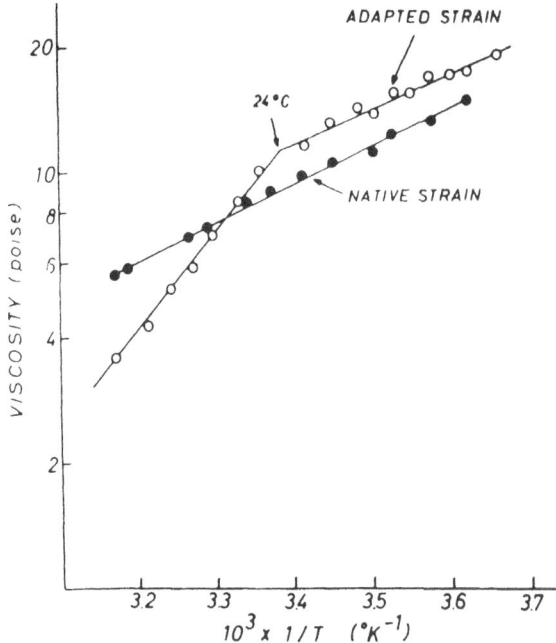

FIGURE 14. The effect of temperature on the microviscosity of membrane lipids of the native and adapted *M. mycoides* var. *capri* strains as determined by fluorescence measurements of diphenylhexatriene incorporated into the membranes. The growth medium was supplemented with palmitic and oleic acids (5 μg/ml of each) and with 10 μg cholesterol/ml for the native strain and 0.12 μg cholesterol/mg for the adapted strain. (From Rottem *et al.*, 1973.)

ring nucleus, and the introduction of a double bond into ring B, serve to improve the membrane function of the sterol molecule. The removal of the 14-methyl group, and to a lesser extent the 4β-methyl group, are particularly important in this regard, as their presence decreases the planarity of the α and β faces of the sterol ring nucleus and increases the bulk of the sterol molecule, which in turn disrupts the interactions between the sterol ring nucleus and the fatty acyl chains of adjacent membrane glycerolipid molecules.

J. S. Dahl *et al.* (1980) subsequently studied the growth characteristics and membrane lipid organization of *M. capricolum* cultured on cholesterol and lanosterol, again using DPH fluorescence polarization spectroscopy. They found that progressive increases in the cholesterol content, but not in the lanosterol content, of the membrane results in a progressive increase in microviscosity (orientational order) when measured at 37°C, the optimal growth temperature of this organism. Moreover, in cholesterol-rich membranes Arrhenius plots of microviscosity are linear whereas those from lanosterol-rich membranes exhibit pronounced discontinuities in slope at 20 and 25°C. Moreover, the micro-viscosities are considerably higher in cholesterol- than in lanosterol-containing membranes, especially at higher temperatures. In fact, the microviscosity values of lanosterol-enriched membranes and cholesterol-poor membranes are quite similar. These workers conclude that only cholesterol is able to completely abolish the cooperative gel-to-liquid-crystalline lipid phase transition in *M. capricolum* membranes and to order the fluid lipid bilayer at the growth temperature.

In this study, Dahl *et al.* (1980) also found that the provision of exogenous cholesterol allowed *M. capricolum* cells to grow on media containing a wider variety of exogenous fatty acids than did lanosterol, a finding compatible with the greater effectiveness of cholesterol in regulating membrane lipid fluidity. An unexpected finding, however, was that low levels of cholesterol, which on their own are unable to support cell growth, are able to support good growth when high levels of lanosterol are also present, despite the fact that the microviscosity values of the membrane lipid remain low in these circumstances. These workers suggest that lanosterol may fulfill one of the more "primitive" functions of sterols in membranes, namely to separate polar lipid head groups and decrease surface charge density, even though it cannot effectively regulate the fluidity of the lipid hydrocarbon chains. Alternatively, small amounts of cholesterol were suggested to serve other, more specialized cellular functions (Dahl and Dahl, 1983, 1984; Dahl *et al.*, 1983), a suggestion later confirmed by subsequent studies from this laboratory.

Lala *et al.* (1979) investigated the role of the free sterol hydroxyl group of cholesterol on cell growth and membrane lipid organization of *M. capricolum*, again utilizing DPH fluorescence polarization spectroscopy. They found that cholesterol methyl ether or cholesterol acetate are relatively effective at supporting the growth of this organism, although not as effective as cholesterol itself. However, the microviscosity values at 25°C of all three membranes are essentially identical, indicating that a free hydroxyl group at C-3 of the cholesterol molecule is not required for ordering the lipid bilayer in the membrane of this organism. In model membrane systems, however, it has been reported that the free hydroxyl group is required to produce the maximum reduction in phospholipid cross-sectional area and passive permeability associated with cholesterol incorporation (see Demel and de Kruyff, 1976).

Le Grimellec *et al.* (1981) studied the effect of variations in membrane lipid fatty acid composition and cholesterol content on lipid orientational order in *M. gallisepticum*, using DPH fluorescence polarization. In isolated membranes containing high amounts of cholesterol, increasing the chain length of the exogenous saturated fatty acid present in the growth medium increases the apparent "lipid" order parameter almost imperceptibly, while replacing stearic acid with oleic acid decreases order slightly. In all cases orientational order decreases almost linearly with temperature with no abrupt changes in slope (see Figure 15). In cholesterol-

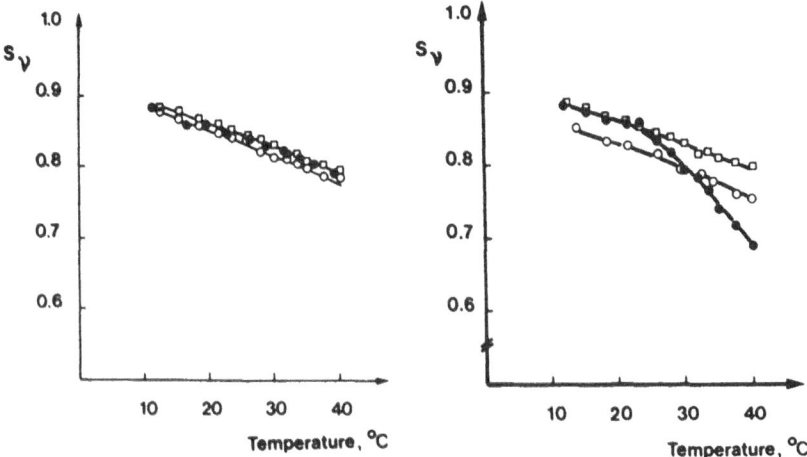

FIGURE 15. (A) Temperature dependence of the lipid order parameter S_v in native strain A5969 of *M. mycoides* subsp. *capri* grown in the presence of the following saturated fatty acids: $C_{14:0}$ (absorbance at 640 nm, 0.26) (○); $C_{16:0}$ (absorbance at 640 nm, 0.20) (●); and $C_{18:0}$ (absorbance at 640 nm, 0.18) (□). (B) Effects of a *cis*-double bond and reduced amounts of cholesterol on the lipid order parameter in membranes enriched with C_{18} acids. Native cells were grown in the presence of 10 μg cholesterol/ml and either 10 μg stearate/ml (□) (absorbance at 640 nm, 0.28) or 10 μg oleate/ml (○) (absorbance at 640 nm, 0.24). Adapted cells (●) were grown with 0.9 μg cholesterol/ml and 10 μg oleate/ml (absorbance at 640 nm, 0.14). (From Le Grimellec *et al.*, 1981.)

poor membranes, changes in fatty acid composition produce larger changes in lipid orientational order and plots of orientational order versus temperature exhibit abrupt changes in slope at temperatures of 28, 25, and 23°C for cells enriched in palmitate, palmitate and oleate, or oleate, respectively (see Figure 16). Although these changes in slopes were interpreted as membrane lipid gel-to-liquid-crystalline phase transitions, it seems highly unlikely that cells containing relatively high levels of oleic acid would melt above 0°C. At low temperatures, low-cholesterol membranes exhibit a higher lipid ordering and at high temperatures, a lower degree of lipid ordering than high-cholesterol membranes. These results confirm earlier findings in other organisms that cholesterol modulates the effect of lipid fatty acid composition variations on membrane fluidity by disordering gel and ordering liquid-crystalline lipid domains so as to produce a state of intermediate fluidity (see Demel and de Kruyff, 1976). Interestingly, the lipid order parameter values obtained for the *M. gallisepticum* membranes are about twice as high as those obtained for *M. mycoides* var. *capri* membranes grown under similar conditions (Rottem *et al.*, 1973). This higher degree of ordering of the *DPH probe* may be due to the presence of larger amounts of protein in the *M.*

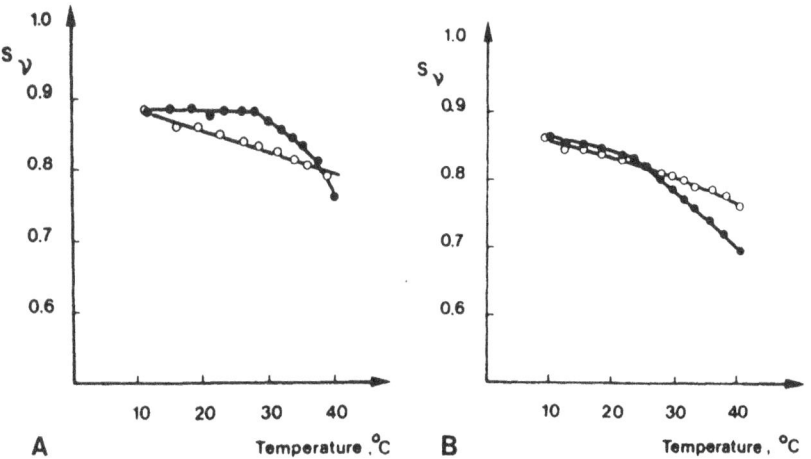

FIGURE 16. (A) Effect of low-cholesterol adaptation on palmitate-enriched cells. ○, native cells (absorbance at 640 nm, 0.20); ●, cells grown in the presence of 0.9 μg exogenous cholesterol/ml (absorbance at 640 nm, 0.16). (B) Effect of low-cholesterol adaptation on cells enriched with oleate and palmitate. Values were obtained for membrane preparations derived from cells harvested at identical absorbance values (absorbance at 640 nm, 0.15). ○, native cells; ●, cells grown in the presence of 0.9 μg exogenous cholesterol/ml. (From Le Grimellec *et al.*, 1981.)

gallisepticum membrane, since the presence of integral, transmembrane proteins has been shown to decrease the degree of wobble (increase the degree of order) of the rigid DPH molecule in model and other biological membranes (see Shinitzky and Barenholz, 1978). Rottem and Verkleij (1982), however, report that the apparent microviscosity (actually orientational order) of *M. gallisepticum* membranes is actually slightly lower than in aqueous dispersions of the membrane lipids over the physiological temperature range, a finding at odds with DPH fluorescence polarization studies in other membrane systems. Clearly, additional experimental work is required here to resolve these apparently contradictory results. However, since other spectroscopic techniques indicate that integral, transmembrane proteins *disorder* the flexible hydrocarbon chains of adjacent membrane lipids in model membrane systems, the results of Le Grimellec *et al.* (1981), even if valid, do not necessarily mean that the *lipid molecules* themselves are more ordered in *M. gallisepticum* membranes.

Rottem (1981) utilized the DPH fluorescence polarization technique to compare lipid microviscosity (orientational order) in dispersions of the total membrane lipids from *M. arginini*, both with their normal high levels of cholesterol and when the cholesterol had been removed chromatographically. When cholesterol was present, Arrhenius plots of microviscosity are linear over the physiological temperature range and lipid microviscosity is relatively higher. When cholesterol was absent, similar Arrhenius plots show changes in slope at 35 and 40°C and lipid microviscosity values are lower, particularly at higher temperatures. The breaks in the Arrhenius plot of the cholesterol-free lipid dispersion correspond to the lower region of the gel-to-liquid-crystalline phase transition detected by DSC only in the cholesterol-free dispersions. These results support the earlier findings that high levels of cholesterol abolish the cooperative chain-melting transition of the membrane lipids and increase order in the liquid-crystalline state. However, in this study cholesterol appears to increase order in the *gel* state as well, a result at odds with previous work on mycoplasma and model membranes (see Demel and de Kruyff, 1976, and previous sections).

Le Grimellec *et al.* (1982) utilized fluorescence polarization and ESR spectroscopy to study the effects of energization on membrane organization in *M. mycoides* var. *capri* and *M. capricolum* and their results were discussed earlier. Briefly, these workers found that cellular energization resulting from the addition of glucose to resting cells did not alter membrane lipid order or dynamics in a significant way, although it did affect the conformation of at least some membrane proteins.

2.8.   Other Physical Techniques

Abramson and Pisetsky (1972) utilized light-scattering measurements to study the phase behavior of isolated *A. laidlawii* membranes and of aqueous dispersions of the total membrane lipids. As the temperature increases, the turbidity values of palmitate- and stearate-enriched membranes and lipid dispersions exhibit a decrease, which correlates reasonably well with the gel-to-liquid-crystalline phase transition detected by calorimetry and other techniques. Oleate-enriched membranes, which do not undergo a phase transition within the physiological temperature range, exhibit a relatively constant degree of light scattering over the temperature range 10–45°C. These authors suggested that this technique may be a convenient and relatively sensitive method to measure phase transitions in biological membranes. Indeed, Macdonald and Cossins (1983) utilized this technique (along with DSC and DPH fluorescence polarization) in their studies of the effects of pressure and alcohols on the lipid phase transition in *A. laidlawii* membranes (discussed earlier). A major disadvantage of thermal-turbidimetric studies, however, is that they provide no direct thermodynamic or structural information about the phenomenon which is monitored.

3.   SUMMARY AND CONCLUSIONS

It should be clear from this summary that we currently know a great deal about the organization and dynamics of the lipids in mycoplasma membranes in general, and in the cell membrane of A. *laidlawii* in particular. In fact, research on mycoplasma membranes has been important in unambiguously establishing the fundamental lipid bilayer structure of all biological membranes and in elucidating some of the major properties of bilayers in biomembranes, such as their thermotropic phase behavior and interactions with cholesterol and membrane proteins. Although a great deal has been learned, a number of issues have not been fully resolved. In particular, the concept of membrane lipid fluidity must be refined and quantitated, and the relationship between orientational order and rates of motion better understood. This will require that the apparent discrepancies between some of the results obtained, for example, by the various spectroscopic techniques, be resolved. In particular, the nature of the boundary lipid surrounding integral membrane proteins will require further study, as will the question of the specificity of lipid–protein interactions. Also, accurate quantitative measurements for the lateral and rotational mobilities of the various lipid components in the mycoplasma membranes have not yet been made. Although not reviewed in this chapter, the related questions of the *in vivo* rate of phospholipid, glycolipid, and cholesterol transverse diffusion (flip-flop), and the possible asymmetric transbilayer distribution of these components in mycoplasma membranes, are still not well understood. Although much remains to be done, particularly with respect to our understanding of protein structure and function in mycoplasma membranes, a solid basis for further advances has now been laid. The many natural advantages of mycoplasma for biochemical and biophysical investigations of membrane structure and function should continue to make these organisms very useful for membrane studies for years to come.

4. REFERENCES

Abramson, M. B., and Pietsky, D., 1972. Thermal-turbidimetric studies of membranes from *Acholeplasma laidlawii*, *Biochim. Biophys. Acta* **282**:80–84.

Askarova, E. A., Kapitanov, A. B., Kol'tover, V. K., and Tatishchev, O. S., 1987, Generation of superoxide radicals and the fluidity of the membrane lipids of *Acholeplasma laidlawii* on aging of the cell culture, *Biophysics* **32**:99–104.

Burke, P. V., Kanki, R., and Wang, H. H., 1985, Effect of positively charged local anesthetics on a membrane-bound phosphatase in *Acholeplasma laidlawii*, *Biochem. Pharmacol.* **34**:1917–1924.

Butler, K. W., Johnson, K. G., and Smith, I. C. P., 1978, *Acholeplasma laidlawii* membranes: An electron spin resonance study of the influence on molecular order of fatty acid composition and cholesterol, *Arch. Biochem. Biophys.* **191**:289–297.

Cameron, D. G., Martin, A., and Mantsch, H. H., 1983, Membrane isolation alters the gel to liquid-crystalline transition of *Acholeplasma laidlawii*, *Science* **219**::180–182.

Cameron, D. G., Martin, A., Moffat, D. J., and Mantsch, H. H., 1985, Infrared spectroscopic study of the gel to liquid-crystal phase transition in live *Acholeplasma laidlawii* cells, *Biochemistry* **24**:4355–4359.

Casal, H. L., and McElhaney, R. N., 1990, Quantitative determination of hydrocarbon chain conformational order in bilayers of saturated phosphatidylcholines of various chain lengths by Fourier transform infrared spectroscopy, *Biochemistry* **29**:5423–5427.

Casal, H. L., and Mantsch, H. H., 1984, Polymorphic phase behavior of phospholipid membranes studied by infrared spectroscopy, *Biochim. Biophys. Acta* **779**:381–401.

Casal, H. L., Smith, I. C. P., Cameron, D. G., and Mantsch, H. H., 1979, Lipid reorganization in biological membranes: A study by Fourier transform infrared difference spectroscopy, *Biochim. Biophys. Acta* **550**:145–149.

Casal, H. L., Cameron, D. G., Smith, I. C. P., and Mantsch, H. H., 1980, *Acholeplasma laidlawii* membranes: A Fourier transform infrared study of the influence of protein on lipid organization and dynamics, *Biochemistry* **19**:444–451.

Casal, H. L., Cameron, D. G., Jarrell, H. C., Smith, I. C. P., and Mantsch, H. H., 1982, Lipid phase transitions in fatty acid-homogeneous membranes of *Acholeplasma laidlawii* B, *Chem. Phys. Lipids* **30**:17–26.

Chapman, D., and Urbina, J., 1971, Phase transitions and bilayer structure of *Mycoplasma laidlawii* B, *FEBS Lett.* **12**:169–172.

Dahl, C. E., Dahl, J. S., and Bloch, K., 1980, Effect of alkyl-substituted precursors of cholesterol on artificial and natural membranes and on the viability of *Mycoplasma capricolum*, *Biochemistry* **19**:1462–1467.

Dahl, C. E., Dahl, J. S., and Bloch, K., 1983, Proteolipid formation in *Mycoplasma capricolum*, *J. Biol. Chem.* **258**:11814–11818.

Dahl, J. S., and Dahl, C. E., 1983, Coordinate regulation of unsaturated phospholipid, RNA, and protein synthesis in *Mycoplasma capricolum* by cholesterol, *Proc. Natl. Acad. Sci. USA* **80**:692–696.

Dahl, J. S., and Dahl, C. E., 1984, Effect of cholesterol on phospholipid, RNA, and protein synthesis in *Mycoplasma capricolum*, *Isr. J. Med. Sci.* **20**:807–811.

Dahl, J. S., Dahl, C. E., and Bloch, K., 1980, Sterols in membranes: Growth characteristics and membrane properties of *Mycoplasma capricolum* cultured on cholesterol and lanosterol, *Biochemistry* **19**:1467–1472.

Davis, J. A., 1983, The description of membrane lipid conformation, order and dynamics by ^{2}H-NMR, *Biochim. Biophys. Acta* **737**:117–171.

Davis, J. H., Bloom, M., Butler, K. W., and Smith, I. C. P., 1980, The temperature dependence of molecular order and the influence of cholesterol in *Acholeplasma laidlawii* membranes, *Biochim. Biophys. Acta* **597**:477–491.

de Kruyff, B., Demel, R. A., and van Deenen, L. L. M., 1972, The effect of cholesterol and epicholesterol incorporation on the permeability and on the phase transition of intact *Acholeplasma laidlawii* cell membranes and derived liposomes, *Biochim. Biophys. Acta* **255**:331–347.

de Kruyff, B., van Dijck, P. W. M., Goldback, R. W., Demel, R. A., and van Deenen, L. L. M., 1973, Influence of fatty acid and sterol composition on the lipid phase transition and activity of membrane-bound enzymes in *Acholeplasma laidlawii*, *Biochim. Biophys. Acta* **330**:269–282.

de Kruyff, B., Cullis, P. R., Radda, G. K., and Richards, R. E., 1976, Phosphorus nuclear magnetic resonance of *Acholeplasma laidlawii* cell membranes and derived liposomes, *Biochim. Biophys. Acta* **419**:411–424.

Demel, R. A., and de Kruyff, B., 1976, The function of sterols in membranes, *Biochim. Biophys. Acta* **457**:109–132.

Engelman, D. M., 1970, X-ray diffraction studies of phase transitions in the membrane of *Mycoplasma laidlawii*, *J. Mol. Biol.* **47**:115–117.

Engelman, D. M., 1971, Lipid bilayer structure in the membrane of *Mycoplasma laidlawii*, *J. Mol. Biol.* **58**:153–165.

Eriksson, P.-O., Rilfors, L., Wieslander, A., Lundberg, A., and Lindblom, G., 1991, Order and dynamics in mixtures of membrane glucolipids from *Acholeplasma laidlawii* studied by ^{2}H-NMR, *Biochemistry* **30**:4916–4924.

Grant, C. W. M., and McConnell, H. M., 1973, Fusion of phospholipid vesicles with viable *Acholeplasma laidlawii*, *Proc. Natl. Acad. Sci. USA* **70**:1238–1240.

Gross, Z., and Rottem, S., 1979, Lipid distribution in *Acholeplasma laidlawii* membrane. A study using lactoperoxidase, *Biochim. Biophys. Acta* **555**:547–552.

Haberer, K., Pfisterer, M., and Galla, H.-J., 1982, Virus capping on mycoplasma cells and its effect on membrane structure, *Biochim. Biophys. Acta* **688**:720–726.

Huang, L., and Haug, A., 1974, Regulation of membrane lipid fluidity in *Acholeplasma laidlawii*: Effect of carotenoid pigment content, *Biochim. Biophys. Acta* **352**:361–370.

Huang, L., Jaquet, D. D., and Huag, A., 1974a, Effect of fatty acyl chain length on some structural and functional parameters of *Acholeplasma* membranes, *Can. J. Biochem.* **52**:483–490.

Huang, L., Lorch, S. K., Smith, G. G., and Haug, A., 1974b, Control of membrane lipid fluidity in *Acholeplasma laidlawii*, *FEBS Lett.* **43**:1–5.

Huang, T., DeSiervo, A. J., and Yang, Q.-X., 1991, Effect of cholesterol and lanosterol on the structure and dynamics of the cell membrane of *Mycoplasma capricolum*. Deuterium nuclear magnetic resonance study, *Biophys. J.* **59**:691–702.

Iwamoto, K., Sunamoto, J., Inoue, K., Endo, T., and Nojima, S., 1982, Liposomal membranes. IV. Importance of surface structure in liposomal membranes of glyceroglycolipids, *Biochim. Biophys. Acta* **691**:44–51.

James, R., and Branton, D., 1973, Lipid- and temperature-dependent structural changes in *Acholeplasma laidlawii* cell membranes, *Biochim. Biophys. Acta* **323**:378–390.

Jarrell, H. C., Butler, K. W., Byrd, A., Deslauriers, R., Ekiel, I., and Smith, I. C. P., 1982, A ^{2}H-NMR study of *Acholeplasma laidlawii* membranes highly enriched in myristic acid, *Biochim. Biophys. Acta* **688**:622–636.

Jarrell, H. C., Tulloch, A. P., and Smith, I. C. P., 1983, Relative roles of cyclo-propane-containing and *cis*-unsaturated fatty acids in determining membrane properties of *Acholeplasma laidlawii*: A deuterium nuclear magnetic resonance study, *Biochemistry* **22**:5611–5619.

Jost, P. C., Griffith, O. H., Capaldi, R. A., and Vanderkooi, G., 1973, Evidence for boundary lipids in membranes, *Proc. Natl. Acad. Sci. USA* **70**:480–486.

Kang, S.-Y., Kinsey, R. A., Rajan, S., Gutowsky, H. S., Gabridge, M. G., and Oldfield, E., 1981, Protein–lipid interactions in biological and model membrane systems. Deuterium NMR of *Acholeplasma laidlawii* B, *Escherichia coli*, and cytochrome oxidase systems containing specifically deuterated lipids, *J. Biol. Chem.* **256**:1155–1159.

Kinsey, R. A., Kintanar, A., Tsai, M.-D., Smith, R. L., Janes, N., and Oldfield, E., 1981, First observation of amino acid side chain dynamics in membrane proteins using high field deuterium nuclear magnetic resonance spectroscopy, *J. Biol. Chem.* **256**:4146–4149.

Lala, A. K., Buttke, T. M., and Bloch, K., 1979, On the role of the sterol hydroxyl group in membranes, *J. Biol. Chem.* **254**:10582–10585.

Langworthy, T. A., 1979, Special features of thermoplasmas, in: *The Mycoplasmas* (M. F. Barile and S. Razin, eds.), Academic Press, New York, pp. 495–513.

Le Grimellec, C., Cardinal, J., Giocondi, M.-C., and Carriere, S., 1981, Control of membrane lipids in *Mycoplasma gallisepticum*: Effect on lipid order, *J. Bacteriol.* **146**:155–162.

Le Grimellec, C., Lajeunesse, D., and Rigaud, J.-L., 1982, Effects of energization on membrane organization in mycoplasma, *Biochim. Biophys. Acta* **687**:281–290.

Leon, O., and Panos, C., 1981, Long-chain fatty acid perturbations in *Mycoplasma pneumoniae*, *J. Bacteriol.* **146**:1124–1134.

Lindblom, G., Brentel, I., Sjolund, M., Wikander, G., and Wieslander, A., 1986, Phase equilibria of membrane lipids from *Acholeplasma laidlawii:* Importance of a single lipid forming non-lamellar phases, *Biochemistry* **25:**7502–7510.

Macdonald, A. G., and Cossins, A. R., 1983, Effects of pressure and pentanol on the phase transition in the membrane of *Acholeplasma laidlawii* B, *Biochim. Biophys. Acta* **730:**239–244.

Macdonald, P. M., McDonough, B., Sykes, B. D., and McElhaney, R. N., 1983, ¹⁹F-nuclear magnetic resonance studies of lipid fatty acyl chain order and dynamics in *Acholeplasma laidlawii* B membranes. The effects of methyl-branch substitution and of *trans*-unsaturation upon membrane acyl chain orientational order, *Biochemistry* **22:**5103–5111.

Macdonald, P. M., Sykes, B. D., and McElhaney, R. N., 1984a, Fatty acyl chain structure, orientational order, and the lipid phase transition in *Acholeplasma laidlawii* B membranes. A review of recent ¹⁹F nuclear magnetic resonance studies, *Can. J. Biochem. Cell Biol.* **62:**1134–1150.

Macdonald, P. M., Sykes, B. D., and McElhaney, R. N., 1984b, ¹⁹F-nuclear magnetic resonance studies of lipid fatty acyl chain order and dynamics in *Acholeplasma laidlawii* B membranes. ¹⁹F-NMR line shape and orientational order in the gel state, *Biochemistry* **23:**4496–4502.

Macdonald, P. M., Sykes, B. D., and McElhaney, R. N., 1984c, Calorimetric and spectroscopic studies of lipid hydrocarbon chain order in the *Acholeplasma laidlawii* membrane, *Isr. J. Med. Sci.* **20:**803–806.

Macdonald, P. M., Sykes, B. D., McElhaney, R. N., and Gunstone, F. D., 1985a, ¹⁹F nuclear magnetic resonance studies of lipid fatty acyl chain order and dynamics in *Acholeplasma laidlawii* B membranes. Orientational order in the presence of a series of positional isomers of *cis*-octadecenoic acid, *Biochemistry* **24:**177–184.

Macdonald, P. M., Sykes, B. D., and McElhaney, R. N., 1985b, ¹⁹F nuclear magnetic resonance studies of lipid fatty acyl chain order and dynamics in *Acholeplasma laidlawii* B membranes. Orientational order in the presence of positional isomers of *trans*-octadecenoic acid, *Biochemistry* **24:**2237–2245.

Macdonald, P. M., Sykes, B. D., and McElhaney, R. N., 1985c, Fluorine-19 nuclear magnetic resonance studies of lipid fatty acyl chain order and dynamics in *Acholeplasma laidlawii* B membranes. Gel-state disorder in the presence of methyl iso- and anteisobranched chain substituents, *Biochemistry* **24:**2412–2419.

Macdonald, P. M., Sykes, B. D., and McElhaney, R. N., 1985d, Fluorine-19 nuclear magnetic resonance studies of lipid fatty acyl chain order and dynamics in *Acholeplasma laidlawii* B membranes. A direct comparison of the effects of *cis* and *trans* cyclopropane ring and double-bond substituents on orientational order, *Biochemistry* **24:**4651–4659.

McDonough, B., Macdonald, P. M., Sykes, B. D., and McElhaney, R. N., 1983, Fluorine-19 nuclear magnetic resonance studies of lipid fatty acyl chain order and dynamics in *Acholeplasma laidlawii* B membranes. A physical, biochemical and biological evaluation of monofluoropalmitic acids as membrane probes, *Biochemistry* **22:**5097–5103.

McElhaney, R. N., 1974a, The effect of membrane lipid phase transitions on membrane structure and on the growth of *Acholeplasma laidlawii* B, *J. Supramol. Struct.* **2:**617–628.

McElhancy, R. N., 1974b, The effect of alterations in the physical state of the membrane lipids on the ability of *Acholeplasma laidlawii* B to grow at various temperatures, *J. Mol. Biol.* **84:**145–157.

McElhaney, R. N., 1982, The use of differential scanning calorimetry and differential thermal analysis in studies of model and biological membranes, *Chem. Phys. Lipids* **30:**229–259.

McElhaney, R. N., 1984a, The relationship between membrane lipid fluidity and phase state and the ability of bacteria and mycoplasmas to grow and survive at various temperatures, in: *Biomembranes,* Volume 12 (M. Kates and L. Masen, eds.), Academic Press, New York, pp. 249–278.

McElhaney, R. N., 1984b, The structure and function of the *Acholeplasma laidlawii* plasma membrane, *Biochim. Biophys. Acta* **779:**1–42.

McElhaney, R. N., 1985, Membrane lipid fluidity, phase state and membrane function in prokaryotic microorganisms, in: *Membrane Fluidity in Biology,* Volume 4 (R. A. Aloia and J. M. Boggs, eds.), Academic Press, New York.

McElhaney, R. N., 1989, The influence of membrane lipid composition and physical properties on membrane structure and function in *Acholeplasma laidlawii, CRC Crit. Rev. Microbiol.* **17:**1–32.

Mannock, D. A., Lewis, R. N. A. H., Sen, A., and McElhaney, R.N., 1988, The physical properties of glycosyldiacylglycerols. Calorimetric studies of a homologous series of 1,2-di-O-acyl-3-O-(β-D-glucopyranosyl)-sn-glycerols, *Biochemistry* **27:**6852–6859.

Mannock, D. A., Lewis, R. N. A. H., and McElhaney, R. N., 1990, The physical properties of glycosyl diacylglycerols. I. Calorimetric studies of a homologous series of 1,2-di-O-acyl-3-O-(α-D-glucopyranosyl)-sn-glycerols, *Biochemistry* **29:**7790–7799.

Mantsch, H. H., Yang, P. W., Martin, A., and Cameron, D. G., 1988, Infrared spectroscopic studies of *Acholeplasma laidlawii* B membranes. Comparison of the gel to liquid-crystal phase transition in intact cells and isolated membranes, *Eur. J. Biochem.* **178:**335–341.

Marsh, D., 1985, ESR spin label studies of lipid–protein interactions, in: *Progress in Protein–Lipid Interactions,* Volume 1 (A. Watts and J. J. H. H. M. de Pont, eds.), Elsevier, Amsterdam, pp. 143–172.

Melchior, D. L., Morowitz, H. J., Sturtevant, J. M., and Tsong, T. Y., 1970, Characterization of the plasma membrane of *Mycoplasma laidlawii.* VII. Phase transitions of membrane lipids, *Biochim. Biophys. Acta* **219:**114–122.

Melchior, D. L., Scavitto, F. J., Walsh, M. T., and Steim, J. M., 1977, Thermal techniques in biomembrane and lipoprotein research, *Thermochim. Acta* **18:**43–71.

Mendelsohn, R., Davies, M. A., Brauner, J. W., Schuster, H. F., and Dluhy, R. A., 1989, Quantitative determination of conformational disorder in the acyl chains of phospholipid bilayers by infrared spectroscopy, *Biochemistry* **28:**8934–8939.

Metcalfe, J. C., Metcalfe, S. M., and Engelman, D. M., 1971, Structural comparisons of native and reaggregated membranes from *Mycoplasma laidlawii* and erythrocytes by X-ray diffraction and nuclear magnetic resonance techniques, *Biochim. Biophys. Acta* **241:**412–421.

Metcalfe, J. C., Birdsall, N. J. M., and Lee, A. G., 1972, ^{13}C NMR spectra of *Acholeplasma* membranes containing ^{13}C labelled phospholipids, *FEBS Lett.* **21:**335–340.

Metcalfe, S. M., Metcalfe, J. C., and Engelman, D. M., 1971, Structural comparisons of native and reaggregated membranes from *Mycoplasma laidlawii* and erythrocytes using a fluorescence probe, *Biochim. Biophys. Acta* **241:**422–430.

Oldfield, E., Chapman, D., and Derbyshire, W., 1972, Lipid mobility in *Acholeplasma* membranes using deuteron magnetic resonance, *Chem. Phys. Lipids* **9:**69–81.

Pink, D. A., and Zuckerman, M. J., 1980, Lipid chain order in *Acholeplasma laidlawii* membranes. What does ^{2}H-NMR tell us? *FEBS Lett.* **109:**5–8.

Rance, M., Jeffrey, K. R., Tulloch, A. P., Butler, K. W., and Smith, I. C. P., 1980, Orientational order of unsaturated lipids in the membranes of *Acholeplasma laidlawii* as observed by ^{2}H-NMR, *Biochim. Biophys. Acta* **600:**245–262.

Rance, M., Jeffrey, K. R., Tulloch, A. P., Butler, K. W., and Smith, I. C. P., 1982, Effects of cholesterol on the conformational order of unsaturated lipids in the membranes of *Acholeplasma laidlawii, Biochim. Biophys. Acta* **688:**191–200.

Rance, M., Smith, I. C. P., and Jarrell, H.C., 1983, The effect of headgroup class on the conformation of membrane lipids in *Acholeplasma laidlawii:* A ^{2}H-NMR study, *Chem. Phys. Lipids* **32:**57–71.

Razin, S., 1979a, Membrane proteins, in: *The Mycoplasmas* (M. F. Barile and S. Razin, eds.), Academic Press, New York, pp. 289–322.

Razin, S., 1979b, Isolation and characterization of mycoplasma membranes, in: *The Mycoplasmas* (M. F. Barile and S. Razin, eds.), Academic Press, New York, pp. 213–229.

Razin, S., 1982, The mycoplasma membrane, in: *Organization of Prokaryotic Cell Membranes* (B. K. Ghosh, ed.), CRC Press, Boca Raton, Fla., pp. 165–250.

Reinert, J. C., and Steim, J. M., 1970, Calorimetric detection of a membrane-lipid phase transition in living cells, *Science* **168**:1580–1582.

Rottem, S., 1975, Heterogeneity in the physical state of the exterior and interior regions of mycoplasma membrane lipids, *Biochem. Biophys. Res. Commun.* **64**:7–12.

Rottem, S., 1979, Molecular organization of membrane lipids, in: *The Mycoplasmas* (M. F. Barile and S. Razin, eds.), Academic Press, New York, pp. 259–288.

Rottem, S., 1981, Cholesterol is required to prevent crystallization of *Mycoplasma arginini* phospholipids at physiological temperatures, *FEBS Lett.* **133**:161–164.

Rottem, S., and Markowitz, O., 1979, Carotenoids as reinforcers of the *Acholeplasma laidlawii* lipid bilayer, *J. Bacteriol.* **140**:944–948.

Rottem, S., and Samuni, A., 1973, Effects of proteins on the motion of spin-labeled fatty acids in mycoplasma membranes, *Biochim. Biophys. Acta* **298**:32–38.

Rottem, S., and Verkleij, A. J., 1982, Possible association of segregated lipid domains of *Mycoplasma gallisepticum* membranes with cell resistance to osmotic lysis, *J. Bacteriol.* **149**:338–345.

Rottem, S., Hubbell, W. L., Hayflick, L., and McConnell, H. M., 1970, Motion of fatty acid spin labels in the plasma membrane of mycoplasma, *Biochim. Biophys. Acta* **219**:104–113.

Rottem, S., Cirillo, V. P., de Kruyff, B., Shinitzky, M., and Razin, S., 1973, Cholesterol in mycoplasma membranes. Correlation of enzymic and transport activities with physical state of lipids in membranes of *Mycoplasma mycoides* var. *capri* adapted to grow with low cholesterol concentrations, *Biochim. Biophys. Acta* **323**:509–519.

Schreier, S., Polnaszek, C. F., and Smith, I.C.P., 1978, Spin labels in membranes. Problems in practice, *Biochim. Biophys. Acta* **515**:395–436.

Seelig, J., and Macdonald, P. M., 1987, Phospholipids and proteins in biological membranes. ^{2}H NMR as a method to study structure, dynamics, and interactions, *Acc. Chem. Res.* **20**:221–228.

Seelig, J., and Seelig, A., 1980, Lipid conformation in model membranes and biological membranes, *Q. Rev. Biophys.* **13**:19–61.

Seguin, C., Lewis, R. N. A. H., Mantsch, H. H., and McElhaney, R. N., 1987, Calorimetric studies of the thermotropic phase behavior of cells, membranes and lipids from fatty acid-homogenous *Acholeplasma laidlawii* B, *Isr. J. Med. Sci.* **23**:403–407.

Sen, A., Hui, S.-W., Mannock, D. A., Lewis, R. N. A. H., and McElhaney, R. N., 1990, The physical properties of glycosyl diacylglycerols. 2. X-ray diffraction studies of a homologous series of 1,2-di-O-acyl-3-O-(α-D-glycopyranosyl)-*sn*-glycerols, *Biochemistry* **29**:7799–7804.

Shinitzky, M., and Barenholz, Y., 1978, Fluidity parameters determined by fluorescence polarization, *Biochim. Biophys. Acta* **515**:367–394.

Silvius, J. R., and McElhaney, R. N., 1978, Growth and membrane lipid properties of *Acholeplasma laidlawii* B lacking fatty acid heterogeneity, *Nature* **272**:645–646.

Silvius, J. R., Mak, N., and McElhaney, R. N., 1980, Lipid and protein composition and thermotropic lipid phase transitions in fatty acid-homogeneous membranes of *Acholeplasma laidlawii* B, *Biochim. Biophys. Acta* **597**:199–215.

Singer, S. J., and Nicolson, G. L., 1972, The fluid mosaic model of the structure of cell membranes, *Science* **175**:720–731.

Smith, I. C. P., and Jarrell, H. C., 1983, Deuterium and phosphorus NMR of microbial membranes, *Acc. Chem. Res.* **16**:266–272.

Smith, I. C. P., Butler, K. W., Tulloch, A. P., Davis, J. H., and Bloom, M., 1979, The properties of gel state lipid in membranes of *Acholeplasma laidlawii* as observed by ^{2}H NMR, *FEBS Lett.* **100**:57–61.

Steim, J. M., Tourtellotte, M. E., Reinert, J. C., McElhaney, R. N., and Rader, R. L., 1969, Calorimetric evidence for the liquid-crystalline state of lipids in a biomembrane, *Proc. Natl. Acad. Sci. USA* **63**:104–109.

Stockton, G. W., Johnson, K. G., Butler, K. W., Ponaszek, C. F., Cyr, R., and Smith, I.C.P., 1975, Molecular order in *Acholeplasma laidlawii* membranes as determined by deuterium magnetic resonance of biosynthetically-incorporated specifically-labelled lipids, *Biochim. Biophys. Acta* **401**:535–539.

Stockton, G. W., Johnson, K. G., Butler, K. W., Tulloch, A. P., Boulanger, Y., Smith, I. C. P., Davis, J. H., and Bloom, M., 1977, Deuterium NMR study of lipid organization in *Acholeplasma laidlawii* membranes, *Nature* **269**:267–268.

Surewicz, W. K., and Mantsch, H. H., 1988, New insight into protein secondary structure from resolution-enhanced infrared spectra, *Biochim. Biophys. Acta* **952**:115–130.

Tourtellotte, M. E., Branton, D., and Keith, A., 1970, Membrane structure: Spin labeling and freeze-etching of *Mycoplasma laidlawii*, *Proc. Natl. Acad. Sci. USA* **66**:909–916.

Wieslander, A., Rilfors, L., Johansson, L.B.-A., and Lindblom, G., 1981, Reversed cubic phase with membrane glucolipids from *Acholeplasma laidlawii:* ^{1}H, ^{2}H, and diffusion nuclear magnetic resonance measurements, *Biochemistry* **20**:730–735.

Wong, P.T.T., Siminovitch, D. J., and Mantsch, H. H., 1988, Structure and properties of model membranes: New knowledge from high-pressure vibrational spectroscopy, *Biochim. Biophys. Acta* **947**:139–171.

Chapter 4

Regulation and Physicochemical Properties of the Polar Lipids in *Acholeplasma laidlawii*

Leif Rilfors, Åke Wieslander, and Göran Lindblom

Although *Acholeplasma laidlawii* was originally isolated from sewage, its continuous isolation from different mammalian, plant, and insect hosts clearly points to its broad parasitic and pathogenic character (Tully *et al.*, 1990). In addition, *A. laidlawii* is one of the mycoplasma species commonly isolated as contaminants in eukaryotic cell cultures (McGarrity *et al.*, 1985). Most mycoplasmas, including *A. laidlawii*, are usually found in close association with the membrane surface of their hosts (Neupert and Sterba, 1983; McGarrity *et al.*, 1985). This association, in combination with the depletion of important metabolites consumed by the mycoplasmas, has deleterious effects for the host cells. Toxic effects, such as that of H_2O_2 from the metabolism of fermentative mycoplasmas, and less well characterized immunological signals and interactions, are also involved (Razin, 1981; Chowdhury *et al.*, 1990). The close association of the parasites and hosts is perhaps best illustrated by the observed exchanges of membrane protein and lipid constituents between the mycoplasma and the eu-

Leif Rilfors and Göran Lindblom   Department of Physical Chemistry, University of Umeå, S-901 87 Umeå, Sweden.   **Åke Wieslander**   Department of Biochemistry, University of Umeå, S-901 87 Umeå, Sweden.

Subcellular Biochemistry, Volume 20: Mycoplasma Cell Membranes, edited by Shlomo Rottem and Itzhak Kahane. Plenum Press, New York, 1993.

karyotic host cell membranes (Powell *et al.*, 1976; Wise *et al.*, 1978; Tarshis *et al.*, 1981). Several mycoplasmas were recently also shown to be able to fuse with a membrane-surrounded animal virus (Citovsky *et al.*, 1988).

1. GENERAL FEATURES OF *A. LAIDLAWII* AND ITS MEMBRANE

Mycoplasmas possess three unique properties which are of utmost importance to their survival potential and which are fundamental to a discussion about membrane physiology and function. These are: (1) the lack of a bacterial cell wall and outer membrane; (2) the smallest size of all free-living cells; and (3) a correspondingly small genome size. Like all membranes, those of mycoplasmas are permeable to H_2O and small organic molecules (De Gier *et al.*, 1971). The lack of a cell wall prevents the generation of a large osmotic pressure in the cell. Hence, the large turgor pressure of other gram-positive bacteria (Csonka, 1989), which is possible due to the mechanical strength and support of the cagelike peptidoglycan mesh, is absent. The lack of an outer membrane, which in gram-negative bacteria is resistant to the detergent properties of bile salts, makes mycoplasmas sensitive to several lytic agents (Razin and Argaman, 1962). The genome of *A. laidlawii* has approximately 1600 kbp (Weisburg *et al.*, 1989), compared with 4700 kbp for the *Escherichia coli* genome (Kröger *et al.*, 1991). Most likely, this means that the fraction of basal housekeeping genes is larger in *A. laidlawii* and that it has a substantially lower capacity to respond or adapt to different conditions. This is supported by its complex nutritional demands (Tourtellotte *et al.*, 1964).

A. laidlawii has a fermentative metabolism and is indifferent to O_2. It lacks the tricarboxylic acid cycle (Manolukas *et al.*, 1988), cytochromes, and quinones (Pollack *et al.*, 1981). The glycolysis and the hexose monophosphate shunt (Desantis *et al.*, 1989) yield lactate, pyruvate, and acetate as end products (Beaman and Pollack, 1981). Hence, all ATP is probably obtained from substrate-level phosphorylations although the excretion of lactate may theoretically create a transmembrane proton gradient (Driessen and Konings, 1990). However, no (or a minor) ΔpH has been found, which probably relates to the alkaline conditions in the growth medium (Clementz *et al.*, 1986). The transmembrane electrical gradient ($\Delta \psi$) is approximately -50 mV (Clementz *et al.*, 1986), which together with the minor ΔpH yield a proton-motive force substantially lower than that in common eubacteria. The intracellular concentration of K^+ can reach 190 mM at an extracellular concentration of 3 mM (Clementz *et al.*, 1986), and the intracellular concentration of Na^+ can be maintained at low levels (15 mM) at high external concentrations because of the outward pumping of Na^+ by the membrane-bound Na^+/Mg^{2+} −ATPase (Mahajan *et al.*, 1988). These two ions are probably the main determinants of the $\Delta \psi$, which probably is created and maintained by the expense of ATP (Clementz *et al.*, 1987).

2. BASIC PROPERTIES OF THE *A. LAIDLAWII* MEMBRANE

A. laidlawii has been a tool for a large number of investigations of the physicochemical properties of biological membranes. The reason for this is the combination of mainly two features: (1) the ability to introduce controlled changes in membrane acyl chain and sterol composition (including cholesterol) and (2) the ease with which pure membranes free from contaminants can be obtained. Several of the findings from these investigations were incorporated into the "fluid mosaic" model of biological membranes by Singer and Nicolson (1972). In many aspects this model still is the framework for our understanding of the structure of biological membranes. However, several important revisions of this simple model have been made since 1972 as is also shown in this review.

A bilayer structure for the lipids in the intact *A. laidlawii* membrane, previously observed for synthetic and purified membrane lipids (Luzzati, 1968), was inferred from low- and wide-angle x-ray diffraction studies (Engelman, 1971; Wilkins *et al.*, 1971). The thickness of the bilayer varied in concordance with the lengths of different selectively supplied fatty acids incorporated as membrane acyl chains. A double-track appearance of several biological membranes, including that of *A. laidlawii* (Razin *et al.*, 1965), had been observed before by electron microscopy. However, the molecular interpretation is uncertain since the same pattern is also observed after extraction of all *A. laidlawii* membrane lipids with organic solvents (Weibull *et al.*, 1983). A study by Rilfors and Weibull (1985) also indicated that the trilaminar appearance of the cytoplasmic membrane of *A. laidlawii* is mainly caused by the membrane proteins in combination with a staining agent. Support for a lipid bilayer structure was obtained from differential scanning calorimetry. A reversible endothermic transition, typical for the melting of the acyl chains during the gel-to-liquid-crystalline transition of a lamellar phase of phospholipids in water (Luzzati, 1968), was observed in cells, membranes as well as protein-free lipids from *A. laidlawii* (Steim *et al.*, 1969). The membrane proteins had small effects on this transition. However, the midpoint of the transition temperature interval was affected in a predictable manner by the extent of acyl chain unsaturation. This transition was verified by x-ray diffraction investigations (Engelman, 1970). The presence of membrane lipids in the gel state, the extent of which is determined by the acyl chain length and saturation, had a strong influence on the minimum temperature at which the cells could grow (McElhaney, 1974). It seems that *A. laidlawii* cannot grow when more than approximately 90% of the lipids are in the gel state (reviewed by McElhaney, 1984). Optimum and maximum temperatures are also affected by the acyl chain properties but to lesser extents.

The intact membrane of *A. laidlawii* was the first for which an orientational order profile for the acyl chains of the membrane lipids was recorded. This profile is very similar to the corresponding profiles determined for the acyl chains in lamellar phases of synthetic phospholipids (Seelig and Seelig, 1980). In a series of papers, Smith and colleagues have characterized the orientational properties of saturated, unsaturated, and cyclopropane acyl chains in *A. laidlawii* membranes and purified lipids by ^{2}H-NMR (summarized in Smith and Jarrell, 1983; Smith, 1984). Both cholesterol and low temperature increase the molecular ordering of especially the plateau region and extend the length of this region, whereas the presence of membrane proteins has a smaller influence. Furthermore, the *sn*-1 acyl chain on the lipid glycerol backbones projects directly downward into the bilayer whereas the *sn*-2 chain begins parallel to the bilayer surface before bending down into the bilayer. A similar orientation has been observed in crystals as well as in lamellar phases of synthetic phosphatidylcholines (PC) and phosphatidylethanolamines (PE) (Hitchcock *et al.*, 1974; Büldt *et al.*, 1978; Hauser *et al.*, 1981). Recently, the acyl chain orientational order of the two major *A. laidlawii* membrane lipids, preferring lamellar and nonlamellar phases, respectively, was shown to differ analogously to the differences in order between PC and PE (Eriksson *et al.*, 1991; see Section 5.1.1). These results highlight the similarities in physicochemical properties between well-characterized synthetic phospholipids and the *A. laidlawii* membrane lipids. The similarities have been extended to other cell membranes which contain completely different polar lipids, for example that of *E. coli*. For further details and analyses of *A. laidlawii* with other techniques (such as ESR and FT-IR), the reader is referred to Chapter 3.

Approximately 85% of the membrane proteins in *A. laidlawii* are of the integral type, anchored by hydrophobic interactions, and cannot be washed off the membrane (Ne'eman *et al.*, 1971). A majority of the proteins are exposed on the intracellular side of the membrane (Amar *et al.*, 1974; Johansson and Hjertén, 1974). At least 15% (on a mass basis) of the more than 200 membrane proteins are covalently modified with preferentially C_{14} and C_{16} saturated acyl chains of endogenous origin (Nyström *et al.*, 1992). These acyl proteins are enriched in hydrophilic amino acid residues and have lower pI values than the average for the membrane. A few other membrane proteins are also covalently modified with isoprenoid chains (Nyström and Wieslander, 1992). These two modifications are in several aspects similar to modifications occurring among eukaryotic membrane proteins (Nyström and Wieslander, 1992). Freeze-fracture electron microscopy revealed that a large fraction of the membrane proteins are deeply embedded into, or traversing, the membrane bilayer (James and Branton, 1973). These proteins are more evenly distributed in the lateral plane of the membrane when the lipids are in a liquid-crystalline state, whereas patching of the proteins is induced by the presence of lipids in a gel state (James and Branton, 1973). A low pH value can also increase the patching (Copps *et al.*, 1976). This may depend on a diminished charge repulsion between the proteins, which is corroborated by the low average pI values of all of the *A. laidlawii* membrane proteins (Nyström *et al.*, 1992). The membrane lipids are negatively charged at physiological pH values as shown *in vitro* (Christiansson *et al.*, 1985) as well as in intact cells and membranes (Schiefer *et al.*, 1976).

The large transmembrane electrochemical proton gradients in mitochondria and certain eubacteria, as well as the large osmotic gradient in many of the latter, are likely to affect the molecular organization of both proteins and lipids in the membrane (Cevc, 1990). So far, this has been best visualized for certain membrane transport and channel proteins responding to ionic and voltage gradients. In *A. laidlawii* the accessibility of the major membrane phospholipid phosphatidylglycerol (PG) to degradation by exogenous phospholipase A_2 depends on the energy status of the cells. More PG was accessible after glucose starvation or exposure of the cells to certain ionophores (Bevers *et al.*, 1978b). In addition, PG was protected from degradation in relation to the fraction of membrane lipids in the gel state (Bevers *et al.*, 1978a). The accessibility of the membrane proteins to iodination was greater in energized cells (Amar *et al.*, 1978). However, the disposition of the proteins showed no obvious correlation to different properties of the lipid matrix (Amar *et al.*, 1979). Although these properties and changes were very evident, a molecular explanation is hard to give since the proposed effects of the ionophores and uncouplers used are at variance with recorded changes in the transmembrane potential (Clementz *et al.*, 1986, 1987). One possibility is that the observed effects depend on changes in the intracellular concentrations of K^+, Na^+, or H^+ (Clementz *et al.*, 1986; see Section 1).

3. BIOCHEMICAL PATHWAYS AND CHEMICAL STRUCTURE OF THE POLAR LIPIDS

Understanding of the mechanisms for the maintenance of the physicochemical properties of the membrane is intimately associated with the synthesis and physiological regulation of the lipids. In *A. laidlawii* all of the normally occurring polar and neutral lipids are derived from metabolites produced by the glycolysis (Figure 1). An unsaturated fatty acid precursor, and occasionally cholesterol, must be obtained from exogenous sources (Smith, 1979).

Saturated fatty acids are synthesized from acetyl CoA by the common malonyl coA pathway (Rottem and Panos, 1970), also involving acyl carrier protein (Rottem *et al.*, 1973). Unsaturated fatty acids cannot be synthesized because of the lack of one specific enzyme. The genes for the pyruvate dehydrogenase complex, i.e., the enzymes connecting the glycolysis with acetyl CoA, have been identified and sequenced recently (Wallbrandt *et al.*, 1992). At least 95% of the processed pyruvate, ending up in the macromolecules, is found in the membrane lipid fraction. A thorough extraction of the undefined growth medium components with organic solvents withdraws pantetheine, a necessary precursor to CoA and acyl carrier protein, and thereby inhibits the synthesis of fatty acids

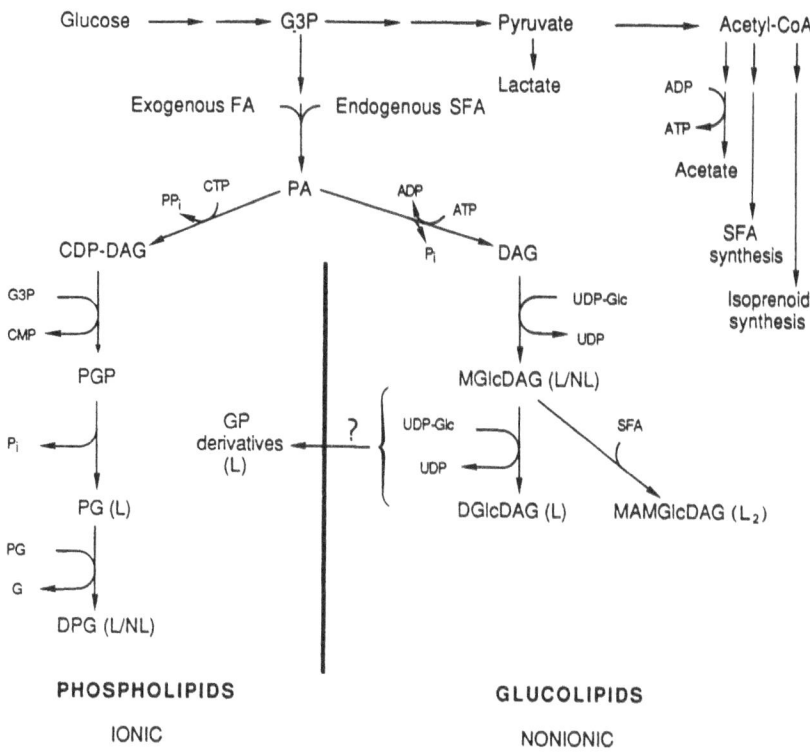

FIGURE 1. Tentative pathways for biosynthesis of polar membrane lipids in *Acholeplasma laidlawii* A-EF22. Abbreviations used: P_i, inorganic orthophosphate; PP_i, inorganic pyrophosphate; G, glycerol; GP, glycerophosphoryl; G3P, *sn*-glycero-3-phosphate, PGP, phosphatidylglycerophosphate; FA, fatty acids; SFA, saturated fatty acids; L, lamellar phase; NL, nonlamellar phase; ?, suggested pathway. For the other abbreviations, see List of Abbreviations. Endogenous SFA and exogenous FA are incorporated from their acyl carrier protein and CoA derivatives, respectively. (Modified from Smith, 1979, and Wieslander *et al.*, 1980.)

(Christiansson and Wieslander, 1980). A partial inhibition is obtained by the addition of avidin, a protein which binds strongly to the essential cofactor biotin in the growth medium (Silvius and McElhaney, 1978; Nyström *et al.*, 1992). This synthesis is also partially inhibited by the presence of exogenous saturated fatty acids, probably by an end-product inhibition mechanism (Christiansson and Wieslander, 1980; Christiansson, 1981). Cholesterol cannot be synthesized by any mycoplasma organism but isoprenoyl chains are synthesized by *A. laidlawii* along the 3-hydroxy-3-methylglutaryl-CoA pathway, which has similarities to the one in yeast (Smith, 1968, 1979; Nyström and Wieslander, 1992).

The polar lipids are most likely synthesized from the glycolysis intermediate glycerol-3-phosphate by the action of acyltransferases, adding two acyl chains and yielding phosphatidic acid (PA) (Romijn *et al.*, 1972; Rottem and Greenberg, 1975; Smith, 1979; Pieringer, 1989). From this lipid, PG and diphosphatidylglycerol (DPG) are synthesized according to the common pathway in most bacteria (Smith, 1979; Pieringer, 1989). An alanyl derivative of PG can be synthesized during certain conditions in old cultures (Koostra and Smith, 1969; Å. Wieslander, unpublished observation). The relative proportions of the anionic lipids vary between different strains. PG is the dominating anionic lipid in the commonly used strains A-EF22 (Wieslander and Rilfors, 1977; Christiansson and Wieslander, 1978) and B-PG9 (Saito and McElhaney, 1977; Silvius *et al.*, 1980), while DPG is synthesized in largest amounts in many other *A. laidlawii* strains (Efrati *et al.*, 1986). It has also been observed that strains which are sensitive to, or persistently infected with, acholeplasma virus synthesize significant amounts of DPG, while resistant strains contain only trace amounts of this lipid (Steinick *et al.*, 1980). The ionic lipid fraction usually makes up between 15 and 45 mole% of the lipids in the A and B strains mentioned above (e.g., Wieslander *et al.*, 1980; Johansson *et al.*, 1981; Rilfors *et al.*, 1987; Saito and McElhaney, 1977). Two glucolipids constitute the major fraction of the membrane lipids in both the A and B strains of *A. laidlawii* under most environmental conditions (Wieslander *et al.*, 1986; Bhakoo and McElhaney, 1988; and references therein). They are synthesized by a two-step glucosylation of diacylglycerol (DAG) (Smith, 1969; Dahlqvist *et al.*, 1992), which in turn probably is derived from PA by a dephosphorylation step similar to that found in streptococci (Pieringer, 1989). A chemical analysis (Shaw *et al.*, 1968) revealed that the structure of the two glucolipids, according to the present *sn* nomenclature (Strickland, 1973), should correspond to 1,2-diacyl-3-*O*-(α-D-glucopyranosyl)-*sn*-glycerol (MGlcDAG) and 1,2-diacyl-3-*O*-[α-D-glucopyranosyl-(1→2)-*O*-α-D-glucopyranosyl]-*sn*-glycerol (DGlcDAG). This is supported by an optical analysis of native MGlcDAG from the A-EF22 strain where an optical rotatory dispersion analysis revealed the carbohydrate moiety to be an α-anomer and circular dichroism showed that the ester-linked fatty acid residues have the 1,2-diacyl-*sn*-glycero configuration (Michelsen, 1985).

The chemical structure of a glycerophosphoryl derivative of DGlcDAG was determined to be 1,2-diacyl-3-*O*-[glycero-3-phosphoryl-6-*O*-(α-D-glucopyranosyl-(1→2)-*O*-α-D-glucopyranosyl)]-*sn*-glycerol (GPDGlcDAG) (Shaw *et al.*, 1972). A phosphatidyl derivative of DGlcDAG can also appear in old cultures of *A. laidlawii* strain B (Smith, 1972). The existence of a glycerophosphoryl derivative of MGlcDAG (GPMGlcDAG) in strain A-EF22 has been asserted (Wieslander and Rilfors, 1977). However, a chemical and enzymatic analysis gave GPDGlcDAG as one, and lyso-GPDGlcDAG as the other, of the two phosphoglucolipids in the A-EF22 strain (Smith, 1986). On the other hand, ^{13}C-NMR spectra recorded from the lipid claimed by Smith to be lyso-GPDGlcDAG revealed two separate carbonyl-carbon signals of equal intensity, which strongly indicates a lipid structure with two acyl chains (Rilfors *et al.*, 1993b); a lysolipid would need to have a 50:50 isomerization between positions *sn*-1 and *sn*-2 to give rise to two signals of equal intensity. Furthermore, this lipid forms a lamellar phase with 95 wt% water (Rilfors *et al.*, 1993b) instead of a micellar solution, which one would expect a lysolipid to form at high water contents. In addition to MGlcDAG and DGlcDAG, a third glucolipid is synthesized in A-EF22 when large amounts of saturated, straight-chain fatty acids are incorporated into the lipids (Wieslander and Rilfors, 1977; Christiansson and Wieslander, 1978, 1980; Wieslander *et al.*, 1979; Christiansson, 1981; Rilfors, 1985). The structure was determined to be 1,2-diacyl-3-*O*-[3-*O*-acyl-(α-D-glucopyranosyl)]-*sn*-glycerol (MAMGlcDAG) (Smith, 1986). The latter lipid is different from the 2-*O*-acyl-1-*O*-polyprenyl-α-D-glucopyranoside (MAPGlc), which is synthesized by the B strain in media containing large amounts of high-melting fatty acids, such as straight-chain saturated acids, methyl-branched iso acids, and ω-cyclohexyl acids (Bhakoo *et al.*, 1987; Lewis *et al.*, 1990b). Synthesis of this polyprenylated lipid is also dependent on the type of semidefined growth medium ingredients used (Wieslander *et al.*, 1993b).

4. PHASE EQUILIBRIA AND STRUCTURAL POLYMORPHISM OF LIPIDS

Amphiphilic molecules can be divided into two groups depending on their behavior in water: (1) water-soluble, micelle-forming amphiphiles and (2) water-insoluble, swelling amphiphiles. Detergents and lysolipids belong to the first group and form micelles at high water contents, while the common membrane lipids belong to the second group and form liquid-crystalline phases. There are, however, not always sharp confines between the two groups, since a lipid that is insoluble at room temperature may be soluble at a higher temperature. As will be seen below, amphiphiles can form a large number of different liquid-crystalline phases having, for example, lamellar, hexagonal, and cubic structures.

Phase diagrams have been collected for some biological membrane lipid–water systems (e.g., see Lindblom and Rilfors, 1989). Determination of phase diagrams is usually very tedious work using classical methods such as x-ray diffraction and polarized light microscopy, and the need for more rapid methods led to the development of NMR techniques for such investigations (Ulmius *et al.*, 1977; Arvidson *et al.*, 1985). The most abundant phases found in the lipid–water systems of interest here are: (1) the micellar solutions with normal (L_1) or reversed (L_2) aggregate structures; (2) lamellar gel (L_β) and liquid-crystalline (L_α) phases; (3) normal (H_I) and reversed (H_{II}) hexagonal liquid-crystalline phases; and (4) different cubic liquid-crystalline phases. The structures of the aggregates in the first three groups of phases are well established (Figure 2A–C), whereas the structures of some of the cubic phases are still under debate (Lindblom and Rilfors, 1989; Figure 2D,E). L_β and L_α phases exhibit a one-

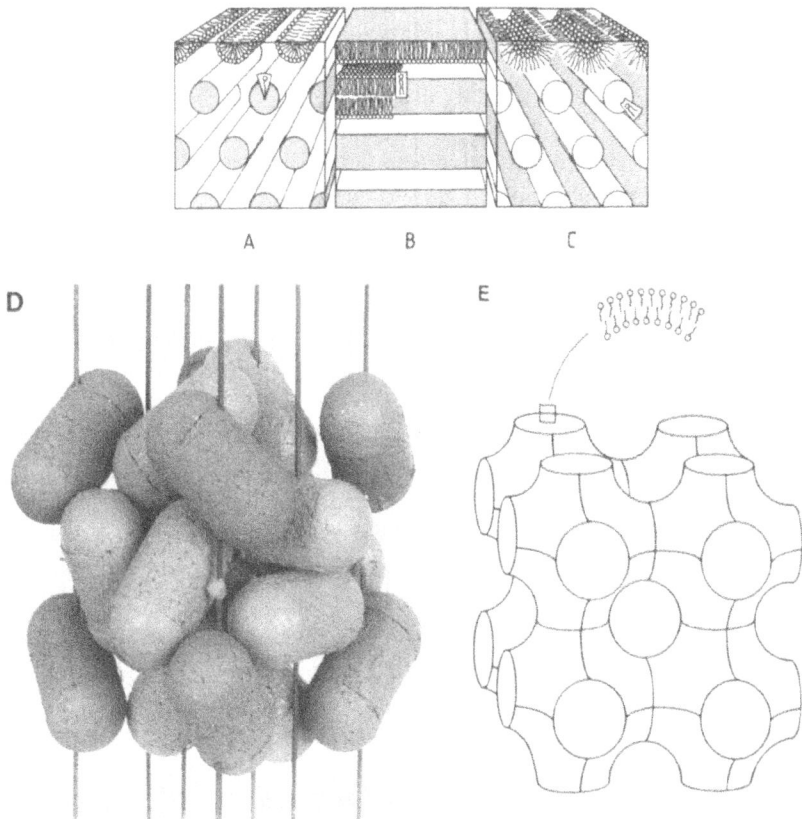

FIGURE 2. Structure of different liquid-crystalline phases of membrane lipids. (A) Normal hexagonal (H$_I$) phase; (B) lamellar (L$_\alpha$) phase; (C) reversed hexagonal (H$_{II}$) phase. (Panels A–C from Rilfors *et al.*, 1984.) (D) Cubic phase built up of short rodlike micelles with an axial ratio of about 2. (From Eriksson *et al.*, 1985.) (E) Bicontinuous cubic phase built up of lamellar aggregates. The surface shown is an infinite periodic minimal surface that represents the midplane of the lamellar aggregates. (From Lindblom and Rilfors, 1989).

dimensional periodicity with lamellar units of infinite extension stacked regularly (Figure 2B); in the L$_\beta$ phase almost all of the carbon–carbon bonds in the hydrocarbon chains have a *trans* conformation, while the L$_\alpha$ phase is characterized by a "liquid" state of the hydrocarbon region with several *trans–gauche* conformations in the hydrocarbon chains. The hexagonal liquid-crystalline phases exhibit a two-dimensional periodicity with rodlike aggregates of infinite length packed into a hexagonal lattice (Figure 2A,C). A cubic liquid-crystalline phase is one in which the lipid aggregate units form a three-dimensional lattice. The aggregates can have different shapes like closed spherical or nonspherical units, rods, or lamellae (Figure 2D,E). The cubic phases are optically isotropic, while the lamellar and hexagonal phases are optically anisotropic. However, the cubic phases, like other liquid-crystalline phases, have no short-range order, i.e., the hydrocarbon chains are partially disordered.

The positions in the phase diagram of the various phases occurring in an amphiphile–water system often give very helpful information about the physicochemical properties of the phases and, in particular for the cubic phases, about the structure of the aggregates building up these phases. Cubic phases have been observed between any two of the main phases in the phase diagram, and they are therefore often called "intermediary phases" (e.g., see Tiddy, 1980). The cubic phase regions are usually very narrow (considering either concentration or temperature), but systems exist where a large part of the phase diagram is occupied by a cubic phase (Lindblom and Rilfors, 1989). Schemes for the phase sequences have been given in several publications (Ekwall, 1975; Tiddy, 1980). Cubic phases of membrane lipids often occur between the L_α and H_{II} phases. It has also been observed that for some systems, isotropic phases, probably of cubic structure, form as metastable intermediates at the thermal transition between the L_α and H_{II} phases (Siegel, 1986; Lindblom and Rilfors, 1989).

The cubic phases can be divided into two fundamentally different groups: the bicontinuous structures having regions which are continuous with respect to both polar (water) and nonpolar (hydrocarbon) components (Figure 2E), and the structures built up either of discontinuous hydrocarbon regions but with continuous water regions (e.g., close-packed normal micelles; Figure 2D), or of discontinuous water regions but with continuous hydrocarbon regions (e.g., close-packed reversed micelles). Most of the cubic phases formed by membrane lipids have been shown to be bicontinuous (Rilfors *et al.*, 1986). However, lysophosphatidylcholines with certain saturated acyl chains form a cubic phase built up of normal micelles (Eriksson *et al.*, 1985; Lindblom and Rilfors, 1989), and a cubic phase built up of reversed micelles was recently reported (Seddon, 1990a). Further details about the cubic and reversed hexagonal phases can be found in the recent reviews by Lindblom and Rilfors (1989) and Seddon (1990b), respectively.

5. PHYSICOCHEMICAL PROPERTIES OF THE POLAR LIPIDS IN *A. LAIDLAWII*

5.1. Glucolipids

Several investigations of the physicochemical properties of native MGlcDAG and DGlcDAG isolated from *A. laidlawii* membranes have been performed (de Kruijff *et al.*, 1973; Wieslander *et al.*, 1978, 1981a; Silvius *et al.*,

1980; Khan *et al.*, 1981; Rance *et al.*, 1983; Lindblom *et al.*, 1986; Eriksson *et al.*, 1991). Because of recent improvements in the syntheses of glycosylglycerolipids (Mannock *et al.*, 1987, 1990b), more systematic studies of synthetic lipids having one or two glucosyl groups have been performed over the last 10 years. Up to now, the greatest attention has been paid to the β-anomer of monoglucosyl (MGlc) lipids with straight, saturated hydrocarbon chains linked to the glycerol with ether bonds (Endo *et al.*, 1982, 1983; Hinz *et al.*, 1985, 1991; Blöcher *et al.*, 1985; Jarrell *et al.*, 1986, 1987a,b; Koynova *et al.*, 1988; Auger *et al.*, 1991; Winsborrow *et al.*, 1991); one work included the corresponding α-anomer as well (Jarrell *et al.*, 1987b). However, both the α- and the β-anomer of MGlc lipids with straight, saturated acyl chains (ester bonds) have been intensively investigated over the last 5 years (Mannock *et al.*, 1988, 1990a; Asgharian *et al.*, 1989; Sen *et al.*, 1990; Lewis *et al.*, 1990a). Four works dealing with β-anomers of diglucosyl (DGlc) lipids containing straight, saturated alkyl chains have been published (Endo *et al.*, 1982, 1983; Iwamoto *et al.*, 1982; Hinz *et al.*, 1991). Several physicochemical properties of these lipids have been studied, including phase equilibria and phase structures; hydration; monolayer behavior; conformation, order, and dynamics of different segments of the molecules; and miscibility properties. A summary of these results follows below. Comparisons are often made between the glucolipids on the one hand, and PCs and PEs on the other, for three reasons: (1) PC and PE occur frequently in biological membranes; (2) PC and PE are the most thoroughly investigated membrane lipids; and (3) PC form lamellar phases under most conditions, while PE has a pronounced ability to form nonlamellar phases of the reversed type.

5.1.1. Monoglucosyl Lipids

The temperature for the transition between an L_β and an L_α phase (T_m) is 17–24°C higher for the β/ether and β/ester derivatives of synthetic MGlc lipids than for the corresponding PCs, but 2–5°C lower than for the corresponding PEs (Endo *et al.*, 1982; Hinz *et al.*, 1985; Blöcher *et al.*, 1985; Jarrell *et al.*, 1986; Mannock *et al.*, 1988; Koynova *et al.*, 1988); the difference in the T_m values decreases as the length of the hydrocarbon chains increases. The corresponding differences in the T_m values for the synthetic MGlcDAG lipids (α/ester) are 13–16 and 5–10°C, respectively (Mannock *et al.*, 1990a). Thus, the T_m values for the α-anomers are 3–5°C lower than for the β-anomers. The enthalpy change associated with the L_β-to-L_α phase transition is higher for the MGlc lipids than for the comparable PEs, and this change is higher for the α-linked MGlcDAG compounds than for the corresponding β-anomers (Mannock *et al.*, 1990a).

MGlcDAG has a pronounced ability to form reversed cubic (I_{II}) and H_{II} phases besides the lamellar phases. A phase diagram has been determined in the composition range 0.5–15 moles water/mole lipid at temperatures between -20

and 50°C for dioleoyl-MGlcDAG isolated from *A. laidlawii* (Figure 3; Lindblom *et al.*, 1986). Above 10°C, only I_{II} and H_{II} phases exist; below 10°C is a region with an L_α phase. The I_{II} phase is bicontinuous and belongs to the space group *Ia3d*. The H_{II} phase stands in equilibrium with excess water, and the maximum hydration of this phase is about 11 moles water/mole lipid. An L_β phase is formed at about -15°C. It is interesting to note that dioleoyl-PE (DOPE) transforms from an L_α to an H_{II} phase at about 10°C (Tilcock and Cullis, 1982).

The lamellar–nonlamellar phase transitions for MGlcDAG and the corresponding β-anomers containing straight, saturated acyl chains with 10 to 20 carbon atoms ($N = 10$–20) have been studied by McElhaney and colleagues (Mannock *et al.*, 1988, 1990a; Sen *et al.*, 1990). The nature of the lamellar–nonlamellar phase transition is affected by the length of the acyl chains and the anomeric linkage. The α-anomers with $N = 14$–16 form an I_{II} phase, while the compounds with $N = 17$–20 form an H_{II} phase; the temperature at which the nonlamellar phases form decreases from 105°C for $N = 14$ to 77°C for $N = 20$. The β-linked anomers are more prone to form nonlamellar phases. An I_{II} phase is formed when $N = 12$–15; the temperature for the formation of this phase *increases* from 58°C for $N = 12$ to 73°C for $N = 15$. An H_{II} phase is formed between 75 and 80°C for the compounds with $N = 16$–20. Both MGlcDAG and its β-linked counterpart with straight, saturated acyl chains are much more prone to form nonlamellar phases than the corresponding PEs (Mannock *et al.*, 1990a).

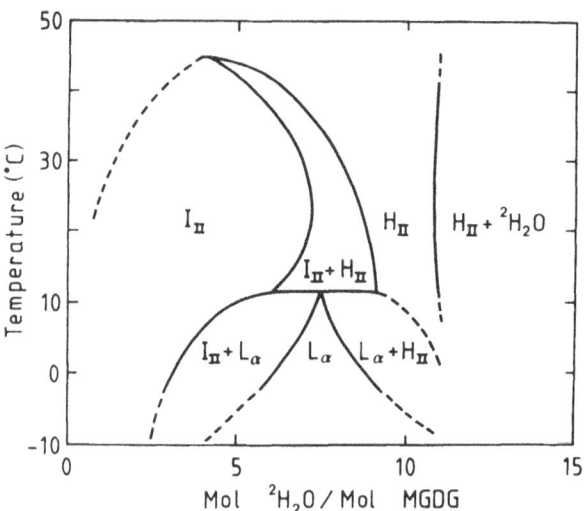

FIGURE 3. Phase diagram of the system dioleoyl-MGlcDAG–^2H_2O. The phase equilibria were deduced by NMR spectroscopy and polarized light microscopy. I_{II}, reversed cubic liquid-crystalline phase. Other notations according to legend of Figure 2. (From Lindblom *et al.*, 1986.)

The phase equilibria of the synthetic MGlcDAG compounds can be compared with those of MGlcDAG's isolated form *A. laidlawii*. Three preparations of MGlcDAG have been found to form an H_{II} phase at maximum hydration and physiological temperatures; the acyl chain composition of these preparations were: (1) 97 mole% oleoyl chains (18:1c); (2) 92 mole% elaidoyl chains (18:1t); and (3) 52 mole% palmitoyl chains (16:0) and 46 mole% 18:1c, respectively (Wieslander *et al.*, 1978; Lindblom *et al.*, 1986). However, an MGlcDAG preparation containing 30 mole% 16:0, 47 mole% 18:1c, and 22 mole% short saturated acyl chains ($N = 12-15$) forms a mixture of L_β and L_α phases between 25 and 47°C; above 47°C a mixture of L_α and I_{II} phases is obtained, and a pure I_{II} phase is formed at about 70°C (Eriksson *et al.*, 1991). Thus, when the average length of the saturated acyl chains in the third preparation above is shortened, MGlcDAG no longer forms an H_{II} phase, but instead lamellar and l_{II} phases. These observations are in line with the results obtained from the synthetic α-anomers (Mannock *et al.*, 1990a). Consequently, it can be concluded that the ability of MGlcDAG to form H_{II} and I_{II} phases increases with an increasing length and degree of unsaturation of the acyl chains.

MGlcDAG takes up smaller amounts of water than both PC and PE with corresponding acyl chains. The maximum hydration of MGlcDAG in the H_{II} phase at 35°C is approximately 7, 8, and 11 moles water/mole lipid for di-elaidoyl-, palmitoyl/oleoyl-, and dioleoyl-MGlcDAG, respectively (Lindblom *et al.*, 1986); an increase in the fraction of acyl chains with *cis* unsaturation thus increases the water binding capacity. The maximum hydration also increases with increasing temperature and in the presence of divalent cations (Wieslander *et al.*, 1978). ^{2}H-NMR investigations of MGlcDAG–^{2}H$_2$O mixtures have indicated that the water molecules bind to several sites on the lipid, probably by hydrogen bonding to the hydroxyl groups of the sugar ring (Lindblom *et al.*, 1986). The maximum hydration of dioleoyl-PC (DOPC) at 2 and 25°C is 30–34 moles water/mole lipid (Gutman *et al.*, 1984; Sjölund *et al.*, 1987; Bergenståhl and Stenius, 1987; Gruner *et al.*, 1988). DOPE takes up about 18 moles water/mole lipid at 2°C (Gruner *et al.*, 1988), while PE enriched in anteiso and iso methyl-branched, saturated acyl chains takes up approximately 14 and 8 moles water/mole lipid at 26°C, respectively (Rilfors *et al.*, 1982, 1993a). Thus, the maximum hydration of PE seems to be much more sensitive to the acyl chain composition than MGlcDAG, which indicates that the hydrogen bonding is stronger between the head groups of MGlcDAG.

Monolayer studies at the air/water interface of dipalmitoyl-MGlcDAG and its β-linked counterpart have shown that both the condensed and the expanded areas of the α-anomers are greater than those of the β-anomers. However, both anomers occupy smaller areas than dipalmitoyl-PC but larger areas than dipalmitoyl-PE (Asgharian *et al.*, 1989). From x-ray diffraction studies of an L_α phase formed by MGlcDAG isolated from *A. laidlawii*, the area per molecule was calculated to be 61 and 63 Å^2 at 20 and 47°C, respectively; the acyl chain composition of this MGlcDAG was 30 mole% of 16:0, 47 mole% of 18:1c, and 22 mole% of short acyl chains ($N = 12$–15) (Eriksson *et al.*, 1991).

The polar head groups of the α- and β-anomers of a MGlc lipid with tetradecyl chains have been studied with ^{2}H-NMR by using specifically ^{2}H-labeled glucose molecules (Jarrell *et al.*, 1986, 1987a,b). In order to determine the molecular ordering of the head groups with respect to the bilayer normal, two assumptions were made: the sugar ring is rigid and there is a rapid rotation of the sugar ring about the axis to which an order parameter, S_{mol}, can be defined (this has to be done since no "natural" axis of high symmetry exists in the sugar group). From the quadrupole splittings of three C–^{2}H bonds, which showed surprisingly similar values for the two anomers, S_{mol} was calculated to be equal to 0.56 and 0.45 for the α- and β-anomers, respectively. From the ^{2}H-NMR data obtained, Jarrell and colleagues concluded that the orientation of the glucopyranose ring differs greatly for the two anomers. The sugar ring of the β-anomer is essentially fully extended away from the bilayer surface, while the pyranose ring of the α-anomer is almost parallel to the bilayer surface. Moreover, from measurements of ^{2}H-NMR spin-relaxation times it was found that the rate of the head-group motion is slower in the α-anomer than in the β-anomer, which in turn has a slower head-group motion than that of PC and PG. Finally, it is interesting to note that the C-2–C-3 bond of the glycerol backbone of the β-anomer was estimated to be tilted away from the bilayer normal by about 3° (Jarrell *et al.*, 1987a); the ordering of the glycerol backbone is consequently very similar to that reported for various phospholipids (Wohlgemuth *et al.*, 1980; Browning and Seelig, 1980; Gally *et al.*, 1981; Strenk *et al.*, 1985).

The order parameter profile and the transverse spin-relaxation rate have been determined for an L_α phase formed by MGlcDAG containing a small fraction of biosynthetically incorporated perdeuterated 16:0 (Eriksson *et al.*, 1991). The order parameter profile is similar to those obtained for phospholipids (Figure 4) (Seelig and Seelig, 1980). Interestingly, the ordering of the acyl chains is higher in MGlcDAG than in DGlcDAG; DGlcDAG, in contrast to MGlcDAG, only forms lamellar phases (see Section 5.1.2). The transverse spin-relaxation data indicate the presence of slow reorientational motions such as bilayer fluctuations, or lipid lateral diffusion over a curved bilayer surface, the amplitude of

---→

FIGURE 4. (A) The order parameter profile for DGlcDAG containing perdeuterated 16:0 (DGlcDAG-d_{31}), with varying amounts of MGlcDAG added. ○, pure DGlcDAG-d_{31}; □, DGlcDAG-d_{31}/MGlcDAG (71:29, mole/mole); △, DGlcDAG-d_{31}/MGlcDAG (24:76, mole/mole). (B) The order parameter profile for MGlcDAG containing perdeuterated 16:0 (MGlcDAG-d_{31}), with varying amounts of DGlcDAG added. ○, pure MGlcDAG-d_{31}; △, MGlcDAG-d_{31}/DGlcDAG (77:23, mole/mole); □, MGlcDAG-d_{31}/DGlcDAG (25:75, mole/mole). The water content was 9 moles ^{1}H$_2$O/mole lipid and the temperature was 48°C. (From Eriksson *et al.*, 1991.)

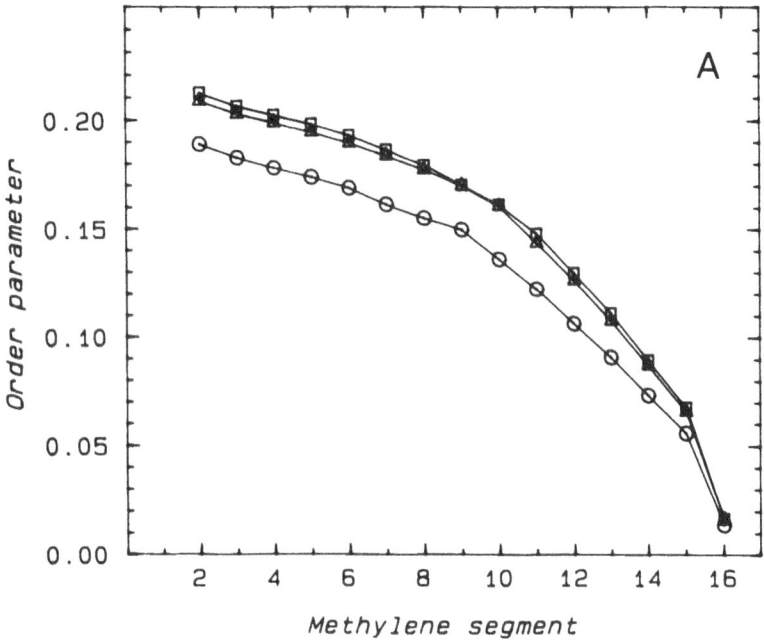

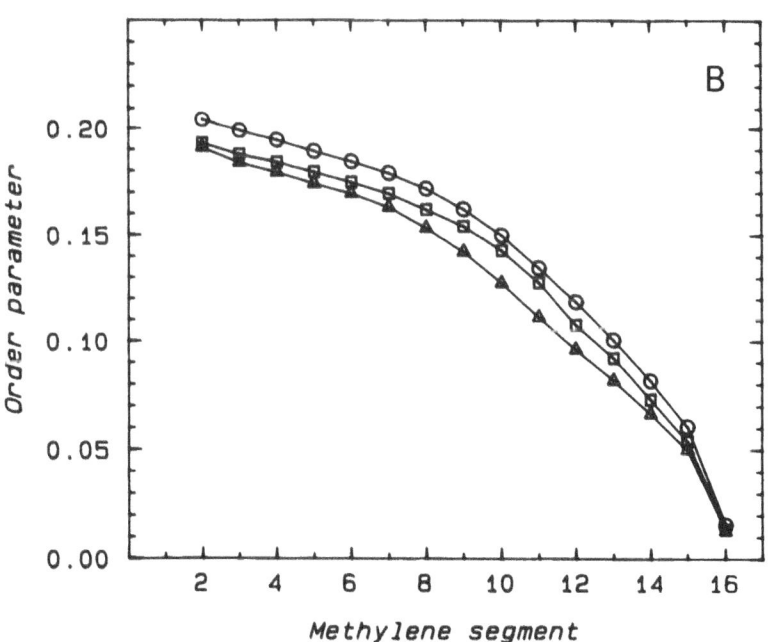

which is directly related to the relative amount of MGlcDAG in binary mixtures with DGlcDAG (Eriksson *et al.*, 1991). However, recently Halle (1991) showed that there is no need to invoke collective reorientation mechanisms, like bilayer fluctuations, to account for the relaxation rates in question. Analogous results of molecular ordering have also been obtained from investigations of DOPC and DOPE and mixtures of these lipids; the ordering of the hydrocarbon chains is smaller in DOPC and is increased by the addition of DOPE (Cullis *et al.*, 1986; Lafleur *et al*, 1990). Consequently, two membrane lipids with the ability to form nonlamellar phases of the reversed type *increase* the order of the acyl chains in a bilayer.

Several authors have discussed the possibility of intermolecular hydrogen bonding between the polar head groups of glycolipids (Hinz *et al.*, 1985; Lindblom *et al.*, 1986; Curatolo, 1987; Boggs, 1987; Mannock *et al.*, 1988, 1990a). In crystals of glycolipids the molecules are packed in such a way that intermolecular hydrogen bonding can occur (Moews and Knox, 1976; Pascher and Sundell, 1977), but no experimental evidence is available demonstrating the existence of such bonds in the presence of water (Boggs, 1987). However, the possible occurrence of intermolecular hydrogen bonding has been used to explain several of the characteristic properties of glycolipids: the high packing density in monolayer films, the high T_m values, the ability to form nonlamellar phases of the reversed type, and the low water binding capacity (Hinz *et al.*, 1985; Curatolo, 1987; Boggs, 1987; Lindblom *et al.*, 1991).

MGlcDAG has similar, but not identical, physicochemical properties as PE. MGlcDAG has a larger area per molecule in a monolayer at the air/water interface; has a lower maximum hydration; is much more prone to form nonlamellar phases when it contains saturated acyl chains; has slightly lower T_m values; and has a larger enthalpy change associated with the L_β-to-L_α phase transition.

5.1.2. Diglucosyl Lipids

Somewhat less information is available for DGlc lipids than for MGlc lipids. Hinz *et al.*, (1991) have performed the most extensive study of synthetic DGlc lipids to date. They examined the phase behavior of the β-anomer of maltosyl derivatives ($\alpha[1\rightarrow4]$ glucoside linkage). For these derivatives, the temperature for the disappearance of the L_β phase (to an L_α or an H_{II} phase) is 6–11°C lower than for the corresponding MGlc lipids but 11–13°C higher than for the corresponding PC species (Hinz *et al.*, 1985; Blöcher *et al.*, 1985). Endo *et al.* (1982) and Iwamoto *et al.* (1982) also investigated the β-anomer of a cellobiosyl compound ($\beta[1\rightarrow4]$ glucoside linkage) with hexadecyl chains. The T_m value was determined to be 54°C (probably underestimated by 3–4°C; see Hinz *et al.*, 1991), which is nearly identical to the value for the corresponding maltosyl species. Hinz *et al.* (1991) extended their examination to the β-anomer of a maltotriosyl compound (three glucopyranosyl residues with $\alpha[1{\to}4]$ linkages) and found that the T_m value of this lipid was further decreased by 9°C relative to the corresponding DGlc lipids. Hinz *et al.* (1991) calculated the transition entropies for the L_β-to-L_α transition for lipids with one, two, and three glucopyranosyl residues in the polar head group; they concluded that an increase in the size of the carbohydrate moiety is associated with a favorable molar entropy gain in this transition. With a successively larger head group, hydrocarbon chains with an increased number of vibrational and rotational degrees of freedom are probably obtained after the melting. In this context it is interesting to note that a very small alteration in the chemical structure of the polar head group of the cellobiosyl derivative causes a surprisingly large change in the T_m value. When the terminal saccharide moiety is galactose (the C_4 epimer of glucose) rather than glucose, the T_m value of the ditetradecyl compound is 66°C (Renou *et al.*, 1989), which is about 25°C above the T_m value of the cellobiosyl compound.

A rather complex picture emerges from the studies of the T_m values of DGlcDAG isolated from *A. laidlawii*. de Kruijff *et al.* (1973) obtained a T_m midpoint value of 53°C for a DGlcDAG preparation containing 86 mole% saturated acyl chains with an average length of 15.2 carbon atoms, and 5 mole% 18:1c and linoleoyl chains (18:2c). This T_m value seems to be in rather good agreement with the values determined for the synthetic DGlcDAG compounds (see above). MGlcDAG and DGlcDAG with 92 and 95 mole% 18:1t, respectively, both have T_m values of 30–35°C (Wieslander *et al.*, 1978). The values were estimated from the disappearance of the quadrupole splitting in ^{2}H-NMR spectra recorded from lipid–^{2}H$_2$O mixtures. It is, however, difficult to make an exact determination of the T_m value with this method. The method was also used by Lindblom *et al.* (1986) to estimate the T_m values of MGlcDAG and DGlcDAG with 97 mole% 18:1c. The values are -15 and -20°C for MGlcDAG and DGlcDAG, respectively. Finally, Silvius *et al.* (1980) used differential thermal analysis to investigate these glucolipids containing 99 mole% isopalmitoyl chains. When heated, DGlcDAG showed three endothermic transitions centered at 26.5, 31.5, and 35°C, while on subsequent cooling the lipid exhibited a single exothermic transition at 15.5°C. In contrast, MGlcDAG showed two completely reversible transitions centered at 22.5 and 25°C.

The system dioleoyl-DGlcDAG–^{2}H$_2$O forms an L_α phase between 4 and 10 moles water/mole lipid and from -20°C up to at least 50°C (Lindblom *et al.*, 1986); ^{2}H-NMR spectra were not recorded at higher temperatures. Below -20°C this lipid–water system forms an L_β phase. A DGlcDAG preparation containing 16 mole% 16:0, 77 mole% 18:1c, and 6 mole% short saturated acyl chains ($N = 12$–15) forms an L_α phase up to at least 70°C with 9 moles water/mole lipid (Eriksson *et al.*, 1991), which is near the maximum hydration (see below). L_β and L_α phases are also formed by DGlcDAG preparations with 95 mole% 18:1t, or with 43 and 56 mole% 16:0 and 18:1c, respectively (Wieslander *et al.*, 1978).

Hinz *et al.* (1991) conclude from their study of synthetic β-anomers with one, two, and three glucopyranosyl residues in the polar head group that the introduction of a di- or a trisaccharide moiety completely suppresses the ability to form nonlamellar phases. Thus, the synthetic DGlc lipids and the native DGlcDAG preparations investigated hitherto have *not* been observed to form any kind of nonlamellar phase under any condition. The phase equilibria are therefore shifted toward a lamellar phase when DGlcDAG is added to a nonlamellar phase formed by MGlcDAG (Wieslander *et al.*, 1981a; Khan *et al.*, 1981; Lindblom *et al.*, 1986; Eriksson *et al.*, 1991).

The maximum hydration at 35°C is approximately 7, 9, and 11 moles water/mole lipid for dielaidoyl-, palmitoyl/oleoyl-, and dioleoyl-DGlcDAG, respectively (Lindblom *et al.*, 1986). Although DGlcDAG has two glucopyranosyl molecules as polar head group, its ability to take up water is the same as that of MGlcDAG (see Section 5.1.1). A partial explanation for this might be that the α[1→2] glucoside linkage of the DGlcDAG occurring in *A. laidlawii* provides a compact structure of the head group, as indicated by a CPK-molecular model of the lipid (Iwamoto *et al.*, 1982). Thus, some of the hydroxyl groups may be shielded from contact with the surrounding aqueous phase. Iwamoto *et al.* (1982) used fluorescence spectroscopy to study the "micropolarity" at the hydrocarbon–water interface of liposomes made of PC and DGlc lipids with maltosyl and cellobiosyl head groups. They concluded that the interface of the glucolipid liposomes is more hydrophobic, and thus less hydrated, than that of the PC liposomes. A plausible explanation for the much lower ability of the glucolipids to take up water as compared with PC might be that there are no strong repulsive forces, resulting from long-range coulombic interactions, between the glucolipids. Furthermore, short-range repulsive forces are also much weaker for the nonionic glucolipids than for PC (Marra, 1986; Lindblom *et al.*, 1986). Israelachvili and Wennerström (1990) gave a possible explanation for this behavior. They discuss the effect of steric forces between amphiphilic surfaces, and most probably the steric or protrusion force between the glucolipid aggregates is diminished by the decreased mobility of the sugar head groups resulting from intermolecular hydrogen bonding (A. Holmgren, G. Lindblom, and L. Rilfors, unpublished results). In contrast, the polar head group of PC moves rapidly, which increases the strength of the protrusion force and the capacity to store water between the bilayers.

The area per DGlcDAG molecule in an L_α phase has been determined by x-ray diffraction. It was calculated to be 71 and 74 Å^2 at 20 and 47°C, respectively (Eriksson *et al.*, 1991). The acyl chain composition of this DGlcDAG was 16 mole% 16:0, 77 mole% 18:1c, and 6 mole% short saturated acyl chains ($N = 12-15$). The area per molecule for this DGlcDAG preparation is thus comparable to the area per DOPC molecule (Rand and Parsegian, 1989).

The order parameter profile and the transverse spin-relaxation rate have been determined by ^{2}H-NMR for an L_α phase formed by DGlcDAG (Eriksson *et al.*, 1991). The order parameter profile is similar to those obtained for phospholipids (Figure 4) (Seelig and Seelig, 1980). For a comparison of the order and dynamics in bilayers of DGlcDAG and MGlcDAG, see Section 5.1.1. The order parameter of the hydrocarbon chains in an L_α phase of PC and a DGlc lipid with maltosyl and cellobiosyl head groups has been studied with fluorescence spectroscopy, and it was concluded that the order parameter is larger in the glucolipid bilayers (Iwamoto *et al.*, 1982). However, the lateral diffusion coefficient of DGlcDAG is about five times larger than that of the corresponding PC species (Wieslander *et al.*, 1981a; Lindblom *et al.*, 1981, 1986).

The order and dynamics of the polar head group of the β-anomer of lactosylditetradecylglycerol have been studied (Renou *et al.*, 1989); the chemical structure of the lactosyl group is galactopyranosyl-β[1→4]-glucopyranosyl. It was concluded that the lactosyl head group is extended away from the membrane surface into the aqueous phase, and that the head group fluctuates about the bilayer normal as a rigid unit with no substantial segmental motion about the disaccharide linkage. The values of the order parameter, defined for the sugar group in a similar way as discussed in Section 5.1.1 (Jarrell *et al.*, 1987a,b), are 0.51 and 0.53 for the galactopyranosyl and glucopyranosyl groups, respectively, which can be compared with the value of 0.45 for the corresponding monoglucosyl species (Jarrell *et al.*, 1986). The addition of the second carbohydrate moiety thus attenuates the amplitude of the overall head-group motion (Renou *et al.*, 1989).

It can be concluded that DGlcDAG resembles PC in two respects: both lipids have a very strong preference for forming lamellar phases and occupy about the same area per molecule in a bilayer. In other respects they differ; DGlcDAG has higher T_m values, has a much lower maximum hydration, and has a higher lateral diffusion coefficient.

5.1.3. MAPGlc

The phase behavior of MAPGlc is markedly different in the heating and cooling modes and, in the former, depends also on the thermal history of the lipid–water sample (Lewis *et al.*, 1990b). Lipid dispersions that have been fully equilibrated at temperatures below 15°C exhibit a conversion from a crystal-like lamellar gel phase to an H_{II} phase near 65°C. However, upon cooling from high temperatures, the H_{II} phase persists to temperatures near 40°C; a mixture of cubic and H_{II} phases seems to be present between 35 and 40°C, and an L_α phase is formed between 33 and 35°C; below 33°C the lipid dispersion transforms to a metastable lamellar gel phase. When this metastable gel phase is heated an L_α phase forms at 33°C and an H_{II} phase forms at 39°C. The phase behavior of the metastable gel phase is considered to reflect the physical properties of MAPGlc when present in the *A. laidlawii* membrane (Lewis *et al.*, 1990b). Since the corresponding MGlcDAG compound transforms from a lamellar to a cubic phase at 79°C (Mannock *et al.*, 1990a), the newly discovered glucolipid is more prone to form nonlamellar phases.

5.1.4. MAMGlcDAG

Studies of the phase equilibria of MAMGlcDAG by various NMR methods (^{2}H-NMR, translational diffusion measurements with the NMR pulsed field gradient method, moment calculations of ^{1}H-NMR spectra) show that a gel or crystalline phase exists up to temperatures around 80°C for samples having 5 and 10 moles water/mole lipid. An L_2 phase is formed when the gel or crystalline phase melts (Lindblom *et al.*, 1993). The fraction of saturated acyl chains (predominantly 16:0) in this MAMGlcDAG preparation is about 95 mole%, the rest being 18:1c. In conformity with MAPGlc, MAMGlcDAG also seems to have somewhat greater tendencies than MGlcDAG to form aggregate structures with negative curvatures since synthetic dipalmitoyl-MGlcDAG transforms to a cubic phase at 79°C (Mannock *et al.*, 1990a). This conclusion seems reasonable considering that MAMGlcDAG has three hydrocarbon chains. When MAMGlcDAG is mixed with DGlcDAG, it increases the order parameter of the acyl chains (Lindblom *et al.*, 1993), which is analogous to the results obtained with MGlcDAG and PE (Section 5.1.1; Killian *et al.*, 1992).

5.2. Phospholipids and Phosphoglucolipids

5.2.1. PG

Dioleoyl-PG forms an L_α phase with 10–98 wt% of water at temperatures between 25 and up to at least 55°C (Lindblom *et al.*, 1991). At physiological pH values, the T_m values of the PG compounds are nearly identical to those of the corresponding PC compounds (Findlay and Barton, 1978). Due to the net negative charge of anionic lipids at physiological pH values, the interaction with monovalent and divalent cations, and changes of the pH value in these lipid–water systems, can affect the phase equilibria. When the pH is changed from neutral to acidic values, the T_m value for PG is increased by approximately 20–25°C (Watts *et al.*, 1978; Findlay and Barton, 1978; Van Dijck *et al.*, 1978). The addition of Ca^{2+} or Mg^{2+} to PG–water dispersions increases the T_m values by 50–80°C depending on the ion and the acyl chain length (Verkleij *et al.*, 1974; Van Dijck *et al.*, 1975, 1978). These effects are apparently due to a decrease in the electrostatic repulsion between the negatively charged head groups. PG forms an L_α phase, or a dehydrated precipitate, in the presence of Ca^{2+} (Farren and Cullis, 1980). Consequently, PG seems to have *no* tendencies to form non-lamellar phases.

5.2.2. DPG

Few studies of synthetic DPG compounds have been performed. However, if the results obtained with tetramyristoyl-DPG (Sankaram *et al.*, 1989) can be generalized, the T_m values for the DPG compounds are about 10–15°C higher than the corresponding PC compounds. Bovine heart DPG forms an L_α phase with 20–95 wt% water at temperatures between 25 and 55°C (Rilfors *et al.*, 1986; Lindblom *et al.*, 1991). At temperatures above 55°C, DPG with 20 and 30 wt% water forms a mixture of L_α, cubic, and H_{II} phases. DPG with 10 wt% water probably forms a mixture of an L_α phase and at least one nonlamellar phase at 25 and 35°C, and an H_{II} phase at 45°C and higher temperatures (Rand and Sengupta, 1972; Lindblom *et al.*, 1991). However, DPG is able to form nonlamellar phases also at high water contents under certain conditions. The addition of cations or cationic local anesthetics induces the formation of such phases: (1) an H_{II} phase in the presence of high concentrations of Na^+ (Seddon *et al.*, 1983; Sankaram *et al.*, 1989); (2) a cubic or an H_{II} phase by the addition of Ca^{2+}, Mg^{2+}, Mn^{2+}, or Ba^{2+} (Rand and Sengupta, 1972; Cullis *et al.*, 1978; Vail and Stollery, 1979; Vasilenko *et al.*, 1982; de Kruijff *et al.*, 1982); and (3) a cubic or an H_{II} phase in the presence of dibucaine or chlorpromazine (Cullis *et al.*, 1978; Rilfors *et al.*, 1986). Finally, an H_{II} phase is formed at acidic pH values (Seddon *et al.*, 1983). The acyl chain composition of DPG affects the phase equilibria in an expected manner (Vasilenko *et al.*, 1982).

5.2.3. GPMGlcDAG and GPDGlcDAG

In the membrane of *A. laidlawii* strain A-EF22, the fraction of GPDGlcDAG is usually larger than that of the lipid proposed to be GPMGlcDAG (Wieslander and Rilfors, 1977; Christiansson and Wieslander, 1978), while the opposite condition is valid for many other strains (Efrati *et al.*, 1986). Both lipids form lamellar phases as revealed by x-ray diffraction and ^{31}P-NMR (Wieslander *et al.*, 1978; Lindblom *et al.*, 1986; Rilfors *et al.*, 1993b). Moreover, the quadrupole splitting obtained from ω-d_3-16:0-labeled GPDGlcDAG is of the same magnitude as those obtained from the corresponding DGlcDAG and PG species, but twice as large as that obtained from MGlcDAG (Wieslander *et al.*, 1978), which further supports the conclusion that GPDGlcDAG forms a lamellar phase. The T_m values for GPDGlcDAG seem to be somewhat higher than for MGlcDAG and DGlcDAG (Wieslander *et al.*, 1978). The chemical shift anisotropy obtained by ^{31}P-NMR for dioleoyl-GPMGlcDAG is significantly smaller than that for dioleoyl-PG, which indicates that the molecular ordering of the head group in GPMGlcDAG is lower. The reason for this is probably that the phosphate group on the glucose residue of GPMGlcDAG is located much farther out in the water region than that of PG (Lindblom *et al.*, 1986).

6.   REGULATION OF THE MEMBRANE LIPID COMPOSITION

The membrane lipid composition has been studied in several strains of *A. laidlawii*, and these strains alter the lipid composition in somewhat different ways when they are grown under various conditions. Therefore, the results obtained with the different strains are summarized separately.

6.1.   *A. laidlawii* A

6.1.1.   Incorporation of Fatty Acids

When fatty acids supplied in the growth medium, or synthesized by the organism, are incorporated into the membrane lipids, the relative distribution of the polar head groups of the lipids is adapted to the chemical structure, and thus the physicochemical properties, of the fatty acids. In principle, three kinds of changes in the head-group composition have been observed to occur: (1) the proportion between the glucolipids MGlcDAG and DGlcDAG; these lipids usually make up between 55 and 75 mole% of the membrane lipids in strain A; (2) the proportion between the nonionic glucolipids and the three anionic lipids PG, GPMGlcDAG, and GPDGlcDAG; and (3) the proportion of the third glucolipid, MAMGlcDAG. The different chemical structures of the fatty acids that have been investigated are: type and degree of unsaturation, methyl-branching, and length of the hydrocarbon chain.

Three key features of the regulation of the polar head-group composition were observed early when strain A-EF22 was grown on 18:1c and an equimolar mixture of 16:0 and 18:1c: a lower value of the ratio MGlcDAG/DGlcDAG, a larger fraction of anionic lipids, and just trace amounts of MAMGlcDAG, were obtained with the 18:1c supplement (Wieslander and Rilfors, 1977). These features have been substantiated repeatedly in experiments where the proportion of 16:0 and 18:1c was systematically varied over a wide range (Figure 5) (Christiansson and Wieslander, 1978; Wieslander *et al.*, 1979, 1980; Christiansson *et al.*, 1985; Lindblom *et al.*, 1986; Rilfors *et al.*, 1987; Wieslander and Selstam, 1987). These studies, together with unpublished results, have also shown that a substantial synthesis of MAMGlcDAG only occurs when the value of the ratio 16:0/18:1c is around 1 or higher; MAMGlcDAG can at the most make up 15–20 mole% of the membrane lipids (Wieslander *et al.*, 1979, 1993a). Moreover, the

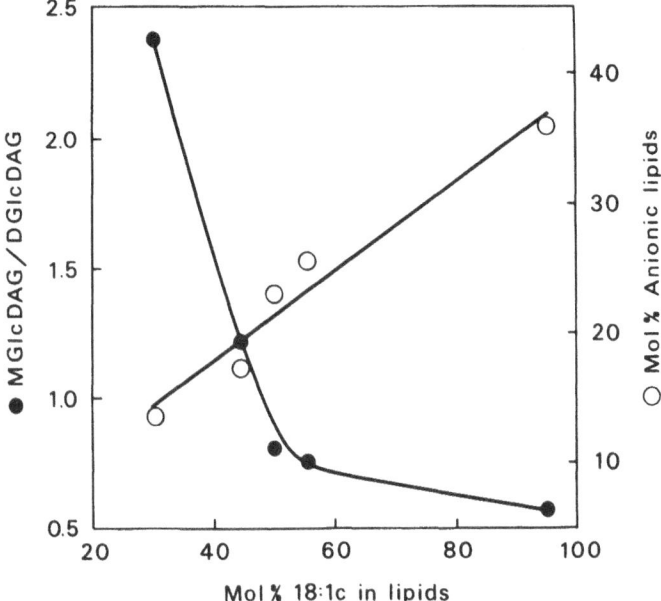

FIGURE 5. The ratio MGlcDAG/DGlcDAG and the fraction of anionic lipids in membranes from *A. laidlawii* strain A-EF22 grown for 18 hr at 37°C in media supplemented with different proportions of 16:0 and 18:1c. The glucolipid ratio and the fraction of anionic lipids are shown as a function of the molar fraction of 18:1c in the membrane lipids. (From Wieslander *et al.*, 1980.)

key features mentioned above are exhibited also when saturated fatty acids, synthesized by the cells themselves, are incorporated into the membrane lipids instead of the exogenously supplied 16:0 (Christiansson and Wieslander, 1980). Incorporation of 18:1t gives a higher value of the ratio MGlcDAG/DGlcDAG, and a smaller fraction of anionic lipids, as compared with 18:1c (Wieslander and Rilfors, 1977), and incorporation of 18:2c further decreases the glucolipid ratio, and raises the fraction of anionic lipids, as compared with 18:1c (Rilfors *et al.*, 1987; Wieslander and Selstam, 1987). In addition to the effects on the membrane polar head-group composition, the incorporation of different fatty acids promotes growth differently and affects the osmotic fragility of the cells (e.g., see Rottem and Panos, 1969).

Several genera of gram-positive and gram-negative bacteria have membrane lipids which contain mainly iso and anteiso methyl-branched saturated acyl chains (Kaneda, 1977). When these fatty acids are incorporated into the membrane lipids of *A. laidlawii* strain A-EF22, a lower value of the ratio MGlcDAG/DGlcDAG, and a larger fraction of the anionic lipids, are obtained with anteiso acids than with iso acids of the same length (Rilfors, 1985). The glucolipid ratio is further decreased, and the fraction of anionic lipids increased, when the lipids contain acyl chains with the methyl-branching point one position above that of anteiso acids (Rilfors *et al.*, 1993a). MAMGlcDAG makes up merely 1–4 mole% of the membrane lipids when the organism is fed with the branched-chain fatty acids (Rilfors, 1985).

The length of the fatty acids also influences the polar head-group composition of the membrane lipids. The value of the ratio MGlcDAG/DGlcDAG is directly related to the average acyl chain length of the lipids. Short chains give a high value of the ratio; the value gradually decreases up to a chain length of approximately 18 carbon atoms, from which a further increase in length yields no, or a minor, change in the ratio (Wieslander *et al.*, 1993a). The chain length dependence has been shown to occur with straight and branched-chain saturated fatty acids, and with *cis*-monounsaturated fatty acids (Wieslander and Rilfors, 1977; Rilfors, 1985; Wieslander *et al.*, 1993a). Likewise, the extent of the MAMGlcDAG synthesis is related to the acyl chain length. Maximum amounts are made with a high percentage of saturated acyl chains with 16 to 18 carbon atoms. Shorter, and especially longer chains strongly reduce the fraction of MAMGlcDAG made (Wieslander *et al.*, 1993a). The fraction of the anionic lipid PG also seems to correlate with the acyl chain length, with a larger fraction associated with longer chains. Furthermore, PG is the dominating anionic lipid in membranes with long chains, whereas for short chains the phosphoglucolipids dominate this subfraction. The amounts of GPMGlcDAG and GPDGlcDAG seem to correlate with the synthesized amounts of MGlcDAG and DGlcDAG, respectively (Wieslander *et al.*, 1993a), which might relate to the proposed biosynthetic pathways for these phosphoglucolipids (Figure 1).

The overall acyl chain composition of the membrane lipids in *A. laidlawii* A-EF22 usually reflects the fatty acid composition of the growth medium. However, when two or more fatty acids are added to the medium, these acids are incorporated in different proportions into the individual lipids. When strain A-EF22 cells are grown on mixtures of 16:0 and 18:1c, the fraction of 18:1c in MGlcDAG is approximately 60–90% of that in DGlcDAG, while PG has the same, or a slightly higher, degree of unsaturation than DGlcDAG; MAMGlcDAG is always highly enriched in saturated acyl chains (Wieslander and Rilfors, 1977; Christiansson and Wieslander, 1978, 1980; Wieslander *et al.*, 1978; Eriksson *et al.*, 1991). The position of a methyl-branching point in the hydrocarbon chain, and the hydrocarbon chain length, affect the incorporation of fatty acids into the individual lipids in a rather complex way (Rilfors, 1985). For example, if the cells are fed with 15-methylhexadecanoic acid (iso-C_{17}) together with 14-methylhexadecanoic acid (anteiso-C_{17}) or 13-methyltetradecanoic acid (iso-C_{15}), the fraction of iso-C_{17} is extremely high in MAMGlcDAG, and this acid is also preferentially directed into PG and MGlcDAG. One common feature in the results obtained with the unsaturated and the branched-chain fatty acids is that

MAMGlcDAG is enriched in acids with high melting points, i.e., acids which support poor growth, or no growth at all, of *A. laidlawii* strain A-EF22. It can be concluded that each of the lipid-synthesizing enzymes selects proper species from the substrate pool, which results in the maintenance of a controlled acyl chain composition of the individual lipids. However, it should be mentioned that the covalently bound acyl chains of the acyl proteins in this A strain are much more carefully selected with respect to chain length and degree of unsaturation (Nyström *et al.*, 1992).

6.1.2. Growth Temperature

If cells of strain A-EF22 incorporate two or more exogenously supplied fatty acids into the membrane lipids, or if the cells are able to synthesize saturated fatty acids by themselves, the *acyl chain* composition is regulated when the growth temperature is altered. This ability is most clearly observed by applying a temperature-shift technique (Christiansson and Wieslander, 1978). Cells grown in an equimolar mixture of 16:0 and 18:1c incorporate a larger fraction of 18:1c when shifted from 37 to 17°C. This change takes different courses in the different lipids. The most rapid change occurs in MGlcDAG, and 4 hr after the shift the fraction of 18:1c is highest in this lipid. However, 7 hr after the shift the fraction of 18:1c has increased in all lipids except DGlcDAG (Christiansson and Wieslander, 1978). If the organism is grown in a medium containing equimolar concentrations of 12-methyltetradecanoic acid (anteiso-C_{15}), anteiso-C_{17}, iso-C_{15}, and iso-C_{17}, the value of the ratio iso/anteiso acyl chains is decreased when the temperature is changed from 37 to 17°C. Moreover, when the cells are allowed to synthesize long-chain iso and anteiso acids from the short-chain precursors 2-methylbutanoic acid (anteiso-C_5) and 3-methylbutanoic acid (iso-C_5), the value of the ratio iso/anteiso acyl chains is lower, and the average chain length is shorter, at 27°C than at 37°C (Rilfors, 1985).

It is uncertain if these features are valid for all A strains of *A. laidlawii*. For an oral (A) strain the percentage of 18:1c increased, and the average chain length decreased, upon growth at lower temperatures as expected for a regulatory mechanism (Rottem *et al.*, 1970; Huang *et al.*, 1974). However, this was not observed in an earlier investigation (Rottem and Panos, 1969). Possible explanations for these contradictory results may be that *A. laidlawii* is a cluster of various strains with different extents of relatedness (Stephens *et al.*, 1983), or that crucial, as yet unknown, experimental conditions vary between different laboratories.

Changes in the growth temperature influence also the *polar head-group* composition of the membrane lipids in strain A-EF22. With the temperature-shift technique it was shown that the value of the ratio MGlcDAG/DGlcDAG is profoundly increased when the cells are shifted from 37 to 17°C. This change is exhibited when the organism is fed with an equimolar mixture of 16:0 and 18:1c or with 18:1c alone (Christiansson and Wieslander, 1978; Wieslander *et al.*, 1980); in the latter case the acyl chain composition of the lipids is forced to be constant under conditions of very low endogenous fatty acid synthesis. The value of the glucolipid ratio is also markedly higher at 27°C than at 37°C when the cells are grown on long-chain iso and anteiso fatty acids (Rilfors, 1985). The fraction of anionic lipids is decreased by lowering the growth temperature when the cells are fed with an equimolar mixture of 16:0 and 18:1c, but virtually no change occurs when the cells are fed with 18:1c alone (Wieslander *et al.*, 1980).

6.1.3. Incorporation of Foreign Molecules into the Membrane

Organisms belonging to the family Acholeplasmataceae do not require sterols for growth (see Freundt and Edward, 1979). However, several sterols are incorporated without chemical modifications into the *A. laidlawii* membrane when they are supplied in the growth medium. Similarly, numerous other hydrophobic and amphiphilic molecules, like nonpolar organic solvents, alcohols, detergents, anesthetics, and chlorophyll, can be incorporated into the cell membrane.

The effect on the *acyl chain* composition has been studied with cholesterol and two anesthetics, tetracaine and diethyl ether. When the cells are grown on an equimolar mixture of 16:0 and 18:1c, the fraction of 18:1c in the lipids increases when cholesterol is present in the membrane (Christiansson and Wieslander, 1978), while no effect on the acyl chain composition is obtained with the anesthetics (Christiansson *et al.*, 1981).

The influence of foreign molecules on the *polar head-group* composition has been much more intensively examined. Cholesterol decreases the value of the ratio MGlcDAG/DGlcDAG when the cells are grown on 18:2c, 18:1c, and an equimolar mixture of 16:0 and 18:1c, but leaves the glucolipid ratio essentially unaltered when the cells are grown on a 4/1 (mole/mole) mixture of 16:0 and 18:1c (Christiansson and Wieslander, 1978; Wieslander *et al.*, 1979; Rilfors *et al.*, 1987; Wieslander and Selstam, 1987). The influence of cholesterol on the composition of lipids enriched in iso and anteiso acyl chains is dependent on the acyl chain length; the glucolipid ratio is profoundly decreased when the lipids contain C_{15} chains, while it is slightly increased when the lipids contain C_{17} chains (Rilfors *et al.*, 1987). Cholesterol raises the fraction of anionic lipids when supplied in combination with 18:2c, 18:1c, and branched-chain fatty acids, but has practically no effect when the membrane lipids are enriched in straight saturated acyl chains (Christiansson and Wieslander, 1978; Wieslander *et al.*, 1979; Rilfors, 1982; Wieslander and Selstam, 1987). Linolenic acid (18:3c) does not promote growth of *A. laidlawii* strain A-EF22, but when the acid is supplied together with cholesterol, the cells are able to grow (Wieslander and Selstam, 1987). It has been noticed that the molar fraction of cholesterol, as well as of several other sterols, in the membrane increases with an increasing extent of acyl chain *cis*-unsaturation in the lipids (Razin, 1978; Wieslander and Selstam, 1987; Rilfors *et al.*, 1987).

A study of the effects on the polar head-group composition of the incorporation of 12 sterols, in combination with different fatty acids, has been carried out (Figure 6) (Rilfors *et al.*, 1987). When the sterols were supplied together with a 4/1 (mole/mole) mixture of 16:0 and 18:1c, one sterol (lanosterol) raised the value of the ratio MGlcDAG/DGlcDAG, seven sterols did not alter this ratio, and four sterols (epicoprostanol and three sterols with a keto group instead of a hydroxyl group in position 3) lowered the ratio (Figure 6A). Three out of these latter four sterols lysed the cells when supplied in combination with 18:1c. In this case the shift technique was applied in order to examine the influence of these sterols on the glucolipid ratio; all of the sterols drastically reduced this ratio within 6 hr after their addition. All of the nonlytic sterols also decreased the glucolipid ratio in combination with 18:1c (Figure 6B), and the magnitude of this decrease can be related to the chemical structure of the sterols. The same three sterols were lytic when supplied together with 18:2c, and the effect of the sterol structure on the value of the ratio MGlcDAG/DGlcDAG was similar to the effect obtained with 18:1c. Generally, however, the glucolipid ratio was less affected by the sterols in the membranes enriched in 18:2c (Figure 6C).

The regulation of the polar head-group composition has also been investigated when strain A-EF22 was grown in the presence of three classes of hydrophobic and amphiphilic molecules which the organism does not meet in its natural habitats: nonpolar organic solvents, alcohols, and detergents (Wieslander *et al.*, 1986). The only fatty acid supplied to the growth medium was 18:1c. Thus, 18:1c made up ≥ 95 mole% of the lipid acyl chains in all experiments. The nonpolar organic solvents used (three hydrocarbons and diethyl ether) partition more or less deeply into the hydrophobic interior of the bilayer, and all of them decrease the value of the ratio MGlcDAG/DGlcDAG (Figure 7). Diethyl ether is the least potent of the four molecules (see also Christiansson *et al.*, 1981), which probably is explained by the higher value of its dielectric constant relative to the hydrocarbons.

The alcohols were represented by the *n*-alcohols with an even number of carbon atoms from ethanol to hexadecanol and by phenethyl alcohol. These molecules are anchored with the hydroxyl group near the hydrocarbon–water interface, and the hydrocarbon chain extends into the hydrocarbon region of the lipid bilayer. The effect exerted by the incorporation of the alcohols into the cell membrane is dependent on the hydrocarbon chain length (Figure 8). Ethanol increases the glucolipid ratio, while 1-butanol causes practically no change in it. The largest decrease in the lipid ratio is achieved by 1-hexanol, 1-octanol, and phenethyl alcohol. This decrease becomes gradually smaller when the chain length increases from 1-decanol to 1-hexadecanol. The latter effect probably

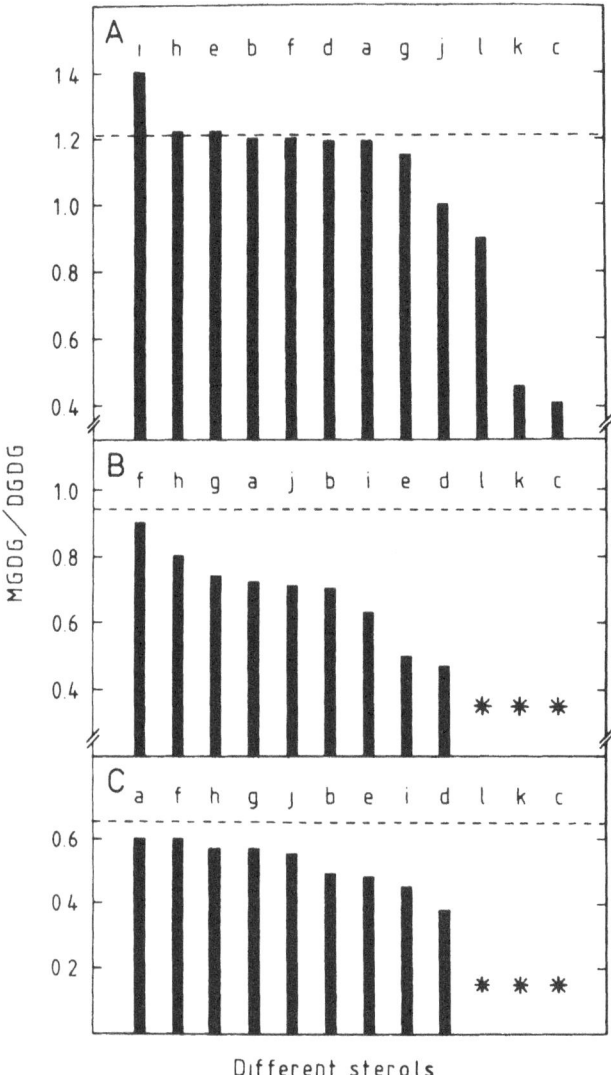

FIGURE 6. MGlcDAG/DGlcDAG ratios in membranes from *A. laidlawii* strain A-EF22 obtained after incorporation of various sterols. Cells were grown for 20 hr at 30°C with a sterol (20 μM) and 120 μM 16:0 plus 30 μM 18:1c (A), 150 μM 18:1c (B), or 150 μM 18:2c (C). The horizontal dashed lines show the MGlcDAG/DGlcDAG ratios in control cultures lacking sterols. The asterisks indicate sterol supplements that inhibited growth due to cell lysis. a, androstanol; b, cholestanol; c, epicoprostanol; d, cholesterol; e, 7-dehydrocholesterol; f, ergosterol; g, β-sitosterol; h, stigmasterol; i, lanosterol; j, cholestanone; k, cholest-4-en-3-one; l, cholest-5-en-3-one. For further details, see Rilfors *et al.* (1987).

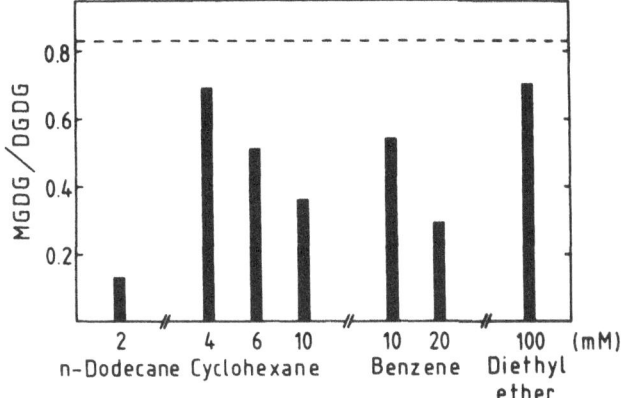

FIGURE 7. The effect of nonpolar organic molecules on the regulation of the ratio MGlcDAG/DGlcDAG in membranes from *A. laidlawii* strain A-EF22. Cells were grown at 30°C and the membrane lipids contained ≥ 95 mole% 18:1c. The horizontal dashed line shows the glucolipid ratio in a control culture devoid of organic molecules. For further details, see Wieslander *et al.* (1986).

depends on the fact that the solubility in the growth medium is about 2000 times lower for 1-hexadecanol than for 1-decanol, while the lipid–buffer partition coefficient is just about six times higher. Thus, the fraction of the long-chain alcohols in the cell membrane is gradually diminished with increasing chain length.

Poly(oxyethylene) alkyl ether detergents were chosen with hydrocarbon chain lengths of 12 and 16 and with a varying number of oxyethylene units in the

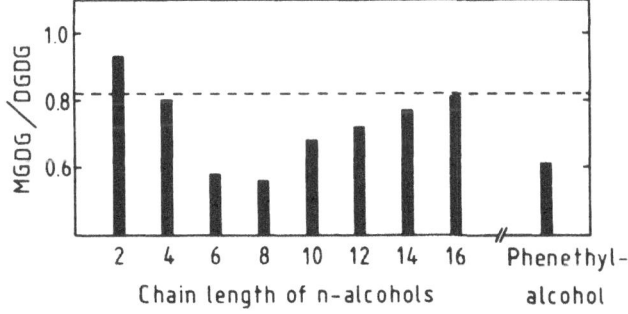

FIGURE 8. The effect of different alcohols on the regulation of the ratio MGlcDAG/DGlcDAG in membranes from *A. laidlawii* strain A-EF22. Cells were grown at 30°C and the membrane lipids contained ≥ 95 mole% 18:1c. The horizontal dashed line shows the glucolipid ratio in a control culture devoid of alcohols. For further details, see Wieslander *et al.* (1986).

head group (Figure 9). When these molecules are present in a bilayer, the oxyethylene units are located in the hydrocarbon–water interfacial region and the alkyl chain extends into the hydrocarbon region. In both surfactant series, the species with the smallest head group decrease the value of the ratio MGlcDAG/DGlcDAG; the ratio then gradually increases when the number of oxyethylene units increases from three to eight.

Besides the general anesthetic diethyl ether, an investigation of the local anesthetic tetracaine has been carried out with the shift technique (Christiansson *et al.*, 1981). The value of the ratio MGlcDAG/DGlcDAG decreases slightly immediately after the addition of the molecule to the growth medium, but 6 hr after the shift and onwards the value of the glucolipid ratio is somewhat higher in the tetracaine-supplemented culture than in the control culture. Incorporation of this cationic anesthetic into the cell membrane raises the relative amount of the anionic lipids by approximately 20%.

Finally, strain A-EF22 has been grown on chlorophyll a together with 18:3c, 18:2c, 18:1c, and a 4/1 (mole/mole) mixture of 16:0 and 18:1c (Wieslander and Selstam, 1987). In comparison with the control culture, chlorophyll causes a small increase in the value of the ratio MGlcDAG/DGlcDAG in combination with 18:2c and the 16:0/18:1c mixture, but a small decrease of the lipid ratio in combination with 18:1c. In conformity with cholesterol, the fraction of anionic lipids is raised by chlorophyll in membranes enriched in 18:1c and 18:2c, but remains unaltered in membranes enriched in the 16:0/18:1c mixture; chlorophyll promotes growth of strain A-EF22 on 18:3c; and chlorophyll is incorporated into the membrane in drastically increasing amounts with an increasing degree of acyl chain unsaturation of the lipids.

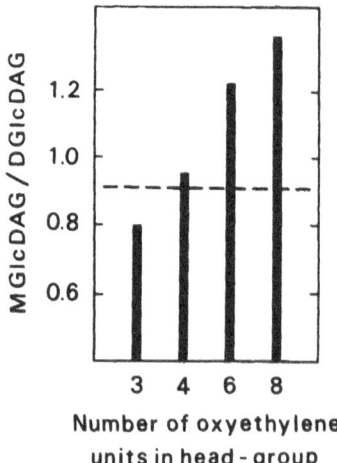

FIGURE 9. The ratio MGlcDAG/DGlcDAG in membranes from *A. laidlawii* strain A-EF22 after incorporation of poly(oxyethylene) alkyl ether detergents with an alkyl chain of 16 carbon atoms and with a varying number of oxyethylene units in the polar head group. Cells were grown at 30°C and the membrane lipids contained ≥ 95 mole% 18:1c. The horizontal dashed line shows the glucolipid ratio in a control culture devoid of detergents. For further details, see Wieslander *et al.* (1986).

6.1.4. Membrane Surface Potential

The activity and kinetic properties of membrane-bound enzymes can be altered when the surface charge density of the membrane is changed (Rilfors *et al.*, 1984; Clementz *et al.*, 1988, and references therein), and it is therefore conceivable that the surface charge density is actively maintained at a certain level by an organism. The latter question has been investigated in strain A-EF22 (Christiansson *et al.*, 1985). The cells were grown in media with different proportions between 16:0 and 18:1c, and the fraction of anionic lipids increased from 30 to 40 mole% when the fraction of 18:1c in the lipids increased from 33 to 95 mole%. It was shown by force–area measurements on monolayers of total lipid extracts that the mean molecular area of the lipids increased concomitantly from 56 to 70 Å² at a surface pressure of 35 mN/m. The surface charge density is thereby kept approximately constant at a value of one elementary charge per 175 ± 6 (S.D.) Å², assuming complete dissociation of the protons of the phosphate groups and the ionic strength prevailing in the growth medium. Measurements of the surface potential revealed that this quantity is also kept fairly constant at a value of −35 mV under the assumptions mentioned above.

A regulation of the fraction of anionic lipids is also achieved by growing the organism in media with different concentrations of NaCl (Christiansson *et al.*, 1985). If the NaCl concentration is raised from 85 to 420 mM, the fraction of anionic lipids increases from 32 to 45 mole%, and from 21 to 41 mole%, when the cells are fed with 18:1c and an equimolar mixture of 16:0 and 18:1c, respectively (Figure 10A). With model systems consisting of liposomes made of dioleoyl-PG and dioleoyl-DGlcDAG mixed in different proportions, it was shown that (Figure 10B): (1) the surface potential increases with an increasing fraction of PG, and decreases with an increasing concentration of NaCl in the surrounding water phase; (2) the largest changes in the surface potential are observed when the fraction of anionic lipids is varied between 0 and 50 mole%; and (3) the surface potential is about −35 mV at physiological ionic strengths and lipid surface charge densities. From these studies it was concluded that the membrane surface potential is an important regulatory factor for the synthesis of anionic lipids in strain A-EF22 when the acyl chain composition and the monovalent cation concentrations are varied.

6.2. *A. laidlawii* B

6.2.1. Incorporation of Fatty Acids

As is the case with the A strains, three kinds of changes in the polar head-group composition have been observed to occur when B strains are grown in

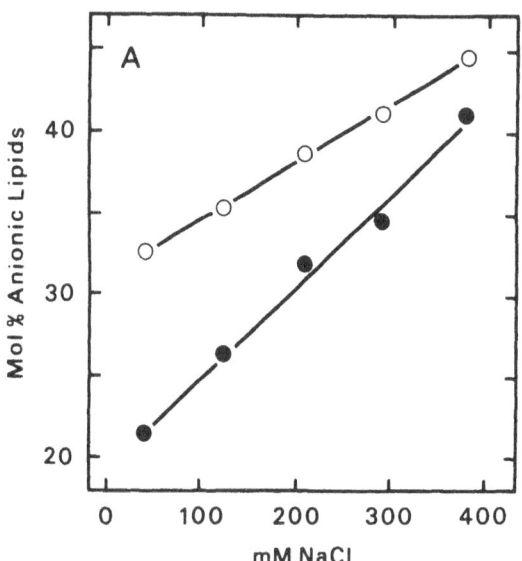

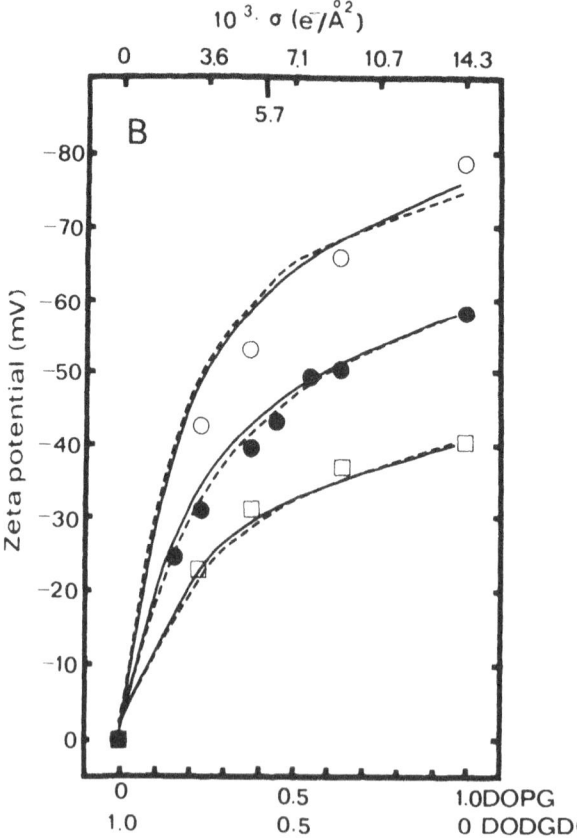

media supplemented with different fatty acids: (1) the proportion between MGlcDAG and DGlcDAG; (2) the proportion between the glucolipids and PG; and (3) the proportion of MAPGlc.

In strain B-PG9 the value of the ratio MGlcDAG/DGlcDAG decreases progressively in the order 16:0 > 18:1t > 18:1c-enriched cells (Bhakoo and McElhaney, 1988). The glucolipid ratio is also lowered when the cells are grown on anteiso acids instead of iso acids, and when the chain length of the fatty acids is increased (Silvius *et al.*, 1980). In the latter investigation, no clear correlation was observed between the fraction of PG and the chemical structure of the fatty acid incorporated into the membrane lipids; the fraction of PG varied over a rather narrow range. On the other hand, Bhakoo and McElhaney (1988) found that the fraction of PG is larger in cells grown on 16:0 than in cells grown on 18:1t and 18:1c, which is the opposite of what is observed in strain A-EF22. We have observed that strain B-PG9 has a lower capacity for incorporation of 18:1c and an enhanced endogenous synthesis of saturated fatty acids, especially in a lipid-depleted bovine serum albumin (BSA)–heart infusion broth–peptone–yeast extract growth medium, as compared with the strain A-EF22 (Wieslander *et al.*, 1993b). This yields more saturated acyl chains and a slightly shorter average acyl chain length in the B strain. A third glucolipid, MAPGlc, is synthesized by the B strain when the cells are fed with high-melting fatty acids, such as straight-chain saturated acids, methyl-branched iso acids, and ω-cyclohexyl acids (Bhakoo *et al.*, 1987; Lewis *et al.*, 1990b).

The regulation of the polar head-group composition in strain B-JU has been examined (Johansson *et al.*, 1981; Clementz *et al.*, 1987). The cells were grown in the same lipid-depleted BSA–tryptose medium (Rilfors, 1985) used to grow strain A-EF22, and the medium was supplemented with PPLO serum fraction or with 16:0 and 18:1c in different proportions. The value of the ratio MGlcDAG/DGlcDAG, as well as the fraction of anionic lipids, varied in the same way as that found for the A strain. Moreover, the acyl chain composition of the individual lipids in strain B-JU grown under these conditions differs in a

---

FIGURE 10. (A) Regulation of the fraction of anionic lipids in membranes from *A. laidlawii* strain A-EF22. Cells were grown for 20 hr at 37°C in media with different concentrations of NaCl (in *addition* to 43 mM NaCl in the basal growth medium). ●, medium containing 75 μM each of 16:0 and 18:1c; ○, medium containing 150 μM 18:1c. (B) Experimental and simulated ζ-potentials in mixtures of dioleoyl-PG and dioleoyl-DGlcDAG isolated from *A. laidlawii* membranes. *x* axis, fraction of dioleoyl-PG and dioleoyl-DGlcDAG. σ, surface charge density assuming an area per lipid molecule of 70 Å^2, complete miscibility, and no transbilayer asymmetry of the two lipids. ○, 50; ●, 100; and □, 200 mM NaCl in a suspending buffer containing 1 mM Tris (pH 7.5) and 0.1 mM EDTA. Dashed and solid curves, simulation of the ζ-potential curves with the Gouy–Chapman–Langmuir equations by assuming a plane of shear of 2 Å and a Na$^+$-binding constant of 1.2 M^{-1}, or 3 Å and 0.6 M^{-1}, respectively. (From Christiansson *et al.*, 1985.)

similar way as in the A strain; the fraction of 18:1c is approximately the same in PG and DGlcDAG, while it is much lower in MGlcDAG (Christiansson, 1981). Saito and McElhaney (1977) studied the acyl chain composition of the individual lipids in strain B-PG9. However, they could discern no regular relationship between the chemical structure of the exogenously added fatty acids and their distribution in the different lipids.

6.2.2. Growth Temperature

The influence of the growth temperature on the *acyl chain* composition of the membrane lipids in strain B-PG9 has been studied in various ways by McElhaney and colleagues. When no exogenous fatty acids were added to the growth medium, the average length of the synthesized acyl chains was slightly decreased with a reduction of the temperature (McElhaney, 1974). However, no systematic effect of the temperature was obtained when the cells were grown on the short-chain precursors 2-methylpropionic acid (iso-C_4) and iso-C_5, on several medium-chain precursors, or on 18:1c (McElhaney, 1974; Saito *et al.*, 1977, 1978). *In vitro* experiments with membranes and derived lipids of this B strain show a temperature-dependent uptake of 16:0 versus 18:1c with increasing temperature that is consistent with a regulatory mechanism, but the process was only analyzed for very short time periods (Melchior and Steim, 1977). During longer time periods *in vivo*, the fraction of exogenously supplied 18:1c in the lipids does not vary with the temperature (McElhaney, 1974).

The effect of the growth temperature on the *polar head-group* composition of the membrane lipids in strain B-PG9 is also limited (Bhakoo and McElhaney, 1988). The value of the ratio MGlcDAG/DGlcDAG is raised by decreasing the temperature; this increase is very marked for cells enriched in 16:0, is modest for cells enriched in 18:1c, and is slight or nonexistent for cells supplemented with 18:1t. A small decrease in the fraction of PG is observed for all fatty acid supplements when the temperature is reduced, but the change may not be statistically significant (Bhakoo and McElhaney, 1988).

6.2.3. Incorporation of Cholesterol into the Membrane

No systematic variations in the *acyl chain* composition of the membrane lipids are obtained when strain B-PG9 is cultured in the presence of cholesterol together with the short-chain precursors iso-C_4 and iso-C_5, or together with several medium-chain precursors (McElhaney *et al.*, 1973; Saito *et al.*, 1977, 1978). However, incorporation of cholesterol affects the *polar head-group* composition of the lipids (Bhakoo and McElhaney, 1988). The value of the ratio MGlcDAG/DGlcDAG is reduced markedly, moderately, and only slightly when the organism is grown on 16:0, 18:1t, and 18:1c, respectively. Thus, at a high level of cholesterol incorporation, the glucolipid ratio of the membranes enriched in 16:0 falls to a value below that of membranes enriched in 18:1t and 18:1c. These results are the opposite of those reported for the A-EF22 strain. This is also the case for the influence of cholesterol on the fraction of PG in the B strain; cholesterol substantially increases the fraction of PG in cells grown on 16:0, while cells grown on the unsaturated fatty acids show only minor alterations of this quantity (Bhakoo and McElhaney, 1988). In conformity with the A strain, the molar fraction of cholesterol in the membrane of the B strain increases with an increasing extent of acyl chain unsaturation in the lipids (Bhakoo and McElhaney, 1988).

6.2.4. Transmembrane Potential

Cells of strain B-JU containing membrane lipids enriched in 18:1c have a transmembrane electrical potential of approximately -50 mV (inside negative) when grown in a lipid-depleted BSA–tryptose medium. The relationship between the transmembrane potential and the membrane lipid composition has been examined (Clementz et al., 1986, 1987). In a first series of experiments the cells were grown on 18:1c, and this fatty acid thus constituted ≥ 95 mole% of lipid acyl chains. The transmembrane electrical potential was changed by adding the ionophores valinomycin, monensin, and nigericin, and the lipophilic tetraphenylphosphonium cation (TPP$^+$). Upon addition of valinomycin to growing cells, a rapid and substantial decrease of the value of the ratio MGlcDAG/DGlcDAG was obtained (Figure 11). The change could be detected in less than 10 min after the addition and at a concentration of 3 nM, which corresponds at the most to about one valinomycin molecule per 2000 lipid molecules. The intracellular concentration of K$^+$ decreased from 190 to 45 mM with 1 µM valinomycin and this ionophore consequently causes a hyperpolarization of the cell membrane. The valinomycin-induced effects on the glucolipid ratio and the membrane hyperpolarization were gradually diminished by adding increasing amounts of K$^+$ to the growth medium (Figure 11). The possibility that the ionophore affects the glucolipid ratio directly by disturbing the lipid packing is therefore excluded. When TPP$^+$ was added to growing cells, the membrane was depolarized, and the glucolipid ratio was increased, in a dose-dependent manner. The concentration of TPP$^+$ in the membrane is at the most about one molecule per 30 lipid molecules, and a direct disturbance of the ionophore on the lipid packing cannot be ruled out. Finally, the addition of monensin or nigericin to a growing cell culture resulted in a lowering of the glucolipid ratio accompanied by an increase in the transmembrane electrical potential.

The effect of valinomycin in combination with different fatty acid supplements was studied in a second series of experiments (Clementz et al., 1987). When the cells were grown on fatty acid supplements ranging from 18:2c to a 4/1

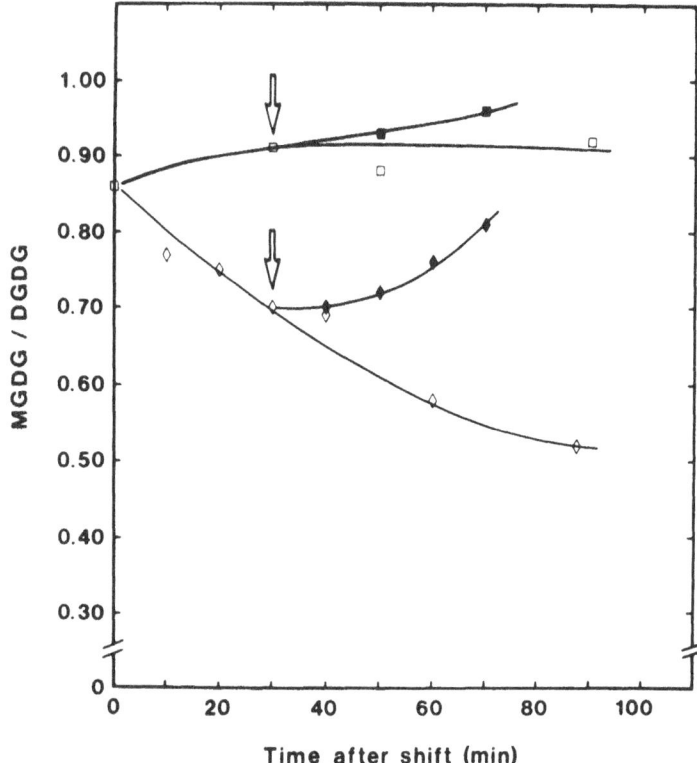

FIGURE 11. Effect of valinomycin and KCl on the ratio MGlcDAG/DGlcDAG in membranes from *A. laidlawii* strain B-JU. Cells were grown at 30°C and the membrane lipids contained ≥ 95 mole% 18:1c. Arrows indicate the time at which KCl was added. □, control; ■, control + 50 mM KCl; ◇, 7.5nM valinomycin; ◆, 7.5 nM valinomycin + 50 mM KCl. For further details, see Clementz *et al.* (1986).

(mole/mole) mixture of 16:0 and 18:1c, the transmembrane electrical potential varied between −52 and −62 mV. However, there was no correlation between the magnitude of the potential and the degree of unsaturation of the lipid acyl chains. The addition of valinomycin caused a hyperpolarization of the membrane and a decreased value of the ratio MGlcDAG/DGlcDAG with all fatty acid supplements, but no correlation was found between the degree of unsaturation of the lipid acyl chains and the magnitude of the hyperpolarization or the magnitude of the decrease of the glucolipid ratio. However, for the cells supplemented with the 16:0/18:1c mixtures, a strong correlation was observed between the two latter quantities, i.e., the larger the hyperpolarization caused by valinomycin, the larger the reduction of the glucolipid ratio; cells enriched in 18:2c did not show this correlation. These experiments suggest that the transmembrane potential, directly or indirectly, affects the lipid composition and regulation in the membrane of *A. laidlawii* strain B-JU, although it cannot be excluded that the observed effects on the glucolipid ratio may be attributed to other factors such as the transmembrane distributions of K^+ and Na^+ ions (Clementz *et al.*, 1986, 1987). However, for large unilamellar vesicles made from *A. laidlawii* lipids containing 95 mole% of 18:1c, the permeability of glucose is affected by the value of the ratio MGlcDAG/DGlcDAG and by the presence of an imposed artificial transmembrane electrical potential (T. Clementz and Å. Wieslander, unpublished results). This indicates that the glucolipid ratio, i.e., the packing of the lipid molecules (see Section 7), and the membrane potential can affect such a crucial property of the membrane as the permeability.

6.2.5. Lipid/Protein Ratio

A correlation has been observed between the chemical structure of the fatty acid incorporated into the membrane lipids and the total amount of membrane protein relative to lipid (Silvius *et al.*, 1980). The lipid/protein ratio has a low value in cells of strain B-PG9 grown in an unsupplemented medium, and in media supplemented with *trans*-unsaturated or cyclopropane fatty acids. *Cis*-unsaturated species increase the ratio as much as twofold, while branched-chain fatty acids have an intermediate effect. Similar results were obtained by Johansson *et al.* (1981) by growing the B-JU strain in the lipid-depleted BSA–tryptose medium used to grow strain A-EF22. The value of the lipid/protein ratio was nearly twice as large in cells having membrane lipids with 95 mole% 18:1c than in cells having membrane lipids with 30 mole% 18:1c. However, cells grown on PPLO serum fraction did not show this correlation. No consistent correlation between the lipid/protein ratio and the acyl chain composition of the membrane lipids has been found in strain A-EF22 (Wieslander and Rilfors, 1977). Reasons for a regulation of the lipid/protein ratio might be that saturated acyl chains are better suited structurally to accommodate the membrane proteins, or that saturated acyl chains decrease the passive permeability of the lipid bilayer which saves energy that instead can be used for protein synthesis.

6.3. Other *A. laidlawii* Strains

The capacity of 20 *A. laidlawii* strains to incorporate cholesterol into the membrane was studied by Efrati *et al.* (1986). The cells were grown in media supplemented either with a 7/1 (mole/mole) mixture of 18:1t and 16:0 or with an equimolar mixture of these acids. According to a statistical analysis, the two lipids MGlcDAG and MAMGlcDAG were highly associated with the variation in the cholesterol uptake observed between the strains. These two lipids explained 90% of the cholesterol uptake variation, and the strongest correlation was found for MAMGlcDAG. The capacity of the *A. laidlawii* strains for cholesterol incorporation is inversely proportional to the relative amounts of these glucolipids. The other membrane lipids do not contribute to the predicted value for the cholesterol uptake.

7. PHYSICOCHEMICAL INTERPRETATIONS OF THE LIPID REGULATION MECHANISMS

It is conceivable that several structural and physicochemical properties of a biological membrane must be kept constant, or within certain limits, in order for a cell to function properly. A first basic demand is, of course, that the membrane is intact and able to maintain its barrier properties. Consequently, apart from possible short-lived and highly localized regions of nonbilayer structures, the membrane lipids must form a bilayer structure together with the membrane proteins. The bilayer structure is the only known lipid aggregate structure that can provide the necessary barrier properties, and the lipid composition of a biological membrane must be such that permanent nonbilayer structures cannot form. A second basic requirement can be presented from studies of *A. laidlawii* and *E. coli:* at least 10–50% of the membrane lipids must be in a liquid-crystalline state. Otherwise the activity of membrane-bound enzymes and transport proteins will decrease below a critical level and the organisms cease to grow and divide (McElhaney, 1984). When these two demands are met, it is still conceivable that some structural properties and physicochemical quantities must be regulated and fine-tuned in order to achieve an optimally working membrane: e.g., the thickness and the passive permeability of the bilayer, the order parameter of the hydrocarbon chains, lipid asymmetry, the surface charge density and the surface potential, the transmembrane chemical and electrical potentials, the lateral and rotational diffusion coefficients of the lipids and proteins, and mechanical properties like the elasticity, the area compressibility modulus, and the curvature modulus. Some of these quantities are related to each other.

The contours of the regulation of a few of the above-mentioned structural properties and physicochemical quantities have emerged from the studies of *A. laidlawii.* As will be discussed in detail below, the lipid regulation mechanisms, which the present authors have observed to occur in *A. laidlawii* strain A-EF22, have mainly been interpreted as: (1) a striving of the organism to maintain a bilayer structure which, however, in thermodynamic terms (e.g., temperature) is comparatively "close" to a transition to nonbilayer structures; and (2) a striving to maintain a constant lipid surface charge density and surface potential. In a few cases the physiological results can be interpreted as a regulation of the transition between L_β and L_α phases, or in more specific molecular terms, a regulation of the order parameter of the acyl chains, or the bilayer thickness. When discussing the first statement above, it is important to realize that the transition between bilayer and nonbilayer structures is of biological relevance whether or not the latter structures occur permanently in a biological membrane or elsewhere in a cell. The reason for this is that the physicochemical interactions and forces which induce a transition between bilayer and nonbilayer structures in pure lipid systems are also operative in a biological membrane (Seddon, 1990b, and references therein). This implies that the presence in the bilayer of a non-bilayer-forming lipid may have subtle, yet decisive influences on membrane-related processes. Lipid systems containing a fraction of non-bilayer-forming lipids, but forming a bilayer, exist in a condition characterized by a more or less pronounced internal stress, since the forces shifting the equilibrium toward nonbilayer structures are already present in the bilayer long before the actual transition occurs. The principal factor driving such a transition is exerted by the packing of the lipid molecules, resulting in a tendency for one or both monolayer halves of the bilayer to curl away from a planar structure (Seddon, 1990b).

More than 10 years ago a hypothesis was presented for the physicochemical causes of the regulation of the membrane lipid composition in *A. laidlawii* strain A-EF22 (Wieslander *et al.*, 1980, 1981b). The theory for the self-assembly of amphiphiles, developed by Israelachvili and colleagues (Israelachvili *et al.*, 1976, 1980), was applied to the results from the biochemical studies of the lipid composition, and the investigations of the phase equilibria in lipid–water systems. Israelachvili *et al.* concluded that the effective geometry of the lipid molecules determines the shape of the aggregates that are formed. The packing of the lipids into different aggregates is dependent on the hydrophobic volume, the hydrocarbon–water interfacial area, and the hydrocarbon chain length of the participating molecules. According to this approach, bilayer aggregates are built up of cylindrical-like molecules. A decrease in the interfacial area, an increase in the hydrophobic volume, or a simultaneous shortening and broadening of the hydrophobic region with no volume change, causes the lipid molecules to adopt an effective shape of a truncated cone. This, in turn, will favor the formation of nonbilayer aggregates of the reversed type. The opposite changes of the above-mentioned quantities will yield nonbilayer aggregates of the normal type. The different aggregate structures, such as spheres, rods, and lamellae, then build up different liquid crystalline phase structures (Figure 2). Below follow the lines of reasoning for the physicochemical interpretations of the different groups of results presented in Section 6.

7.1. Incorporation of Fatty Acids

When the strains A-EF22 and B-JU are grown on mixtures of 16:0 and 18:1c, the value of the ratio MGlcDAG/DGlcDAG is decreased, and the fraction of anionic lipids is increased, by increasing the proportion of 18:1c in the growth medium (Figure 5). The change in the glucolipid ratio is interpreted as a regulation of the equilibrium between lipids forming lamellar and nonlamellar phases. An increase in the degree of monounsaturation of the acyl chains shifts the equilibrium toward nonlamellar phases, while a decreasing fraction of MGlcDAG shifts the equilibrium toward lamellar phases. The variation in the glucolipid ratio cannot be interpreted as a regulation of the T_m value or the order parameter of the acyl chains. The T_m value and the order parameter (Seelig and Seelig, 1977) decrease when the degree of monounsaturated acyl chains is increased, and by decreasing the glucolipid ratio these quantities are further decreased, especially since DGlcDAG contains a larger fraction of monounsaturated acyl chains than does MGlcDAG.

The regulation of the equilibrium between lipids forming lamellar and nonlamellar phases has been verified by studying the phase equilibria of total polar lipid mixtures isolated from strain A-EF22 (Figure 12). The lipids contain different proportions of 16:0 and 18:1c. If the polar head-group composition had *not*

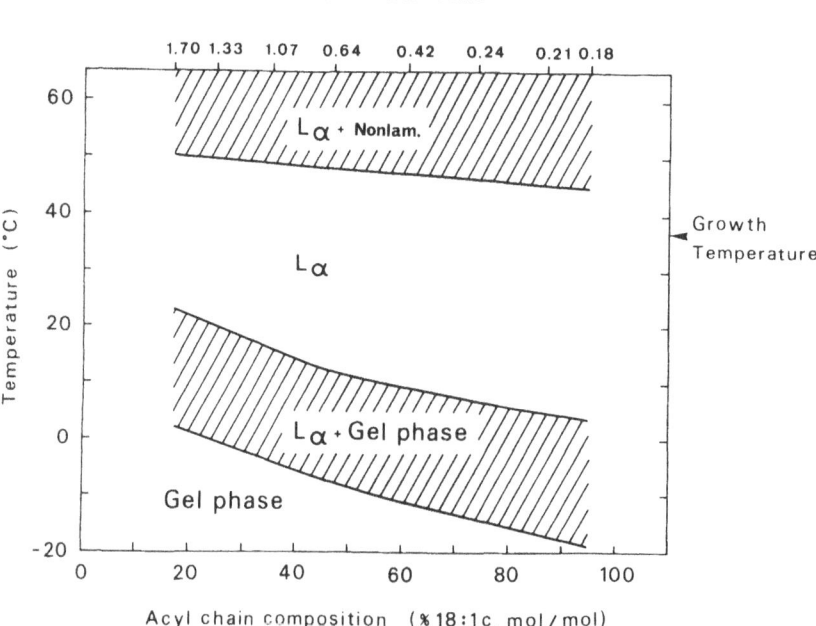

FIGURE 12. Phase equilibria of total polar lipid mixtures, containing different amounts of 16:0 and 18:1c (lower x axis), from membranes of *A. laidlawii* strain A-EF22 grown at 37°C. Water contents were approximately 20 wt%. The upper x axis shows the metabolically obtained MGlcDAG/DGlcDAG ratios. The lower hatched area denotes the gel-to-liquid-crystalline phase transition interval as determined by electron spin resonance. The upper hatched area denotes the appearance of nonlamellar (H_{II} and/or cubic) phases in the lipid mixtures. (Adapted from Lindblom *et al.*, 1986.)

been changed when the fraction of 18:1c is increased, the phase equilibria would be shifted toward cubic and H_{II} phases at the growth temperature (Wieslander *et al.*, 1981; Khan *et al.*, 1981). However, by lowering the value of the ratio MGlcDAG/DGlcDAG, the transition to the nonlamellar phases begins at about 10–15°C above the growth temperature in all of the lipid extracts. As expected, the T_m value of the lipid mixtures varied over a wide range; according to a differential thermal analysis of total lipid mixtures isolated from strain B-PG9 (Silvius *et al.*, 1980), the midpoint T_m value of the transition interval should vary between approximately -10 and 30°C with the acyl chain compositions shown in Figure 12.

The same line of reasoning can be applied to the results obtained when both A and B strains are grown on 18:1t, 18:2c, iso and anteiso acids, and acids with different hydrocarbon chain lengths. The temperature for the transition between an L_β or an L_α phase to an H_{II} phase (T_{LH}) is increased by 18:1t and decreased by 18:2c as compared with 18:1c, and anteiso acids induce a lower T_{LH} value in comparison with iso acids (Rilfors *et al.*, 1982; Lewis *et al.*, 1989). The value of the ratio MGlcDAG/DGlcDAG is adjusted in order to counteract the changes in the lamellar–nonlamellar phase equilibrium brought about by these acids. In the case of acyl chains of different length, the response in the glucolipid ratio is in accordance with the concomitant changes in the phase equilibria observed for monoacylglycerols, PE, and different synthetic glycolipids. For these lipids, the phase areas representing nonlamellar phases in the phase diagrams are decreased, or the temperatures for the transition between lamellar and nonlamellar phases are increased, by a decreased chain length, especially at short lengths (Lutton, 1965; Lewis *et al.*, 1989; Mannock *et al.*, 1990a; Hinz *et al.*, 1991). This has also been observed for MGlcDAG prepared from *A. laidlawii* (Section 5.1.1). The fraction of MGlcDAG is increased, and thus counteracts the shift of the phase equilibria of the membrane lipids toward lamellar phases caused by the decrease in chain length. Since shorter chains yield lower T_m values than longer chains, an increased fraction of MGlcDAG should also counteract the lowering of the T_m value caused by the short acyl chains, but the magnitude of this compensation is probably too small.

With an increasing content of saturated acyl chains of especially medium and short lengths in the *A. laidlawii* lipids, the phase equilibria of MGlcDAG are shifted so much toward the lamellar phases that this lipid probably cannot fulfill its packing functions. Under these conditions, the enzymatic addition in strain A-EF22 of a third acyl chain on MGlcDAG creates a new lipid, MAMGlcDAG, with a small polar head group and a substantially enlarged hydrophobic volume. The latter lipid is more prone to form nonlamellar phases than is MGlcDAG. At extreme proportions of the above-mentioned acyl chains, the fraction of the normally minor precursor lipid DAG (Figure 1) is also markedly increased (Wieslander *et al.*, 1993a). This lipid is very potent in shifting the phase equilibria toward nonlamellar phases (Das and Rand, 1986; Siegel *et al.*, 1989).

Hence, under conditions when MGlcDAG is less efficient to form nonlamellar phases, the cells make two alternative lipids which are more potent in maintaining the appropriate packing conditions. Similar features have also been observed in strain B-PG9. With a large fraction of 16:0 in the membrane lipids, the cells start to produce a lipid, MAPGlc, with a small polar head group and a hydrophobic volume larger than that of MGlcDAG. MAPGlc has, in conformity with MAMGlcDAG, three hydroxyls in the polar head group; however, the former lipid has only two hydrocarbon chains, of which one is more bulky (the polyprenyl chain) than a straight acyl chain. Like MAMGlcDAG, MAPGlc is synthesized mainly at the expense of MGlcDAG. MAPGlc was found to be more potent in forming non-bilayer structures than MGlcDAG, since it undergoes an L_α to H_{II} phase transition at 39°C, while a dipalmitoyl-MGlcDAG species undergoes an L_α to I_{II} phase transition at 79°C (Mannock *et al.*, 1990). It is clear that MAMGlcDAG and MAPGlc play analogous roles in the regulation of the balance between bilayer- and non-bilayer-forming lipids in the cell membrane of *A. laidlawii* strains A-EF22 and B-PG9, despite the fact that MAPGlc is able to form liquid crystalline phases, while MAMGlcDAG can only form reversed micelles when the acyl chains are melted. Consequently, the important property for the cells is to possess the ability to synthesize a second wedge-shaped lipid molecule in order to maintain a slightly negative curvature of the membrane. Together, these findings indicate that the packing conditions in the bilayer probably are sensed at more levels than between MGlcDAG and DGlcDAG (Figure 1).

In strain A-EF22 the fraction of anionic lipids seems to be regulated in a consistent way as a function of the structure of the acyl chains present in the lipids. The incorporation of acyl chains increasing the interfacial area per lipid molecule results in a larger fraction of anionic lipids, and it was shown that the lipid surface charge density and the surface potential are kept fairly constant by this compensating mechanism (Christiansson *et al.*, 1985). The results available from the strains B-JU and B-PG9 are conflicting. When strain B-JU is grown in a lipid-depleted BSA–tryptose medium, it regulates the fraction of anionic lipids in conformity with strain A-EF22. However, when strain B-PG9 is grown in a lipid-depleted BSA–heart infusion broth–peptone–yeast extract medium, it does seemingly not maintain the lipid surface charge density at a constant level; either no clear correlation was found between the fraction of PG and the chemical structure of the acyl chains, or results opposite of those observed in the A strain were obtained.

7.2. Growth Temperature

When cells of strain A-EF22 are grown in media supplemented with one unsaturated fatty acid only, this acid constitutes at least 95 mole% of the acyl chains of the membrane lipids. The cells compensate for a decrease in the growth temperature by raising the value of the ratio MGlcDAG/DGlcDAG. This response is also interpreted as a regulation of the equilibrium between lipids forming lamellar and nonlamellar phases: the equilibrium is shifted toward lamellar phases by decreasing the temperature and is shifted back toward nonlamellar phases by increasing the glucolipid ratio. The response cannot be interpreted as a regulation of the T_m value or the order parameter of the acyl chains. The T_m value is slightly increased by raising the value of the glucolipid ratio; the order parameter increases at lower temperatures (Seelig and Seelig, 1977; Smith, 1984) and this quantity is either left unaltered or further increased by the response in the glucolipid ratio. Strain B-PG9 reacts partly in a different manner than the A strain when the growth temperature is reduced; the value of the glucolipid ratio is increased with 16:0 and 18:1c but not with 18:1t, which results in a higher value of the ratio with 18:1c than with 18:1t at 22°C.

When the cells of strain A-EF22 incorporate two kinds of fatty acids into the membrane lipids, variations of the growth temperature also result in a regulation of the acyl chain composition of the lipids. For example, the value of the ratios 16:0/18:1c and iso/anteiso acyl chains is decreased when the growth temperature is lowered. This response can be interpreted as strivings to avoid a too large fraction of the lipids from entering the gel state and to regulate the balance between lipids forming lamellar and nonlamellar phases. Moreover, the value of the ratio MGlcDAG/DGlcDAG is raised, probably in order to restore the tendency of the lipids to form nonlamellar phases. In the case of growth on 16:0/18:1c, the degree of acyl chain unsaturation becomes *higher* in MGlcDAG than in DGlcDAG (compare Section 6.1.1) after a reduction of the growth temperature, and this further reinforces the ability of MGlcDAG to form nonlamellar phases.

The reduction of the average acyl chain length observed in cells of strain A-EF22 upon a decrease of the growth temperature *cannot* be interpreted as a regulation of the balance between lipids forming lamellar and nonlamellar phases; both of the changes shift the balance toward lamellar phases. The aim of the chain-length regulation in this case may instead be a regulation of the T_m value or the bilayer thickness (Rilfors, 1985).

Finally, by reducing the growth temperature from 37 to 17°C the fraction of anionic lipids is decreased when cells of strain A-EF22 are fed with an equimolar mixture of 16:0 and 18:1c, but virtually no change occurs when the cells are fed with 18:1c alone. In the former case, part of the lipids enter the gel state at 17°C which decreases the average interfacial area per lipid molecule and consequently increases the surface charge density. The cell response counteracts this effect. With membranes containing 18:1c only, 17°C is well above the T_m interval. In strain B-PG9 the fraction of PG is nearly independent of the growth temperature.

7.3. Incorporation of Foreign Molecules into the Membrane

If the approach presented above is of general validity when discussing lipid regulation mechanisms, the origin of the molecules present in the membrane should be of no importance. In a series of experiments the growth medium of *A. laidlawii* strain A-EF22 was supplemented with several hydrophobic and amphiphilic molecules that spontaneously will insert into the cell membrane. These molecules should affect the lipid composition in accordance with their ability to change the effective hydrophobic volume, the hydrocarbon–water interfacial area, and the hydrocarbon chain length, in the membrane of the living cell.

Nonpolar organic solvents like hydrocarbons and diethyl ether all decrease the value of the ratio MGlcDAG/DGlcDAG more or less drastically when present in the cell membrane (Figure 7). These solvents are able to transform lamellar phases into nonlamellar phases of the reversed type, and this effect is counteracted by the change in the glucolipid ratio. It is unlikely that the variation of this ratio can be explained as an effort to regulate the T_m value of the lipids. The solvents slightly decrease the T_m value of model lipid–water systems, and the decrease in the glucolipid ratio further decreases this value.

Figure 8 displays the effect of the presence of various *n*-alcohols on the value of the ratio MGlcDAG/DGlcDAG. This effect can be explained by their influence on the volume of the hydrocarbon region and on the hydrocarbon–water interfacial area. Ethanol raises the T_{LH} value for lipid–water systems and the cells consequently increase the fraction of MGlcDAG in the membrane. 1-Butanol has a small effect on the phase equilibria and the glucolipid ratio is left nearly unaltered. The alcohols from 1-hexanol to 1-hexadecanol decrease the T_{LH} value for lipid–water systems and the cells respond by decreasing the fraction of MGlcDAG. However, the decrease gradually diminishes for the long-chain species which probably depends on their extremely low solubility in water (see Section 6.1.3). The hydrocarbon region of phenethyl alcohol extends like those of 1-pentanol or 1-hexanol and the lipid–buffer partition coefficient is on the same order of magnitude for these alcohols; the value of the glucolipid ratio is also nearly as low for phenethyl alcohol as it is for 1-hexanol. The regulation of the glucolipid ratio cannot be explained as a regulation of the T_m value or the order parameter of the acyl chains. For example, ethanol and 1-hexanol both lower the T_m value and decrease the acyl chain order parameter in a lipid bilayer, but the two alcohols have opposite effects on the glucolipid ratio. Moreover, 1-hexanol and 1-octanol lower the T_m value while 1-tetradecanol raises this value, but the glucolipid ratio is decreased in all cases.

The results obtained when cells of strain A-EF22 are grown in the presence of two different series of poly(oxyethylene) alkyl ether detergents can be interpreted as a regulation of the equilibrium between lamellar and nonlamellar phases (Figure 9). With a growing number of oxyethylene units, the hydrocarbon–water interfacial area of the detergents increases. Concomitantly, the phase equilibria are gradually shifted from an L_α to an L_1 phase in both of the series. In order to compensate for this change, MGlcDAG constitutes a gradually larger fraction of the membrane lipids when the polar head-group size of the detergent increases.

When cells of strain A-EF22 have membrane lipids enriched in one fatty acid like 18:1c or 18:2c, the incorporation of cholesterol into the membrane induces a reduction of the value of the ratio MGlcDAG/DGlcDAG and an increase in the fraction of anionic lipids. These responses counteract two of the effects caused by cholesterol: the ability to induce the formation of nonlamellar phases of the reversed type together with lipids containing unsaturated acyl chains (Rilfors *et al.*, 1984); and the dilution of the surface charge density. If cholesterol is incorporated together with a mixture of 16:0 and 18:1c, a third effect is observed: the fraction of 18:1c increases. This response can be explained by the fact that cholesterol increases the order parameter of the acyl chains for lipids in the liquid-crystalline state (Smith, 1984). If 16:0 constitutes 75 mole% or more of the acyl chains, the influence of cholesterol on the glucolipid ratio and the fraction of anionic lipids is very small; the reason for this may be that the effect of cholesterol is less pronounced in combination with lipids enriched in saturated acyl chains. Qualitatively, cholesterol has the same effects on the glucolipid ratio and the fraction of anionic lipids in strain B-PG9, but quantitatively the effects are reversed, i.e., they are largest for cells grown on 16:0. It is interesting to note that the capacity of 20 *A. laidlawii* strains to incorporate cholesterol into their cell membrane is inversely proportional to the fraction of MGlcDAG and MAMGlcDAG, which in resemblance with cholesterol shift the equilibria of membrane lipids toward nonlamellar phases of the reversed type.

The cells of strain A-EF22 have been grown with a large number of sterols besides cholesterol. The influence of these molecules on the hydrophobic volume, the hydrocarbon–water interfacial area, and the hydrocarbon chain length of the membrane lipids is complex and difficult to predict. It is therefore not possible to give a single plausible physicochemical rationale to the effects of the sterols on the value of the ratio MGlcDAG/DGlcDAG (Figure 6). However, in several cases there is a strong correlation between the ability of the sterols to induce the formation of nonlamellar lipid phases and their ability to reduce the glucolipid ratio. With the information available today, it can also be concluded that it is very unlikely that this lipid ratio is regulated in response to the influence of the sterols on the T_m value of the membrane lipids or the order parameter of the acyl chains.

7.4. Conclusions

For strain A-EF22 it can be stated that the changes of the value of the ratio MGlcDAG/DGlcDAG in *all* cases can be interpreted as a regulation of the balance between lipids forming lamellar and nonlamellar phases; furthermore, in many cases a regulation of the T_m value or the order parameter of the acyl chains can be directly excluded. Concerning the changes of the acyl chain composition of the membrane lipids, they can in one case (degree of unsaturation versus temperature) be seen as a regulation of the lamellar–nonlamellar phase equilibria or the lamellar gel-to-liquid crystalline phase equilibrium. However, in two cases the aim of these changes cannot be to regulate the lamellar–nonlamellar phase equilibria; instead it may be an adjustment of the T_m value, implying an adjustment of the bilayer thickness (acyl chain length versus temperature) and the acyl chain order (incorporation of cholesterol versus degree of unsaturation). The fraction of anionic lipids seems in *all* cases to be varied in order to keep the lipid surface charge density at a constant value. It should, however, be pointed out that a variation of the fraction of anionic lipids also affects other properties of the membrane; an increase in this fraction will probably reduce the T_m value slightly, shift the phase equilibria toward lamellar phases, and probably decrease the order parameter of the acyl chains (Killian *et al.*, 1992). But even with these assumptions it can be excluded that the T_m value or the ordering of the acyl chains is regulated by the fraction of anionic lipids. In conclusion, it should be emphasized that the physiological regulation of lipids forming lamellar and nonlamellar phases, and the regulation of the lipid surface charge density, closely follows the physicochemical principles that govern the behavior of amphiphilic molecules and the aggregates formed by them.

The picture is somewhat more complicated for the B strains. When strain B-JU is grown in a lipid-depleted BSA–tryptose medium, the value of the ratio MGlcDAG/DGlcDAG and the fraction of anionic lipids are regulated in the same way as in strain A-EF22. However, when strain B-PG9 is grown in a lipid-depleted BSA–heart infusion broth–peptone–yeast extract medium, the glucolipid ratio is changed qualitatively as in the A and B-JU strains, and the fraction of anionic lipids is varied in a reversed way, not at all, or qualitatively in the same way as compared with the A and B-JU strains. In strain B-PG9, the acyl chain composition of the membrane lipids seems not to be regulated significantly in response to changes in the environmental conditions. A multivariate data analysis of strains A-EF22 and B-PG9 demonstrates that these strains can be separated into two distinct groups as regards the membrane protein and lipid composition (Wieslander *et al.*, 1993b). However, the regulation of the glucolipid composition as a consequence of changes in the growth and environmental conditions occurs in a similar way in the two strains. The different settings of the glucolipid fraction in the strains are most probably explained by the inherent differences in acyl chain length and unsaturation (Wieslander *et al.*, 1993b). A possible cause of the differences in acyl chain length and unsaturation may be that *A. laidlawii* is a cluster of various strains with different extents of relatedness (Stephens *et al.*, 1983).

The acyltransferases of *A. laidlawii* can apparently accept a broad spectrum of exogenously supplied fatty acids as regards the chemical structure and the chain length. The selectivity of the fatty acids incorporated into the membrane lipids is in most cases rather small; only when the cells are fed with fatty acids having an extreme chain length or chemical structure does the selectivity disfavor these acids. Moreover, *A. laidlawii* has a limited capacity for fatty acid synthesis (Section 3). The limitations in the fatty acid synthesis and the selectivity of the fatty acid incorporation probably necessitate a regulation of the polar head-group composition in order to neutralize the unfavorable properties introduced into the lipid bilayer by the various acyl chains. The essential features of the membrane lipid regulation mechanisms observed in *A. laidlawii* can then be summarized as follows: the polar head groups are responsible for the regulation of the equilibrium between lamellar and nonlamellar phase structures and the regulation of the lipid surface charge density. Important roles of the acyl chains are to ensure that a minimum fraction of the lipids are in a liquid-crystalline state and that the bilayer thickness is held within a certain interval. The limited variation of the acyl chain composition may also partially regulate the equilibria between lamellar and nonlamellar phase structures.

The order parameter of the acyl chains and the lamellar–nonlamellar phase equilibria can together be said to determine the "packing properties" of the lipid molecules in the bilayer. ^{2}H-NMR investigations of mixtures of MGlcDAG and DGlcDAG show that the order parameter of the acyl chains is increased by MGlcDAG, and indicate that the curvature of the bilayer is increased by increasing the fraction of MGlcDAG (Eriksson *et al.*, 1991). It can be speculated that a variation of these packing properties of the lipid molecules affects the lateral pressure exerted by the lipids on the membrane-bound glucolipid synthases. The closeness to a transition to nonbilayer structures might therefore be sensed by the synthases and their activity adjusted in relation to this closeness. The activities of the glucolipid synthases *in vitro* are affected by both the presence and types of the surrounding lipids (Dahlqvist *et al.*, 1992). However, it is also possible that the activity of the lipid synthases is regulated indirectly, i.e., by one or more parameters which in turn are affected by the packing properties of the lipid molecules. In order to differentiate between these alternatives, the lipid synthases must be isolated and their structure and function in different lipid matrices determined.

LIST OF ABBREVIATIONS

Diacylglycerol	DAG
Phosphatidic acid	PA
Phosphatidylglycerol	PG
Diphosphatidylglycerol	DPG
Phosphatidylcholine	PC
Dioleoyl-PC	DOPC
Phosphatidylethanolamine	PE
Dioleoyl-PE	DOPE
Monoglucosyl	MGlc

Diglucosyl   DGlc
1,2-Diacyl-3-*O*-(α-D-glucopyranosyl)-*sn*-glycerol   MGlcDAG
1,2-Diacyl-3-*O*-[α-D-glucopyranosyl-(1→2)-*O*-α-D-   DGlcDAG
  glucopyranosyl]-*sn*-glycerol
Glycerophosphoryl-MGlcDAG   GPMGlc-
                                                                              DAG
1,2-Diacyl-3-*O*-[glycero-3-phosphoryl-6-*O*-(α-D-   GPDGlc-
  glucopyranosyl-(1→2)-*O*-α-D-glucopyranosyl)]-*sn*-   DAG
  glycerol
1,2-Diacyl-3-*O*-[3-*O*-acyl-(α-D-glucopyranosyl)]-*sn*-glycerol   MAMGlc-
                                                                              DAG
2-*O*-Acyl-1-*O*-polyprenyl-α-D-glucopyranoside   MAPGlc
Normal micellar solution L_1
Reversed micellar solution L_2
Lamellar gel phase L_β
Lamellar liquid-crystalline phase L_α
Temperature (interval) for the L_β-to-L_α phase transition T_m
Normal hexagonal liquid-crystalline phase H_I
Reversed hexagonal liquid-crystalline phase H_{II}
Temperature (interval) for the L_β-to-H_{II} or L_α-to-H_{II} phase T_{LH}
  transition
Reversed cubic phase I_{II}
Palmitic acid; palmitoyl chains   16:0
Elaidic acid; elaidoyl chains   18:1t
Oleic acid; oleoyl chains   18:1c
Linoleic acid; linoleoyl chains   18:2c
Linolenic acid; linolenoyl chains   18:3c
2-Methylpropionic acid   Iso-C_4
3-Methylbutanoic acid   Iso-C_5
2-Methylbutanoic acid   Anteiso-C_5
13-Methyltetradecanoic acid   Iso-C_{15}
12-Methyltetradecanoic acid   Anteiso-C_{15}
15-Methylhexadecanoic acid   Iso-C_{17}
14-Methylhexadecanoic acid   Anteiso-C_{17}
Tetraphenylphosphonium ion TPP^+
Bovine serum albumin   BSA

REFERENCES

Amar, A., Rottem, S., and Razin, S., 1974, Characterization of the Mycoplasma membrane proteins. IV. Disposition of proteins in the membrane, *Biochim. Biophys. Acta* **352**:228–244.
Amar, A., Rottem, S., and Razin, S., 1978, Disposition of membrane proteins as affected by

changes in the electrochemical gradient across Mycoplasma membranes, *Biochem. Biophys. Res. Commun.* **84**:306–312.

Amar, A., Rottem, S., and Razin, S., 1979, Is the vertical disposition of Mycoplasma membrane proteins affected by membrane fluidity? *Biochim. Biophys. Acta* **552**:457–467.

Arvidson, G., Brentel, I., Khan, A., Lindblom, G., and Fontell, K., 1985, Phase equilibria in four lysophosphatidylcholine–water systems. Exceptional behavior of 1-palmitoyl-glycerophosphocholine, *Eur. J. Biochem.* **152**:753–759.

Asgharian, B., Cadenhead, D. A., Mannock, D. A., Lewis, R. N. A. H., and McElhaney, R. N., 1989, A comparative monomolecular film study of 1,2-di-*O*-palmitoyl-3-*O*-(α- and β-D-glucopyranosyl)-*sn*-glycerols, *Biochemistry* **28**:7102–7106.

Auger, M., Smith, I. C. P., and Jarrell, H. C., 1991, Slow motions in lipid bilayers. Direct detection by two-dimensional solid-state deuterium nuclear magnetic resonance, *Biophys. J.* **59**:31–38.

Beaman, K. D., and Pollack, J. D., 1981, Adenylate energy charge in *Acholeplasma laidlawii*, *J. Bacteriol.* **146**:1055–1058.

Bergenståhl, B. A., and Stenius, P., 1987, Phase diagrams of dioleoylphosphatidylcholine with formamide, methylformamide, and dimethylformamide, *J. Phys. Chem.* **91**:5944–5948.

Bevers, E. M., Op den Kamp, J. A. F., and van Deenen, L. L. M., 1978a, Physico-chemical properties of phosphatidylglycerol in membranes of *Acholeplasma laidlawii*, *Eur. J. Biochem.* **84**:35–42.

Bevers, E. M., Leblanc, G., Le Grimellec, C., Op den Kamp, J. A. F., and van Deenen, L. L. M., 1978b, Disposition of phosphatidylglycerol in metabolizing cells of *Acholeplasma laidlawii*, *FEBS Lett.* **87**:49–51.

Bhakoo, M., and McElhaney, R. N., 1988, The effects of variations in growth temperature, fatty acid composition and cholesterol content on the lipid polar head-group composition of *Acholeplasma laidlawii* B membranes, *Biochim. Biophys. Acta* **945**:307–314.

Bhakoo, M., Lewis, R. N. A. H., and McElhaney, R. N., 1987, Isolation and characterization of a novel monoacylated glucopyranosyl neutral lipid from the plasma membrane of *Acholeplasma laidlawii* B, *Biochim. Biophys. Acta* **922**:34–45.

Blöcher, D., Six, L., Gutermann, R., Henkel, B., and Ring, K., 1985, Physicochemical characterization of tetraether lipids from *Thermoplasma acidophilum*. Calorimetric studies on miscibility with diether model lipids carrying branched or unbranched alkyl chains, *Biochim. Biophys. Acta* **818**:333–342.

Boggs, J. M., 1987, Lipid intermolecular hydrogen bonding: Influence on structural organization and membrane function, *Biochim. Biophys. Acta* **906**:353–404.

Browning, J. L., and Seelig, J., 1980, Bilayers of phosphatidylserine: A deuterium and phosphorus nuclear magnetic resonance study, *Biochemistry* **19**:1262–1270.

Büldt, G., Gally, H.-U., Seelig, A., Seelig, J., and Zaccai, G., 1978, Neutron diffraction studies on selectively deuterated phospholipid bilayers, *Nature* **271**:182–184.

Cevc, G., 1990, Membrane electrostatics, *Biochim. Biophys. Acta* **1031**:311–382.

Chowdhury, M. I. H., Munakata, T., Koyanagi, Y., Kobayashi, S., Arai, S., and Yamamoto, N., 1990, Mycoplasma can enhance HIV replication in vitro: A possible cofactor responsible for the progression of AIDS, *Biochem. Biophys. Res. Commun.* **170**:1365–1370.

Christiansson, A., 1981, Lipid metabolism in *Acholeplasma laidlawii*, Ph.D. thesis, University of Lund, Lund, Sweden.

Christiansson, A., and Wieslander, Å., 1978, Membrane lipid metabolism in *Acholeplasma laidlawii* A EF 22. Influence of cholesterol and temperature shift-down on incorporation of fatty acids and synthesis of membrane lipid species, *Eur. J. Biochem.* **85**:65–76.

Christiansson, A., and Wieslander, Å., 1980, Control of membrane polar lipid composition in *Acholeplasma laidlawii* A by the extent of saturated fatty acid synthesis, *Biochim. Biophys. Acta* **595**:189–199.

Christiansson, A., Gutman, H., Wieslander, Å., and Lindblom, G., 1981, Effects of anesthetics on water permeability and lipid metabolism in *Acholeplasma laidlawii* membranes, *Biochim. Biophys. Acta* **645**:24–32.

Christiansson, A., Eriksson, L. E. G., Westman, J., Demel, R., and Wieslander, Å., 1985, Involvement of surface potential in regulation of polar membrane lipids in *Acholeplasma laidlawii*, *J. Biol. Chem.* **260**:3984–3990.

Citovsky, V., Rottem, S., Nussbaum, O., Laster, Y., Rott, R., and Loyter, A., 1988, Animal viruses are able to fuse with prokaryotic cells, *J. Biol. Chem.* **263**:461–467.

Clementz, T., Christiansson, A., and Wieslander, Å., 1986, Transmembrane electrical potential affects the lipid composition of *Acholeplasma laidlawii*, *Biochemistry* **25**:823–830.

Clementz, T., Christiansson, A., and Wieslander, Å., 1987, Membrane potential, lipid regulation and adenylate energy change in acyl chain modified *Acholeplasma laidlawii*, *Biochim. Biophys. Acta* **898**:299–307.

Clementz, T., Christiansson, A., and Wieslander, Å., 1988, Effect of electrical potential on membrane organization and function, in: *Physiological Regulation of Membrane Fluidity* (R. C. Aloia, C. C. Curtain, and L. M. Gordon, eds.), Liss, New York, pp. 41–74.

Copps, T. P., Chelack, W. S., and Petkau, A., 1976, Variation in distribution of membrane particles in *Acholeplasma laidlawii* B with pH, *J. Ultrastruct. Res.* **55**:1–3.

Csonka, L. N., 1989, Physiological and genetic responses of bacteria to osmotic stress, *Microbiol. Rev.* **53**:121–147.

Cullis, P. R., Verkleij, A. J., and Ververgaert, P. H. J. T., 1978, Polymorphic phase behavior of cardiolipin as detected by ^{31}P NMR and freeze-fracture techniques, *Biochim. Biophys. Acta* **513**:11–20.

Cullis, P. R., Hope, M. J., and Tilcock, C. P. S., 1986, Lipid polymorphism and the role of lipids in membranes, *Chem. Phys. Lipids* **40**:127–144.

Curatolo, W., 1987, The physical properties of glycolipids, *Biochim. Biophys. Acta* **906**:111–136.

Dahlqvist, A., Andersson, S., and Wieslander, Å., 1992, The enzymatic synthesis of membrane glucolipids in *Acholeplasma laidlawii*, *Biochim. Biophys. Acta* **1105**:131–140.

Das, S., and Rand, R. P., 1986, Modification by diacylglycerol of the structure and interaction of various phospholipid bilayer membranes, *Biochemistry* **25**:2882–2889.

De Gier, J., Mandersloot, J. G., Hupkes, J. V., McElhaney, R. N., and Van Beek, W. P., 1971, On the mechanism of non-electrolyte permeation through lipid bilayers and through biomembranes, *Biochim. Biophys. Acta* **233**:610–618.

de Kruijff, B., Demel, R. A., Slotboom, A. J., Van Deenen, L. L. M., and Rosenthal, A. F., 1973, The effect of the polar headgroup on the lipid–cholesterol interaction: A monolayer and differential scanning calorimetry study, *Biochim. Biophys. Acta* **307**:1–19.

de Kruijff, B., Verkleij, A. J., Leunissen-Bijvelt, J., Van Echteld, C. J. A., Hille, J., and Rijnbout, H., 1982, Further aspects of the Ca^{2+}-dependent polymorphism of bovine heart cardiolipin, *Biochim. Biophys. Acta* **693**:1–12.

Desantis, D., Tryon, V. V., and Pollack, J. D., 1989, Metabolism of Mollicutes: The Embden–Meyerhof–Parnas pathway and the hexose monophosphate shunt, *J. Gen. Microbiol.* **135**:683–691.

Driessen, A. J. M., and Konings, W. N., 1990, Energetic problems of bacterial fermentations: Extrusion of metabolic end products, in: *The Bacteria*, Volume XII (T. A. Krulwich, ed.), Academic Press, New York, pp. 449–478.

Efrati, H., Wax, Y., and Rottem, S., 1986, Cholesterol uptake capacity of *Acholeplasma laidlawii* is affected by the composition and content of membrane glycolipids, *Arch. Biochem. Biophys.* **248**:282–288.

Ekwall, P., 1975, Composition, properties and structures of liquid crystalline phases in systems of amphiphilic compounds, *Adv. Liq. Crystallogr.* **1**:1–142.

Endo, T., Inoue, K., and Nojima, S., 1982, Physical properties and barrier functions of synthetic glyceroglycolipids, *J. Biochem.* **92**:953–960.

Endo, T., Inoue, K., Nojima, S., Sekiya, T., Ohki, K., and Nozawa, Y., 1983, Electron microscopic study on the structures formed by mixtures containing synthetic glyceroglycolipids, *J. Biochem.* **93**:1–6.

Engelman, D. M., 1970, X-ray diffraction studies of phase transitions in the membrane of *Mycoplasma laidlawii*, *J. Mol. Biol.* **47**:115–117.

Engelman, D. M., 1971, Lipid bilayer structure in the membrane of *Mycoplasma laidlawii*, *J. Mol. Biol.* **58**:153–165.

Eriksson, P.-O., Lindblom, G., and Arvidson, G., 1985, NMR studies of 1-palmitoyllysophosphatidylcholine in a cubic liquid crystal with a novel structure, *J. Phys. Chem.* **89**:1050–1053.

Eriksson, P.-O., Rilfors, L., Wieslander, Å., Lundberg, A., and Lindblom, G., 1991, Order and dynamics in mixtures of membrane glucolipids from *Acholeplasma laidlawii* studied by ²H NMR, *Biochemistry* **30**:4916–4924.

Farren, S. B., and Cullis, P. R., 1980, Polymorphism of phosphatidylglycerol–phosphatidylethanolamine model membrane systems: A ³¹P-NMR study, *Biochem. Biophys. Res. Commun.* **97**:182–191.

Findlay, E. J., and Barton, P. G., 1978, Phase behavior of synthetic phosphatidylglycerols and binary mixtures with phosphatidylcholines in the presence and absence of calcium ions, *Biochemistry* **17**:2400–2405.

Freundt, E. A., and Edward, D. G., 1979, Classification and taxonomy, in: *The Mycoplasmas*, Volume I (M. F. Barile and S. Razin, eds.), Academic Press, New York, pp. 1–41.

Gally, H.-U., Pluschke, G., Overath, P., and Seelig, J., 1981, Structure of *Escherichia coli* membranes. Glycerol auxotrophs as a tool for the analysis of the phospholipid headgroup region by deuterium magnetic resonance, *Biochemistry* **20**:1826–1831.

Gruner, S. M., Tate, M. W., Kirk, G. L., So, P. T. C., Turner, D. C., Keane, D. T., Tilcock, C. P. S., and Cullis, P. R., 1988, X-ray diffraction study of the polymorphic behavior of N-methylated dioleoylphosphatidylethanolamine, *Biochemistry* **27**:2853–2866.

Gutman, H., Arvidson, G., Fontell, K., and Lindblom, G., 1984, ³¹P and ²H NMR studies of phase equilibria in the three component system monoolein–dioleoylphosphatidylcholine–water, in: *Surfactants in Solution*, Volume 1 (K. L. Mittal and B. Lindman, eds.), Plenum Press, New York, pp. 143–152.

Halle, B., 1991, ²H NMR relaxation in phospholipid bilayers. Toward a consistent molecular interpretation, *J. Phys. Chem.* **95**:6724–6733.

Hauser, H., Pascher, I., Pearson, R. H., and Sundell, S., 1981, Preferred conformation and molecular packing of phosphatidylethanolamine and phosphatidylcholine, *Biochim. Biophys. Acta* **650**:21–51.

Hinz, H.-J., Six, L., Ruess, K.-P., and Lieflländer, M., 1985, Head-group contributions to bilayer stability: Monolayer and calorimetric studies on synthetic, stereochemically uniform glucolipids, *Biochemistry* **24**:806–813.

Hinz, H.-J., Kuttenreich, H., Meyer, R., Renner, M., Fründ, R., Koynova, R., Boyanov, A. I., and Tenchov, B. G., 1991, Stereochemistry and size of sugar head groups determine structure and phase behavior of glycolipid membranes: Densitometric, calorimetric, and X-ray studies, *Biochemistry* **30**:5125–5138.

Hitchcock, P. B., Mason, R., Thomas, K. M., and Shipley, G. G., 1974, Structural chemistry of 1,2-dilauroyl-DL-phosphatidylethanolamine: Molecular conformation and intermolecular packing of phospholipids, *Proc. Natl. Acad. Sci. USA* **71**:3036–3040.

Huang, L., Lorch, S. K., Smith, G. G., and Haug, A., 1974, Control of membrane lipid fluidity in *Acholeplasma laidlawii*, *FEBS Lett.* **43**:1–5.

Israelachvili, J. N., and Wennerström, H., 1990, Hydration or steric forces between amphiphilic surfaces, *Langmuir* **6**:873–876.

Israelachvili, J. N., Mitchell, D. J., and Ninham, B. W., 1976, Theory of self-assembly of hydrocarbon amphiphiles into micelles and bilayers, *J. Chem. Soc. Faraday Trans. II* **72**:1525–1568.

Israelachvili, J. N., Marcelja, S., and Horn, R. G., 1980, Physical principles of membrane organization, *Q. Rev. Biophys.* **13**:121–200.

Iwamoto, K., Sunamoto, J., Inoue, K., Endo, T., and Nojima, S., 1982, Liposomal membranes. XV. Importance of surface structure in liposomal membranes of glyceroglycolipids, *Biochim. Biophys. Acta* **691**:44–51.

James, R., and Branton, D., 1973, Lipid- and temperature-dependent structural changes in *Acholeplasma laidlawii* cell membranes, *Biochim. Biophys. Acta* **323**:378–390.

Jarrell, H. C., Giziewicz, J. B., and Smith, I. C. P., 1986, Structure and dynamics of a glyceroglycolipid: A ^{2}H NMR study of head group orientation, ordering, and effect on lipid aggregate structure, *Biochemistry* **25**:3950–3957.

Jarrell, H. C., Jovall, P. Å., Giziewicz, J. B., Turner, L. A., and Smith, I. C. P., 1987a, Determination of conformational properties of glycolipid head groups by ^{2}H NMR of oriented multibilayers, *Biochemistry* **26**:1805–1811.

Jarrell, H. C., Wand, A. J., Giziewicz, J. B., and Smith, I. C. P., 1987b, The dependence of glyceroglycolipid orientation and dynamics on head-group structure, *Biochim. Biophys. Acta* **897**:69–82.

Johansson, K.-E., and Hjertén, S., 1974, Localization of the Tween 20-soluble membrane proteins of *Acholeplasma laidlawii* by crossed immunoelectrophoresis, *J. Mol. Biol.* **86**:341–348.

Johansson, K.-E., Jägersten, C., Christiansson, A., and Wieslander, Å., 1981, Protein composition and extractability of lipid-modified membranes from *Acholeplasma laidlawii*, *Biochemistry* **20**:6073–6079.

Kaneda, T., 1977, Fatty acids of the genus *Bacillus:* An example of branched-chain preference, *Microbiol. Rev.* **41**:391–418.

Khan, A., Rilfors, L., Wieslander, Å., and Lindblom, G., 1981, The effect of cholesterol on the phase structure of glucolipids from *Acholeplasma laidlawii* membranes, *Eur. J. Biochem.* **116**:215–220.

Killian, J. A., Fabrie, C. H. J. P., Baart, W., Morein, S., and de Kruijff, B., 1992, Effects of temperature variation and phenethylalcohol addition on acyl chain order and lipid organization in *E. coli* derived membrane systems. A ^{2}H and ^{31}P NMR study, *Biochim. Biophys. Acta* **1105**:253–262.

Koostra, W. L., and Smith, P. F., 1969, D- and L-alanylphosphatidylglycerols from *Mycoplasma laidlawii*, strain B, *Biochemistry* **8**:4794–4806.

Koynova, R. D., Kuttenreich, H. L., Tenchov, B. G., and Hinz, H.-J., 1988, Influence of headgroup interactions on the miscibility of synthetic, stereochemically pure glycolipids and phospholipids, *Biochemistry* **27**:4612–4619.

Kröger, M., Wahl, R., and Rice, P., 1991, Compilation of DNA sequences of *Escherichia coli* (update 1991), *Nucleic Acid Res.* **19**(Suppl):2023–2043.

Lafleur, M., Bloom, M., and Cullis, P. R., 1990, Lipid polymorphism and hydrocarbon order, *Biochem. Cell Biol.* **68**:1–8.

Lewis, R. N. A. H., Mannock, D. A., McElhaney, R. N., Turner, D. C., and Gruner, S. M., 1989, Effect of fatty acyl chain length and structure on the lamellar gel to liquid-crystalline and lamellar to reversed hexagonal phase transitions of aqueous phosphatidylethanolamine dispersions, *Biochemistry* **28**:541–548.

Lewis, R. N. A. H., Mannock, D. A., and McElhaney, R. N., 1990a, Physical properties of glycosyldiacylglycerols: An infrared spectroscopic study of the gel-phase polymorphism of 1,2-di-O-acyl-3-O-(β-D-glucopyranosyl)-sn-glycerols, *Biochemistry* **29**:8933–8943.

Lewis, R. N. A. H., Yue, A. W. B., McElhaney, R. N., Turner, D. C., and Gruner, S. M., 1990b, Thermotropic characterization of the 2-O-acyl,polyprenyl α-D-glucopyranoside isolated from palmitate-enriched *Acholeplasma laidlawii* B membranes, *Biochim. Biophys. Acta* **1026**:21–28.

Lindblom, G., and Rilfors, L., 1989, Cubic phases and isotropic structures formed by membrane lipids—Possible biological relevance, *Biochim. Biophys. Acta* **988**:221–256.

Lindblom, G., Johansson, L. B.-Å., and Arvidson, G., 1981, Effect of cholesterol in membranes. Pulsed nuclear magnetic resonance measurements of lipid lateral diffusion, *Biochemistry* **20**:2204–2207.

Lindblom, G., Brentel, I., Sjölund, M., Wikander, G., and Wieslander, Å., 1986, Phase equilibria of membrane lipids from *Acholeplasma laidlawii:* Importance of a single lipid forming non-lamellar phases, *Biochemistry* **25**:7502–7510.

Lindblom, G., Rilfors, L., Hauksson, J. B., Brentel, I., Sjölund, M., and Bergenståhl, B., 1991, Effect of head-group structure and counterion condensation on phase equilibria in anionic phospholipid–water systems studied by ^{2}H, ^{23}Na, and ^{31}P NMR, and X-ray diffraction, *Biochemistry* **30**:10938–10948.

Lindblom, G., Hauksson, J. B., Rilfors, L., Bergenståhl, B., Wieslander, Å., and Eriksson, P.-O., 1993, Membrane lipid regulation in *Acholeplasma laidlawii* grown with saturated fatty acids. Synthesis of a triacylglucolipid not forming liquid crystalline structures, submitted.

Lutton, E. S., 1965, Phase behavior of aqueous systems of monoglycerides, *J. Am. Oil Chem. Soc.* **42**:1068–1070.

Luzzati, V., 1968, X-ray diffraction studies of lipid–water systems, in: *Biological Membranes* (D. Chapman, ed.), Academic Press, New York, pp. 71–123.

McElhaney, R. N., 1974, The effect of alterations in the physical state of the membrane lipids on the ability of *Acholeplasma laidlawii* B to grow at various temperatures, *J. Mol. Biol.* **84**:145–157.

McElhaney, R. N., 1984, The relationship between membrane lipid fluidity and phase state and the ability of bacteria and mycoplasmas to grow and survive at various temperatures, in: *Biomembranes*, Volume 12 (M. Kates and L. A. Manson, eds.), Plenum Press, New York, pp. 249–278.

McElhaney, R. N., De Gier, J., and Van der Neut-Kok, E. C. M., 1973, The effect of alterations in fatty acid composition and cholesterol content on the nonelectrolyte permeability of *Acholeplasma laidlawii* B cells and derived liposomes, *Biochim. Biophys. Acta* **298**:500–512.

McGarrity, G. J., Sarama, J., and Vanaman, V., 1985, Cell culture techniques, *ASM News* **51**:170–183.

Mahajan, S., Lewis, R. N. A. H., George, R., Sykes, B. D., and McElhaney, R. N., 1988, Characterization of sodium transport in *Acholeplasma laidlawii* B cells and in lipid vesicles containing purified *A. laidlawii* ($Na^+ - Mg^{2+}$)–ATPase by using nuclear magnetic resonance spectroscopy and ^{22}Na tracer techniques, *J. Bacteriol.* **170**:5739–5746.

Mannock, D. A., Lewis, R. N. A. H., and McElhaney, R. N., 1987, An improved procedure for the preparation of 1,2-di-*O*-acyl-3-*O*-(β-D-glucopyranosyl)-*sn*-glycerols, *Chem. Phys. Lipids* **43**:113–127.

Mannock, D. A., Lewis, R. N. A. H., Sen, A., and McElhaney, R. N., 1988, The physical properties of glycosyldiacylglycerols. Calorimetric studies of a homologous series of 1,2-di-*O*-acyl-3-*O*-(β-D-glucopyranosyl)-*sn* glycerols, *Biochemistry* **27**:6852–6859.

Mannock, D. A., Lewis, R. N. A. H., and McElhaney, R. N., 1990a, Physical properties of glycosyl diacylglycerols. 1. Calorimetric studies of a homologous series of 1,2-di-*O*-acyl-3-*O*-(α-D-glucopyranosyl)-*sn*-glycerols, *Biochemistry* **29**:7790–7799.

Mannock, D. A., Lewis, R. N. A. H., and McElhaney, R. N., 1990b, The chemical synthesis and physical characterization of 1,2-di-*O*-acyl-3-*O*-(α-D-glucopyranosyl)-*sn*-glycerols, an important class of membrane glycolipids, *Chem. Phys. Lipids* **55**:309–321.

Manolukas, J. T., Barile, M. F., Chandler, D. K. F., and Pollack, J. D., 1988, Presence of anaplerotic reactions and transamination, and the absence of the tricarboxylic acid cycle in *Mollicutes*, *J. Gen. Microbiol.* **134**:791–800.

Marra, J., 1986, Direct measurements of attractive van der Waals and adhesion forces between uncharged lipid bilayers in aqueous solutions, *J. Colloid Interface Sci.* **109**:11–20.

Melchior, D. L., and Steim, J. M., 1977, Control of fatty acid composition of *Acholeplasma laidlawii* membranes, *Biochim. Biophys. Acta* **466**:148–159.

Michelsen, P., 1985, A facile method for the determination of the absolute configuration of monoglucosyldiacyl-*sn*-glycerol, *Chem. Scr.* **25**:217–218.

Moews, P. C., and Knox, J. R., 1976, The crystal structure of 1-decyl α-D-glucopyranoside: A polar bilayer with a hydrocarbon subcell, *J. Am. Chem. Soc.* **98**:6628–6633.

Ne'eman, Z., Kahane, I., and Razin, S., 1971, Characterization of the Mycoplasma membrane proteins. II. Solubilization and enzymatic activities of *Acholeplasma laidlawii* membrane proteins, *Biochim. Biophys. Acta* **249**:169–176.

Neupert, G., and Sterba, T., 1983, Persistence of *Acholeplasma laidlawii* in an established cell line RL 19, *Exp. Pathol.* **24**:207–211.

Nyström, S., and Wieslander, Å., 1992, Isoprenoid modification of proteins distinct from membrane acyl proteins in the prokaryote *Acholeplasma laidlawii*, *Biochim. Biophys. Acta* **1107**:39–43.

Nyström, S., Wallbrandt, P., and Wieslander, Å., 1992, Membrane protein acylation: Preference for exogenous myristic acid or endogenous saturated chains in *Acholeplasma laidlawii*, *Eur. J. Biochem.* **204**:231–240.

Pascher, I., and Sundell, S., 1977, Molecular arrangements in sphingolipids. The crystal structure of cerebroside, *Chem. Phys. Lipids* **20**:175–191.

Pieringer, R. A., 1989, Biosynthesis of non-terpenoid lipids, in: *Microbial Lipids*, Volume 2 (C. Ratledge and S. G. Wilkinson, eds.), Academic Press, New York, pp. 51–114.

Pollack, J. D., Merola, A. J., and Booth, R. L., Jr., 1981, Respiration-associated components of *Mollicutes*, *J. Bacteriol.* **146**:907–913.

Powell, D. A., Hu, P. C., Wilson, M., Collier, A. M., and Baseman, J. B., 1976, Attachment of *Mycoplasma pneumoniae* to respiratory epithelium, *Infect. Immun.* **13**:959–966.

Rance, M., Smith, I. C. P., and Jarrell, H. C., 1983, The effect of headgroup class on the conformation of membrane lipids in *Acholeplasma laidlawii*: A ^{2}H NMR study, *Chem. Phys. Lipids* **32**:57–71.

Rand, R. P., and Parsegian, V. A., 1989, Hydration forces between phospholipid bilayers, *Biochim. Biophys. Acta* **988**:351–376.

Rand, R. P., and Sengupta, S., 1972, Cardiolipin forms hexagonal structures with divalent cations, *Biochim. Biophys. Acta* **255**:484–492.

Razin, S., 1978, Cholesterol uptake is dependent on membrane fluidity in mycoplasmas, *Biochim. Biophys. Acta* **513**:401–404.

Razin, S., (ed.), 1981, Mycoplasma infections, *Isr. J. Med. Sci.* **17**:509–686.

Razin, S., and Argaman, M., 1962, Susceptibility of *Mycoplasma* (pleuropneumonia-like organisms) and bacterial protoplasts to lysis by various agents, *Nature* **193**:502–503.

Razin, S., Morotwitz, H. J., and Terry, T. M., 1965, Membrane subunits of *Mycoplasma laidlawii* and their assembly to membranelike structures, *Proc. Natl. Acad. Sci. USA* **54**:219–225.

Renou, J.-P., Giziewicz, J. B., Smith, I. C. P., and Jarrell, H. C., 1989, Glycolipid membrane surface structure: Orientation, conformation, and motion of a disaccharide headgroup, *Biochemistry* **28**:1804–1814.

Rilfors, L., 1982, Regulation and physical properties of bacterial membrane lipids, Ph.D. thesis, University of Lund, Lund, Sweden.

Rilfors, L., 1985, Difference in packing properties between iso and anteiso methyl-branched fatty acids as revealed by incorporation into the membrane lipids of *Acholeplasma laidlawii* strain A, *Biochim. Biophys. Acta* **813**:151–160.

Rilfors, L., and Weibull, C., 1985, The consumption of osmium tetroxide by components of the cytoplasmic membrane of *Acholeplasma laidlawii* and its morphological implications, *Micron Microsc. Acta* **16**:77–83.

Rilfors, L., Khan, A., Brentel, I., Wieslander, Å., and Lindblom, G., 1982, Cubic liquid crystalline phase with phosphatidylethanolamine from *Bacillus megaterium* containing branched acyl chains, *FEBS Lett.* **149**:293–298.

Rilfors, L., Lindblom, G., Wieslander, Å., and Christiansson, A., 1984, Lipid bilayer stability in biological membranes, in: *Biomembranes*, Volume 12 (M. Kates and L. A. Manson, eds.), Plenum Press, New York, pp. 205–245.

Rilfors, L., Eriksson, P.-O., Arvidson, G., and Lindblom, G., 1986, Relationship between three-dimensional arrays of "lipidic particles" and bicontinuous cubic lipid phases, *Biochemistry* **25**:7702–7711.

Rilfors, L., Wikander, G., and Wieslander, Å., 1987, Lipid acyl chain-dependent effects of sterols in *Acholeplasma laidlawii* membranes, *J. Bacteriol.* **169**:830–838.

Rilfors, L., Hauksson, J. B., and Lindblom, G., 1993a, Phase equilibria in membrane lipids from *Acholeplasma laidlawii* and *Bacillus megaterium* containing iso and anteiso methyl-branched acyl chains, submitted.

Rilfors, L., Hauksson, J. B., and Lindblom, G., 1993b, The chemical structure of two phosphoglucolipids occurring in the membrane of *Acholeplasma laidlawii* strain A-EF22, submitted.

Romijn, J. C., van Golde, L. M. G., McElhaney, R. N., and van Deenen, L. L. M., 1972, Some studies on the fatty acid composition of total lipids and phosphatidylglycerol from *Acholeplasma laidlawii* B and their relation to the permeability of intact cells of this organism, *Biochim. Biophys. Acta* **280**:22–33.

Rottem, S., and Greenberg, A., 1975, Changes in composition, biosynthesis and physical state of membrane lipids occurring upon aging of *Mycoplasma hominis* cultures, *J. Bacteriol.* **121**:631–639.

Rottem, S., and Panos, C., 1969, The effect of long chain fatty acid isomers on growth, fatty acid composition and osmotic fragility of *Mycoplasma laidlawii* A, *J. Gen. Microbiol.* **59**:317–328.

Rottem, S., and Panos, C., 1970, The synthesis of long-chain fatty acids by a cell-free system from *Mycoplasma laidlawii* A, *Biochemistry* **9**:57–63.

Rottem, S., Hubbell, W. L., Hayflick, L., and McConnell, H. M., 1970, Motion of fatty acid spin labels in the plasma membrane of Mycoplasma, *Biochim. Biophys. Acta* **219**:104–113.

Rottem, S., Muhsam-Peled, O., and Razin, S., 1973, Acyl carrier protein in mycoplasmas, *J. Bacteriol.* **113**:586–591.

Saito, Y., and McElhaney, R. N., 1977, Membrane lipid biosynthesis in *Acholeplasma laidlawii* B: Incorporation of exogenous fatty acids into membrane glyco- and phospholipids by growing cells, *J. Bacteriol.* **132**:485–496.

Saito, Y., Silvius, J. R., and McElhaney, R. N., 1977, Membrane lipid biosynthesis in *Acholeplasma laidlawii* B: De novo biosynthesis of saturated fatty acids by growing cells, *J. Bacteriol.* **132**:497–504.

Saito, Y., Silvius, J. R., and McElhaney, R. N., 1978, Membrane lipid biosynthesis in *Acholeplasma laidlawii* B: Elongation of medium- and long-chain exogenous fatty acids in growing cells, *J. Bacteriol.* **133**:66–74.

Sankaram, M. B., Powell, G. L., and Marsh, D., 1989, Effect of acyl chain composition on salt-induced lamellar to inverted hexagonal phase transitions in cardiolipin, *Biochim. Biophys. Acta* **980**:389–392.

Schiefer, H.-G., Krauss, H., Brunner, H., and Gerhardt, U., 1976, Ultrastructural visualization of anionic sites on Mycoplasma membranes by polycationic ferritin, *J. Bacteriol.* **127**:461–468.

Seddon, J. M., 1990a, An inverse face-centered cubic face formed by diacylglycerol–phosphatidylcholine mixtures, *Biochemistry* **29**:7997–8002.

Seddon, J. M., 1990b, Structure of the inverted hexagonal (H_{II}) phase, and non-lamellar phase transitions of lipids, *Biochim. Biophys. Acta* **1031**:1–69.

Seddon, J. M., Kaye, R. D., and Marsh, D., 1983, Induction of the lamellar-inverted hexagonal phase transition in cardiolipin by protons and monovalent cations, *Biochim. Biophys. Acta* **734**:347–352.

Seelig, A., and Seelig, J., 1977, Effect of a single cis double bond on the structure of a phospholipid bilayer, *Biochemistry* **16**:45–50.

Seelig, J., and Seelig, A., 1980, Lipid conformation in model membranes and biological membranes, *Q. Rev. Biophys.* **13**:19–61.

Sen, A., Hui, S.-W., Mannock, D. A., Lewis, R. N. A. H., and McElhaney, R. N., 1990, Physical properties of glycosyl diacylglycerols. 2. X-ray diffraction studies of a homologous series of 1,2-di-*O*-acyl-3-*O*-(α-D-glucopyranosyl)-*sn*-glycerols, *Biochemistry* **29**:7799–7804.

Shaw, N., Smith, P. F., and Koostra, W. L., 1968, The lipid composition of *Mycoplasma laidlawii* strain B, *Biochem. J.* **107**:329–333.

Shaw, N., Smith, P. F., and Verheij, H. M., 1972, The structure of a glycerylphosphoryl-diglucosyl diglyceride from the lipids of *Acholeplasma laidlawii* strain B, *Biochem. J.* **129**:167–173.

Siegel, D. P., 1986, Inverted micellar intermediates and the transitions between lamellar, cubic, and inverted hexagonal amphiphile phases, *Chem. Phys. Lipids* **42**:279–301.

Siegel, D. P., Banschbach, J., and Yeagle, P. L., 1989, Stabilization of H_{II} phases by low levels of diglycerides and alkanes: An NMR, calorimetric, and X-ray diffraction study, *Biochemistry* **28**:5010–5019.

Silvius, J. R., and McElhaney, R. N., 1978, Lipid compositional manipulation in *Acholeplasma laidlawii* B. Effect of exogenous fatty acids on fatty acid composition and cell growth when endogenous fatty acid production is inhibited, *Can. J. Biochem.* **56**:462–469.

Silvius, J. R., Mak, N., and McElhaney, R. N., 1980, Lipid and protein composition and thermotropic lipid phase transitions in fatty acid-homogeneous membranes of *Acholeplasma laidlawii* B, *Biochim. Biophys. Acta* **597**:199–215.

Singer, S. J., and Nicolson, G. L., 1972, The fluid mosaic model of the structure of cell membranes, *Science* **175**:720–731.

Sjölund, M., Lindblom, G., Rilfors, L., and Arvidson, G., 1987, Hydrophobic molecules in lecithin–water systems. I. Formation of reversed hexagonal phases at high and low water contents, *Biophys. J.* **52**:145–153.

Smith, I. C. P., 1984, Conformational and motional properties of lipids in biological membranes as determined by deuterium magnetic resonance, in: *Biomembranes,* Volume 12 (M. Kates and L. A. Manson, eds.), Plenum Press, New York, pp. 133–168.

Smith, I. C. P., and Jarrell, H. C., 1983, Deuterium and phosphorus NMR of microbial membranes, *Acc. Chem. Res.* **16**:266–272.

Smith, P. F., 1968, The lipids of *Mycoplasma, Adv. Lipid Res.* **6**:69–105.

Smith, P. F., 1969, Biosynthesis of glucosyl diglycerides by *Mycoplasma laidlawii* strain B, *J. Bacteriol.,* **99**:480–486.

Smith, P. F., 1972, A phosphatidyl diglucosyl diglyceride from *Acholeplasma laidlawii* B, *Biochim. Biophys. Acta* **280**:375–382.

Smith, P. F., 1979, The composition of membrane lipids and lipopolysaccharides, in: *the Mycoplasmas,* Volume 1 (M. F. Barile and S. Razin, eds.), Academic Press, New York, pp. 231–257.

Smith, P. F., 1986, Structures of unidentified lipids in *Acholeplasma laidlawii,* strain A-EF 22, *Biochim. Biophys. Acta* **879**:107–112.

Steim, J. M., Tourtellotte, M. E., Reinert, J. C., McElhaney, R. N., and Rader, R. L., 1969, Calorimetric evidence for the liquid-crystalline state of lipids in a biomembrane, *Proc. Natl. Acad. Sci. USA* **63**:104–109.

Steinick, L. E., Wieslander, Å., Johansson, K.-E., and Liss, A., 1980, Membrane composition and virus susceptibility of *Acholeplasma laidlawii, J. Bacteriol.* **143**:1200–1207.

Stephens, E. B., Aulakh, G. S., Rose, D. L., Tully, J. G., and Barile, M. F., 1983, Intraspecies genetic relatedness among strains of *Acholeplasma laidlawii* and of *Acholeplasma axanthum* by nucleic acid hybridization, *J. Gen. Microbiol.* **129**:1929–1934.

Strenk, L. M., Westerman, P. W., and Doane, J. W., 1985, A model of orientational ordering in phosphatidylcholine bilayers based on conformational analysis of the glycerol backbone region, *Biophys. J.* **48**:765–773.

Strickland, K. P., 1973, The chemistry of phospholipids, in: *Form and Function of Phospholipids* (G. B. Ansell, J. N. Hawthorne, and R. M. C. Dawson, eds.), Elsevier, Amsterdam, pp. 9–42.

Tarshis, M. A., Ladygina, V. G., Migoushina, V. L., Klebanov, G. I., and Rakovskaya, I. V., 1981, Interaction of *Acholeplasma laidlawii* cells with mouse spleen lymphocytes, *Zentralbl. Bakteriol. Hyg. Abt. 1 Orig. A.* **250**:153–166.

Tiddy, G. J. T., 1980, Surfactant–water liquid crystal phases, *Phys. Rep.* **57**:1–46.

Tilcock, C. P. S., and Cullis, P. R., 1982, The polymorphic phase behaviour and miscibility properties of synthetic phosphatidylethanolamines, *Biochim. Biophys. Acta* **684**:212–218.

Tourtellotte, M. E., Morowitz, H. J., and Kasimer, P., 1964, Defined medium for *Mycoplasma laidlawii*, *J. Bacteriol.* **88**:11–15.

Tully, J. G., Whitcomb, R. F., Rose, D. L., Hackett, K. J., Clark, E., Henegar, R. B., Carle, P., and Bové, J. M., 1990, Current insight into the host diversity of Acholeplasmas, *Int. J. Med. Microbiol. Suppl.* **20**:461–467.

Ulmius, J., Wennerström, H., Lindblom, G., and Arvidson, G., 1977, Deuteron nuclear magnetic resonance studies of phase equilibria in a lecithin–water system, *Biochemistry* **16**:5742–5745.

Vail, W. J., and Stollery, J. G., 1979, Phase changes of cardiolipin vesicles mediated by divalent cations, *Biochim. Biophys. Acta* **551**:74–84.

Van Dijck, P. W. M., Ververgaert, P. H. J. T., Verkleij, A. J., Van Deenen, L. L. M., and De Gier, J., 1975, Influence of Ca^{2+} and Mg^{2+} on the thermotropic behaviour and permeability properties of liposomes prepared from dimyristoyl phosphatidylglycerol and mixtures of dimyristoyl phosphatidylglycerol and dimyristoyl phosphatidylcholine, *Biochim. Biophys. Acta* **406**:465–478.

Van Dijck, P. W. M., de Kruijff, B., Verkleij, A. J., Van Deenen, L. L. M., and De Gier, J., 1978, Comparative studies on the effects of pH and Ca^{2+} on bilayers of various negatively charged phospholipids and their mixtures with phosphatidylcholine, *Biochim. Biophys. Acta* **512**:84–96.

Vasilenko, I., de Kruijff, B., and Verkleij, A., 1982, Polymorphic phase behaviour of cardiolipin from bovine heart and from *Bacillus subtilis* as detected by ^{31}P NMR and freeze-fracture techniques, *Biochim. Biophys. Acta* **684**:282–286.

Verkleij, A. J., De Kruijff, B., Ververgaert, P. H. J. T., Tocanne, J. F., and Van Deenen, L. L. M., 1974, The influence of pH, Ca^{2+} and protein on the thermotropic behaviour of the negatively charged phospholipid phosphatidylglycerol, *Biochim. Biophys. Acta* **339**:432–437.

Wallbrandt, P., Tegman, V., Jonsson, B.-H., and Wieslander, Å., 1992, Identification and analysis of the genes for the putative pyruvate dehydrogenase enzyme complex in *Acholeplasma laidlawii*, *J. Bacteriol.* **174**:1388–1396.

Watts, A., Harlos, K., Maschke, W., and Marsh, D., 1978, Control of the structure and fluidity of phosphatidylglycerol bilayers by pH titration, *Biochim. Biophys. Acta* **510**:63–74.

Weibull, C., Christiansson, A., and Carlemalm, E., 1983, Extraction of membrane lipids during fixation, dehydration and embedding of *Acholeplasma laidlawii*-cells for electron microscopy, *J. Microsc.* **129**:201–207.

Weisburg, W. G., Tully, J. G., Rose, D. L., Petzel, J. P., Oyaizu, H., Yang, D., Mandelco, L., Sechrest, J., Lawrence, T. G., Van Etten, J., Maniloff, J., and Woese, C. R., 1989, A phylogentic analysis of the Mycoplasmas: Basis for their classification, *J. Bacteriol.* **171**:6455–6476.

Wieslander, Å., and Rilfors, L., 1977, Qualitative and quantitative variations of membrane lipid species in *Acholeplasma laidlawii* A, *Biochim. Biophys. Acta* **466**:336–346.

Wieslander, Å., and Selstam, E., 1987, Acyl-chain-dependent incorporation of chlorophyll and cholesterol in membranes of *Acholeplasma laidlawii*, *Biochim. Biophys. Acta* **901**:250–254.

Wieslander, Å., Ulmius, J., Lindblom, G., and Fontell, K., 1978, Water binding and phase structures for different *Acholeplasma laidlawii* membrane lipids studied by deuteron nuclear magnetic resonance and X-ray diffraction, *Biochim. Biophys. Acta* **512**:241–253.

Wieslander, Å., Christiansson, A., Walter, H., and Weibull, C., 1979, Fractionation of membranes
from *Acholeplasma laidlawii* A on the basis of their surface properties by partition in two-
polymer aqueous phase systems, *Biochim. Biophys. Acta* **550**:1–15.

Wieslander, Å., Christiansson, A., Rilfors, L., and Lindblom, G., 1980, Lipid bilayer stability in
membranes. Regulation of lipid composition in *Acholeplasma laidlawii* as governed by molecular shape, *Biochemistry* **19**:3650–3655.

Wieslander, Å., Rilfors, L., Johansson, L. B.-Å., and Lindblom, G., 1981a, Reversed cubic phase
with membrane glucolipids from *Acholeplasma laidlawii*. ¹H, ²H, and diffusion nuclear magnetic resonance measurements, *Biochemistry* **20**:730–735.

Wieslander, Å., Christiansson, A., Rilfors, L., Johansson, L. B.-Å., Khan, A., and Lindblom, G.,
1981b, Lipid phase structure governs the regulation of lipid composition in membranes of
Acholeplasma laidlawii, FEBS Lett. **124**:273–278.

Wieslander, Å., Rilfors, L., and Lindblom, G., 1986, Metabolic changes of membrane lipid composition in *Acholeplasma laidlawii* by hydrocarbons, alcohols, and detergents: Arguments for
effects on lipid packing, *Biochemistry* **25**:7511–7517.

Wieslander, Å., Dahlqvist, A., Rilfors, L., and Lindblom, G., 1993a, Strong influence of acyl chain
length on the regulation of polar lipid composition in membranes and implications for cell size in
Acholeplasma laidlawii, submitted.

Wieslander, Å., Rilfors, L., Dahlqvist, A., Jonsson, J., Hellberg, S., Rännar, S., Sjöström, M., and
Lindblom, G., 1993b, Similar regulatory mechanisms despite differences in membrane lipid
composition in *Acholeplasma laidlawii* strains A-EF22 and B-PG9: A multivariate data analysis,
submitted.

Wilkins, M. H. F., Blaurock, A. E., and Engelman, D. M., 1971, Bilayer structure in membranes,
Nature New Biol. **230**:72–76.

Winsborrow, B. G., Smith, I. C. P., and Jarrell, H. C., 1991, Dynamics of glycolipids in the liquid-
crystalline state. ²H NMR study, *Biophys. J.* **59**:729–741.

Wise, K. S., Cassell, G. H., and Acton, R. T., 1978, Selective association of murine T lympho-
blastoid cell surface alloantigens with *Mycoplasma hyorhinis, Proc. Natl. Acad. Sci. USA*
75:4479–4483.

Wohlgemuth, R., Waespe-Sarcevic, N., and Seelig, J., 1980, Bilayers of phosphatidylglycerol. A
deuterium and phosphorus nuclear magnetic resonance study of the head-group region, *Biochemistry* **19**:3315–3321.

Chapter 5

The Role of Cholesterol in Mycoplasma Membranes

Jean Dahl

1. INTRODUCTION

Cholesterol is the end product of the sterol biosynthetic pathway in animal cells. The sterol molecule arose necessarily with the advent of an aerobic environment (Bloch, 1983), making it a relative latecomer in the evolutionary development of living cells. Perhaps, as a consequence, sterols are not uniformly distributed among life forms as are, say, amino acids and nucleic acids, but are reserved almost without exception as membrane components of eukaryotic rather than prokaryotic cells. Notable exceptions to the exclusionary rule of sterols in pro-karyotes can be found in a few aerobic bacteria (Bird *et al.*, 1971) and in members of the class Mollicutes. Of the three established families comprising the class Mollicutes, the Mycoplasmataceae and Spiroplasmataceae require choles-terol or a related sterol for growth whereas Acholeplasmataceae do not. None of the mycoplasmas* synthesize cholesterol *de novo*, nor do they generally metabo lize or modify the sterol molecule. Rather, mycoplasmas incorporate sterols unchanged from the environment to levels approaching 50 mole% of the total

---

*The trivial term *mycoplasma* will be used to denote any species included in the family My-coplasmataceae.

Jean Dahl   Department of Pathology, Harvard Medical School, Boston, Massachusetts 02115.

Subcellular Biochemistry, Volume 20: Mycoplasma Cell Membranes, edited by Shlomo Rottem and Itzhak Kahane. Plenum Press, New York, 1993.

membrane lipid. These features coupled with mycoplasmas' inherent fatty acid auxotrophy and single membrane structure assure the mycoplasmas a high rank among the organisms of choice for studies aimed at understanding the role of sterols in natural membranes. The degree of control over membrane lipid content one can achieve with mycoplasmas more closely approaches that obtained with model membranes than with most other natural systems used to study sterol structure–function relationships such as yeast, insects, and fibroblasts. Although studies on model membranes have provided valuable information on the functional consequences of sterol–phospholipid interactions, the principles governing both sterol–phospholipid and sterol–protein interactions in the complex milieu of a natural membrane must ultimately be delineated before a full understanding of the role of cholesterol in mycoplasmas and other cells can be reached.

1.1. Sterol as a Required Growth Factor for Mycoplasma

The fact that cholesterol is a required growth factor for mycoplasma was first noted by Edward and Fitzgerald (1951). Serum, which normally provides enough cholesterol and fatty acids needed to promote optimum cell growth, can be replaced by delipidated bovine serum albumin, fatty acids, sterol, and, when required, phosphatidylcholine and spingomyelin as in the case of *M. gallisepticum* (Rodwell, 1983a). By replacing serum with a series of cholesterol analogues, Smith and colleagues defined the requisite sterol structural features for optimum cell growth, namely a planar nucleus, an equatorial 3β-hydroxyl group,

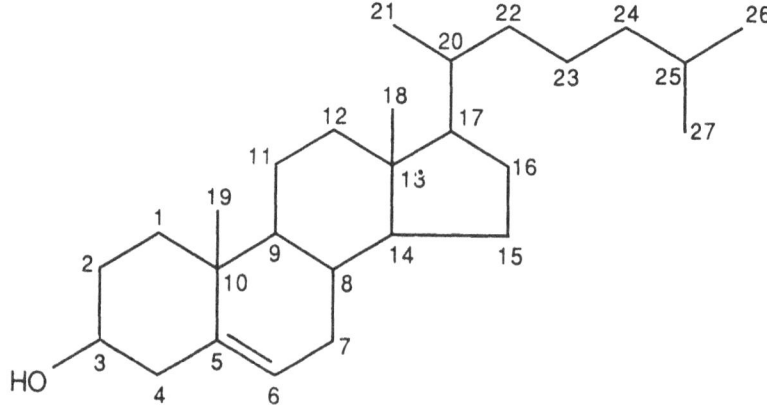

FIGURE 1. Structure of cholesterol indicating the numerical designations of carbon atoms.

and an isooctyl side chain (see Figure 1) (Smith and Lynn, 1958; Smith, 1964; Smith and Rothblat, 1962). These are the same attributes cholesterol must have to modulate the bulk fluidity of the lipid bilayer (Demel and de Kruyff, 1976). Although esterified cholesterol will also be taken up by mycoplasmas (Rottem, 1980), it is not required for growth and appears to form lipid droplets or pockets in the membrane (Melchior and Rottem, 1981).

1.2. Sterol as an Architectural Component of Mycoplasma Membranes

In mycoplasma, cholesterol is localized exclusively in the plasma membrane where it provides architectural integrity. A high membrane cholesterol content correlates with the preservation of cell intactness when mycoplasmas are treated with phospholipase A_2 (Gross and Rottem, 1984). On the other hand, mycoplasmas undergo irreversible lysis when membrane cholesterol is complexed with certain sterol binding agents such as digitonin and streptolysin O (Razin and Argaman, 1963; Bernheimer and Davidson, 1965). Model membrane studies show that cholesterol exerts a profound effect on membrane physical properties such as membrane ordering and permeability, condensation of membrane phospholipids, lateral diffusion of phospholipids, lateral phase separations, and phospholipid phase transition (Yeagle, 1988). Physical studies on membranes of *Mycoplasma* and *Acholeplasma* species have suggested that cholesterol similarly affects the physical characteristics of natural membranes, but more importantly, they have facilitated the exploration of the functional consequences of cholesterol on membrane processes such as ion transport (Le Grimellec and Leblanc, 1978), cell volume control (Romano *et al.*, 1986), control of osmotic fragility (Rottem and Verkleij, 1982), and the activity of membrane-bound enzymes (McElhaney, 1982). The effects of cholesterol on these processes have been attributed principally to cholesterol-mediated alterations in the physical and lipid environment of the respective proteins and not to direct protein–sterol interactions. However, studies in which mycoplasma is grown on pairs of sterols, one major and one minor, indicate that sterol performs a dual role in membranes, one of an architectural or bulk function that influences the membrane physical state, and another of a regulatory nature that may be fulfilled by minor amounts of sterol showing no measurable effect on bulk physical membrane properties (J. S. Dahl *et al.*, 1980). This chapter will review evidence for both roles of cholesterol in mycoplasma membranes as well as remark on the mode of cholesterol uptake by mycoplasmas and its intracellular distribution. In addition to the material presented here, the reader is referred to several recent reviews covering specific aspects of these topics (Bloch, 1983; Yeagle, 1985; Bittman, 1988; Dahl and Dahl, 1988).

2. MODE OF STEROL UPTAKE

2.1. Nature of the Sterol Donor

All axenic cultures of mycoplasma can be maintained in medium supplemented with serum as a supply of cholesterol and other lipid factors such as fatty acids and phospholipids. However, in the case of some of the less fastidious species, lipid-defined or chemically defined media have been developed (Rodwell, 1983b). Here the serum component has been replaced with high- or low-density lipoproteins (Slutzky *et al.*, 1977), defatted serum protein fraction (fraction C) supplemented with sterol and fatty acid (Rodwell, 1969), delipidated bovine serum albumin and ethanolic solutions of sterol and fatty acids (Dahl and Dahl, 1983), or liposomes prepared with phosphatidylcholine and cholesterol along with bovine serum fraction or bovine serum albumin (Kahane and Razin, 1977; Cluss *et al.*, 1983).

2.2. A Receptor-Mediated Mechanism?

In contrast to mammalian cells where cholesterol can be synthesized *de novo* (Bloch, 1965) or taken up by receptor-mediated endocytosis (Brown and Goldstein, 1986), mycoplasmas must absorb all of their cholesterol by processes which take place exclusively at the outer surface of the plasma membrane and which do not involve endocytosis or pinocytosis. The rate and extent of cholesterol uptake is markedly dependent on the cholesterol content of the receptor membrane, and the state of the cell's viability, metabolic activity, and proliferation rate as well as the intactness of its surface proteins (Clejan *et al.*, 1978; Efrati *et al.*, 1981). By adapting *M. capricolum* to grow in serum-poor medium, the membrane cholesterol content can be lowered by a factor of six. Addition of serum-rich medium to a cholesterol-depleted, early logarithmic culture of *M. capricolum* results in a marked increase in the rate of growth accompanied by a fourfold rise in the free cholesterol content (Clejan *et al.*, 1978). Complete inhibition of cell growth attained by treating cholesterol-depleted *M. capricolum* with chloramphenicol or the ionophores valinomycin, nonactin, or gramicidin correlates with a 50% reduction in the level of cellular cholesterol, showing that cholesterol uptake by *M. capricolum* is, at least in part, a growth- and metabolism-dependent phenomenon. Further support for a metabolic determinant in cholesterol dynamics is provided by experiments showing that the translocation of cholesterol from the outer to the inner leaflet of the lipid bilayer is negatively affected by the same set of metabolic inhibitors that stop growth, namely chloramphenicol or the ionophores valinomycin, nonactin, and gramicidin (Clejan *et al.*, 1978).

The possibility of receptor-facilitated cholesterol uptake in mycoplasmas was raised by Efrati *et al.* (1981) when they found that proteolytic digestion of intact cells of *M. capricolum* correlates with a 50% reduction in cholesterol uptake from phospholipid–cholesterol vesicles. However, since the proliferation of *M. capricolum* was also inhibited following trypsin treatment, an alternative explanation is that the cells must be in a dynamic state of growth to drive the accumulation of cholesterol. Stronger support for the notion of receptor-mediated cholesterol uptake in mycoplasmas comes from a study on the transfer of cholesterol from low-density lipoprotein to *M. capricolum* membranes (Efrati *et al.*, 1982). Trypsin treatment decreases both the binding of lipoprotein particles and the transfer of cholesterol to isolated membranes. In this regard it should be pointed out the cholesterol uptake in *A. laidlawii* is not influenced by inhibiting protein synthesis with chloramphenicol (Razin, 1974) or by digesting membrane proteins with trypsin (Efrati *et al.*, 1981), suggesting the lack of protein involvement in the absorption of cholesterol by the non-sterol-requiring Mollicutes.

Evidence disfavoring a receptor-mediated mechanism of cholesterol uptake in mycoplasma derives from studies on sterol exchange. Kinetic data on the exchange of cholesterol between donor and acceptor membranes are thought best to be explained by a mechanism involving transfer of cholesterol through the aqueous phase rather than by direct transfer via transient contact between donor and acceptor moieties (Backer and Dawidowicz, 1981a; McLean and Phillips, 1981; Bittman, 1988). In *M. gallisepticum* agents that increase the solubility of cholesterol in water such as potassium thiocyanate, dimethylsulfoxide, and bovine serum albumin markedly enhance the rate of cholesterol exchange between intact cells and unilamellar lipid vesicles, whereas those which decrease the aqueous solubility of sterol reduce the rate of exchange (Clejan and Bittman, 1984a). Nevertheless, the possibility that an integral membrane cholesterol binding protein functions in cholesterol absorption by mycoplasmas in a manner analogous to that described for fatty acid uptake in *Escherichia coli* (Black *et al.*, 1987) or mammalian cells (Stremmel *et al.*, 1985) cannot be ruled out.

2.3. Transmembrane Distribution of Sterol

Among the various types of eukaryotic membranes, e.g., the plasma membrane, nuclear membranes, mitochondrial membranes, the endoplasmic reticulum, and the Golgi, there is a great disparity in sterol content. At about 45 mole%, the plasma membrane exhibits the highest sterol concentration (Coleman and Finean, 1966), while the rough endoplasmic reticulum at less than 6 mole% is among the lowest (Colbeau *et al.*, 1971). Beneath this layer of complexity are other levels of diversity within a membrane. The sterol content of the respective bilayer halves can differ. Whereas an equal distribution is found in the inner and outer leaflets of the erythrocyte membrane (Blau and Bittman, 1978), a ratio of 80% (inner) to 20% (outer) has been reported for the LM cell plasma membrane

(Schroeder, 1981). Moreover, reports suggesting that coated pits (Montesano *et al.*, 1979) may be depleted of cholesterol compared with the surrounding membrane regions have raised the possibility of heterogeneity in the lateral disposition of cholesterol within a membrane. Despite numerous investigations aimed at understanding the cause of heterogeneity in sterol distribution among various eukaryotic membranes, the underlying mechanism(s) responsible for establishing and maintaining its complexity remains as elusive as its true function.

With mycoplasmas the problem of asymmetric disposition of cholesterol among and within membranes is reduced to one of transmembrane distribution between bilayer halves. The localization of cholesterol in the mycoplasma cytoplasmic membrane has been studied in both *M. gallisepticum* and *M. capricolum* utilizing two different techniques, sterol exchange (Bloj and Zilversmit, 1976) and sterol binding to the polyene antibiotic filipin (Bittman and Rottem, 1976). Sterol exchange between intact cells and lipoproteins or vesicles exhibits biphasic behavior thought to arise from differences in the exchange rates of two cholesterol pools separated by their inner or outer bilayer localization. On the other hand, stopped-flow kinetic measurements of filipin binding compare initial rates obtained with either intact cells or unsealed membranes. The results from both techniques suggest that cholesterol is distributed asymmetrically in *M. capricolum* (two-thirds in the outer monolayer) and symmetrically in *M. gallisepticum* (Bittman and Rottem, 1976). Surprisingly, though, in both species the ability of cholesterol to undergo "flip-flop" is markedly restricted in resting cells. Rates of transmembrane movement in metabolically active cells are characterized by half-times of 2 to 4 hr which expand to about 18 days in resting cells (Rottem *et al.*, 1978; Clejan *et al.*, 1978). By comparison, the transmembrane movement of cholesterol in erythrocytes or phospholipid vesicles is extremely rapid, showing a half-time of less than 1 min (Backer and Dawidowicz, 1981b; Lange *et al.*, 1981).

Although cholesterol may be evenly distributed between the two bilayer halves, analogues of cholesterol may not be. A series of experiments was undertaken to explore the effect of altering certain structural features of the sterol molecule on its transmembrane distribution (Clejan *et al.*, 1981; Clejan and Bittman, 1984b). The sterol analogues used in this study fall into two categories, those which partition between the two bilayer halves like cholesterol and those which reside, preferentially compared with cholesterol, in the outer monolayer. The sterols in the former class resemble cholesterol with respect to their isooctyl side chain whereas the latter category is characterized by sterols bearing alkyl-substituted or unsaturated side chains. In *M. gallisepticum* 52% of the cholesterol is localized to the outer monolayer compared with values of 86% for β-sitosterol, 94% for desmosterol, 93% for *trans*-22-dehydrocholesterol, 91% for *cis*-22-dehydrocholesterol, and 94% for cholesta-5,22*E*,24-trien-3β-ol (Clejan and Bittman, 1984b). In *M. capricolum*, whereas 65% of the cholesterol, cholestanol, or

4,6-cholestadien-3β-ol is found in the outer leaflet of the membrane, 89% of the sitosterol, stigmosterol, or ergosterol resides there (Clejan *et al.*, 1981). Thus, sterols bearing alkyl-substituted or unsaturated side chains are less likely to accumulate in the inner monolayer than those with the isooctyl side chain of cholesterol. Yet to be determined, however, is the extent to which protein–sterol interactions, membrane protein asymmetry, and membrane phospholipid asymmetry contribute to the relationship between sterol structure and transmembrane distribution. Studies along these lines must necessarily await a more detailed knowledge of the proteins that go to make up the mycoplasma membrane as well as the transmembrane distribution of its phospholipids.

The functional significance of the transmembrane distribution of sterols also remains a puzzle. Since in these studies there is no apparent correlation between rates of growth on the various sterols tested and the degree of their respective membrane asymmetries, the functional significance of the transmembrane distribution of cholesterol cannot be evaluated. A comprehensive treatment of the subject of transmembrane distribution of sterols is presented in Chapter 2.

2.4. Effect of Membrane Lipid Composition on Sterol Uptake

It has been suggested that the cholesterol uptake capacity of mycoplasmas is largely dependent on the membrane phospholipid content (Razin, 1974). Whereas the ratio of cholesterol to protein in mycoplasma membranes varies markedly upon cell aging or chloramphenicol treatment, the ratio of cholesterol to phospholipid under the same conditions remains essentially unchanged. A further study by Efrati *et al.* (1981) showed that extensive hydrolysis of *M. capricolum* membrane phospholipids by phospholipase A_2 removes 70% of the polar lipids and results in a 60% decline in the cholesterol uptake capacity. By comparison, similar treatment of *A. laidlawii* membranes results in a 30% decline in polar lipid content and a 55% reduction in cholesterol uptake. The authors suggest that the disproportionate dependence of cholesterol incorporation on phospholipase A_2-sensitive lipids of *A. laidlawii* relative to *M. capricolum* is due to fundamental differences in the lipid profiles of these two organisms. While phosphatidylglycerol and diphosphatidylglycerol comprise about 75% of the polar lipids in *M. capricolum* (Gross *et al.*, 1982), they account for only 35% in *A. laidlawii* where glycolipids are the major polar lipids (Efrati *et al.*, 1986). In *A. laidlawii* the glycolipids remaining after phospholipase treatment have a lower capacity to bind cholesterol than phospholipids.

Support for the idea that the glycolipid content of *A. laidlawii* results in diminished capacity to incorporate cholesterol is provided by a recent study of Efrati *et al.* (1986) which correlates the cholesterol content of 20 strains of *A. laidlawii* with the polar lipid species. As judged by statistical analysis, cholesterol uptake shows a strong negative correlation to glycolipid X and monoglucosyldiglyceride (MGDG). Glycolipid X has been identified as a polyprenol-α-D-glucoside esterified at the 2-hydroxyl group with a long-chain fatty acid (Bhakoo *et al.*, 1987). As pointed out by McElhaney (1989), this new polyprenylated glycolipid shares with MGDG (and cholesterol) the probable propensity for forming nonlamellar lipid phases due to a similar "inverted" wedge shape (smaller at the polar head group). The tendency of cholesterol and MGDG to be mutually exclusive as membrane lipid components of *A. laidlawii* A has been demonstrated convincingly by an extensive series of studies reviewed by Rilfors *et al.* (1984). The investigators have rationalized the inverse relationship between cholesterol and MGDG as compensatory adjustments made by the cell in an effort to maintain a balance between bilayer and reverse hexagonal phases needed to ensure proper membrane stability and function.

3. STEROL STRUCTURE–FUNCTION RELATIONSHIPS

3.1. Effect of Sterol on Membrane Physical Properties and Cell Physiology

Much of our awareness of sterol-induced changes in the physical properties of membranes has necessarily stemmed from studies using artificial bilayers. There is now overwhelming evidence that cholesterol acts to modulate membrane fluidity and minimize membrane permeability to small molecules (Demel and de Kruyff, 1976; Yeagle, 1985). By virtue of its amphipathic character and planar steroid ring system, cholesterol interdigitates between the phospholipids, aligning parallel to the extended fatty acyl chains with the 3β-hydroxyl group pointing toward the water bilayer interface. The ensuing hydrophobic interactions between cholesterol and the phospholipids act to order fatty acyl chains above the phase transition and disorder those below the phase transition, causing the thermal phase transition to disappear (Yeagle, 1985). Thus, cholesterol maintains phospholipid bilayers in an intermediate fluid state. Besides altering the thermodynamic properties of lipid bilayers, cholesterol also acts to reduce the membrane surface area by "condensing" the phospholipid molecules (Hyslop *et al.*, 1990) and serves to alter bilayer thickness (McIntosh, 1978; Ipsen *et al.*, 1990).

In view of the broad and profound effects cholesterol has on the physical properties of artificial bilayers, the need to assess the functional consequences of cholesterol in living cells becomes crucial. In this area mycoplasmas have been particularly useful. Initial experiments along these lines were performed in *A. laidlawii*. This organism, though not dependent on sterol for growth, will nevertheless incorporate it up to 20 mole% of the membrane lipid when it is present in the medium. Consequently, de Kruyff *et al.* (1972) demonstrated in *A.*

laidlawii that cholesterol acts to quench the thermal phase transition of a natural membrane. In *Mycoplasma* species that preferentially incorporate saturated fatty acids into their *de novo*-synthesized phospholipids such as *M. arginini*, cholesterol was deemed necessary to prevent crystallization of the phospholipids at physiological temperatures (Rottem, 1981). Thus, the ability of cholesterol to order the fatty acyl chains of phospholipid above the thermal phase transition and disorder those below also applies to natural membranes.

Studies designed to explore the structural features of cholesterol required to exert its ordering effects in a natural membrane were also undertaken in *A. laidlawii* by de Kruyff *et al.* (1972, 1973). By comparing the effect of cholesterol, epicholesterol, cholestanol, epicholestanol, ergosterol, stigmasterol, and coprostanol on the rate of flux of glycerol and erythritol through the membrane of intact *A. laidlawii*, the investigators showed that only sterols with a 3β-hydroxyl group and a planar steroid nucleus effectively reduce the permeability of *A. laidlawii* membranes.

Although *A. laidlawii* is useful for examining the effect of sterol structure on modulating membrane fluidity, studies on this organism cannot provide insight into the unique roles cholesterol must perform in cells that show an absolute dependency on sterol for growth. For this purpose natural sterol auxotrophs of the *Mycoplasma* species are ideal. Not only do they show an absolute growth requirement for sterol, but the detected sterol-dependencies can be attributed to the sterol molecule *per se* since mycoplasmas do not metabolize cholesterol or other sterols to functional products such as steroid hormones or bile acids as is the case for virtually all eukaryotes. In a few *Mycoplasma* species the amount of sterol required for growth is minimal. For example, a membrane cholesterol concentration as low as 6 mole% is adequate to permit growth of *M. mycoides* subsp. *capri* (Rottem *et al.*, 1973a,b). Presumably, 6 mole% is a threshold level for membrane sterol below which cell proliferation is precluded. Cholesterol-poor *M. mycoides* subsp. *capri* is characterized by a slow growth rate, increased fragility and permeability, and cold sensitivity. Membranes isolated from cholesterol-poor cells are susceptible to thermotropic phase separations and lateral phase separation of membrane proteins. Moreover, the cholesterol-poor *M. mycoides* subsp. *capri* enrich their phospholipids with saturated fatty acids (Rottem *et al.*, 1973b). All of these data support the notion that one role of cholesterol is structural and serves to modulate the physical state of the membrane. However, in *M. mycoides* subsp. *capri* this role is apparently not mandatory for growth but merely optimizes cellular processes. The threshold level of cholesterol remaining in cholesterol-poor cells appears, on the other hand, to serve a function essential to cell proliferation.

Just as the required amount of growth sterol for a few *Mycoplasma* species is minimal, so too are the required structural features. A series of sterols exhibiting structural features far removed from the archetypal 3β-hydroxyl group, planar steroid nucleus, and isooctyl side chain of cholesterol have been found to support the growth of *M. capricolum* (Odriozola *et al.*, 1978). Among this group of sterol analogues is one that is blocked at the 3-hydroxyl by a methyl group (cholesteryl methyl ether), those in which the planarity of the α-face of the ring system has been modified (lanosterol, 4β-methylcholestanol, and 4,4-dimethylcholesterol), and one that lacks the flexible isooctyl side chain and is a pentacyclic triterpene rather than a sterol (β-amyrin). Substantiating the claim of a broad sterol specificity exhibited by certain mycoplasma species is a report by Efrati *et al.* (1980) that cholesteryl betainate serves as a growth sterol for *M. capricolum* and a report by Kannenberg and Poralla (1982) that the hopane diplopterol will moderately support growth of *M. mycoides* subsp. *capri*. Although the efficacies of these cholesterol analogues and hopanoids as essential nutritional factors for *M. capricolum* and *M. mycoides* subsp. *capri* are decidedly less than cholesterol, the analogues nonetheless provide a limited degree of functionality as a sterol substitute.

Clues to this functionality are as follows. NMR spectroscopy has revealed that lanosterol, like cholesterol, causes separation of phospholipid head groups (Yeagle *et al.*, 1977) which may be a general property of amphipathic molecules possessing a rigid polycyclic nucleus. Second, sterol analogues differing in the spatial orientation of methyl groups attached to the sterol nucleus show a progressive effectiveness to modulate membrane fluidity which is paralleled by their serviceability as growth sterols for *M. capricolum* (Dahl *et al.*, 1980a). By both criteria, sterol competence increases in the order lanosterol < 4,4-dimethylcholesterol < 4β-methylcholesterol < 4α-methylcholesterol < cholesterol. The relative ineffectiveness of lanosterol to raise the microviscosity of lecithin vesicles and to support growth of *M. capricolum* was rationalized on the grounds that the methyl group at C-14 projects from the sterol α-plane and interferes with essential hydrophobic interactions (Bloch, 1976). Support for this argument comes from studies on cycloartenol, a 9,19-cyclopropane plant sterol which is isomeric with lanosterol. Although both sterols bear an axial methyl group at C-14, molecular models reveal striking conformational differences between the two molecules. A slight bow in the steroid nucleus caused by the cyclopropyl group on the β-face, appears to bring the C-14 methyl group in alignment with several axial H atoms and thereby restore planarity to the α-face of the molecule. Thus, cycloartenol is substantially more competent as a membrane sterol than lanosterol as judged by its ability to condense fatty acyl chains of phospholipids and support the growth of *M. capricolum* (C. E. Dahl *et al.*, 1980b).

Members of the hopanoid family are pentacyclic sterol-like molecules arising from squalene by an anaerobic route of cyclization. They are principally found in bacterial membranes where they perform a cholesterol-like function (Ourisson *et al.*, 1979). Below the phase transition they act to fluidize phospholipid acyl chains and above the transition temperature they reduce acyl chain mobility, and thereby serve to stabilize membranes by increasing the rigidity of the lipid matrix (Kannenberg *et al.*, 1983).

3.2. Sterol Synergism: Effect of Pairs of Sterols on Growth and Membrane Properties

The foregoing discussion on sterol structure–function relationships serves to reinforce the dogma that cholesterol is an architectural component of membranes that influences the physical state of the resident polar lipids and thereby leads to optimum cell function. A notable outcome of the series of experiments by C. E. Dahl *et al.* (1980a) examining the effectiveness of various methylated sterols as growth factors for *M. capricolum* is the discovery of sterol synergism in mycoplasma (J. S. Dahl *et al.*, 1980). Among the methylated sterols tested as growth factors for *M. capricolum,* lanosterol is the least effective. However, when lanosterol is combined with a small amount of cholesterol (lanosterol:cholesterol, 20:1, by weight), the doubling time is halved. The level of cholesterol used is unable to support growth alone, but with lanosterol is synergistic. The observation of sterol synergism suggests a dual role for sterol in mycoplasma. Cells raised on the lanosterol–cholesterol mixture contain 18 mole% lanosterol and 3 mole% cholesterol. At this level cholesterol shows no measurable effect on the membrane microviscosity. Thus, apart from a bulk function, cholesterol may act in microdomains to regulate specific metabolic processes. A regulatory role for the threshold level of cholesterol remaining in cholesterol-poor *M. mycoides* subsp. *capri* has also been suggested by Razin (1982). What distinguishes the regulatory role of sterol is its structural specificity, whereas the bulk role can be satisfied at least in the caprine mycoplasmas by a broad spectrum of sterol structures.

The phenomenon of sterol synergism in *M. capricolum* is not limited to the sterol pair of lanosterol (10 μg/ml) and cholesterol (0.5 μg/ml), as it has recently been reported by Lelong *et al.* (1988) to occur with a mixture of 7β-OH cholesterol (2.5 μg/ml) and cholesterol (0.5 μg/ml). In this study the bulk sterol function is performed by 7β-OH cholesterol whereas the small amount of cholesterol fulfills a more specialized regulatory role. At equimolar concentrations of cholesterol and 7β-OH cholesterol, the hydroxylated sterol is inhibitory for growth. Whether this inhibition is due to an adverse effect on membrane fluidity or inhibition of a protein-mediated process was not determined, but it was noted that 7β-OH cholesterol correlates with a significant decrease in the diphosphatidylglycerol/phosphatidylglycerol ratio without affecting the total phospholipid ratio. Also worth noting was the remarkable resistance of 7β-OH cholesterol to exchange with lipid vesicles, suggesting strong interaction of this cholesterol analogue with some membrane component.

The phenomenon of sterol synergism has also been reported to occur in insects (Clark and Bloch, 1959) and in yeast (Rodriguez *et al.*, 1982; Ramgopal and Bloch, 1983; Pinto *et al.*, 1983).

4. STEROL-MEDIATED REGULATION

The description of sterol synergism in mycoplasmas has raised the possibility that apart from a bulk role for cholesterol in the control of membrane fluidity, specific cholesterol–protein interactions modulate the function of certain membrane proteins. In fact, the importance of membrane fluidity as the determining factor of cholesterol-mediated effects on membrane enzyme activity has come into question (Carruthers and Melchior, 1984). Although there have been relatively few studies on the interaction of sterols with membrane proteins, several select examples give clear evidence that cholesterol not only binds to membrane proteins but that this binding has functional significance. McNamee *et al.* (1982) have demonstrated that reconstitution of a functionally competent acetylcholine receptor from *Torpedo californica* requires the presence of cholesterol. Cholesterol appears to be excluded from sites at the lipid–protein interface but possibly binds internally between the receptor subunits (Jones and McNamee, 1988). In acetylcholine receptor-rich membranes, two pools of cholesterol occur, one that is easily depleted influencing only the bulk viscosity and a second that is more tightly bound (Liebel *et al.*, 1987). The Ca^{2+}/Mg^{2+}-ATPase from sarcoplasmic reticulum is markedly activated by cholesterol when reconstituted with short-chain phospholipid (Simmonds *et al.*, 1982). Fluorescence quenching studies show that cholesterol binds to nonannular sites on the ATPase possibly between transmembranous α-helices or between dimers (Michelangeli *et al.*, 1990). Band 3, the anion channel in the erythrocyte membrane, possesses a high-affinity sterol binding site which is proposed to be an inhibitory site for anion transport (Schubert and Boss, 1982). Fusion of enveloped viruses and cell membranes is reported to require the presence of cholesterol or other sterols (White *et al.*, 1983). A specific binding of sterol with the fusogenic hydrophobic sequence of the fusion protein has been suggested to facilitate the fusion reaction in Sendai virus (Asano and Asano, 1988).

Due to the relative ease of manipulating the sterol content, both quantitatively and qualitatively, mycoplasmas have provided researchers with a particularly responsive system in which to explore the role of cholesterol in protein-mediated processes. In accordance with the lipophilic nature of the sterol molecule, the processes studied are usually membrane-bound, and as a consequence, have most often been examined in their natural environment, i.e., intact cells or isolated membranes. However, in at least one instance, McElhaney and co-workers (McElhaney, 1989) have pursued these studies to the level of a purified membrane protein, the Na^+/Mg^{2+}-ATPase of *A. laidlawii*, in reconstituted lipid vesicles.

4.1. ATPase

An early study of Rottem *et al.* (1973b) showed that the activity of the Mg^{2+}-dependent ATPase of *My. mycoides* subsp. *capri* is modulated by the cholesterol content of the membrane. In cholesterol-poor membranes the response of ATPase activity to temperature is biphasic and decreases rapidly at temperatures below the breakpoint. The transition temperature is a function of the fatty acid composition of the membrane lipids, shifting from 18–22°C for oleic acid to 30–32°C for elaidic acid. By comparison, the temperature response of the ATPase in cholesterol-rich cells is always linear. The data were interpreted to mean that the cholesterol effects on ATPase are mediated by the corresponding physical properties of the resident bilayer.

The recent purification of the Na^+/Mg^{2+}-ATPase from *A. laidlawii* B has enabled McElhaney and co-workers to study an acholeplasma ATPase in more detail. The enzyme is composed of five subunits ranging in apparent molecular weight from 68,000 to 16,000. Labeling experiments indicate that whereas portions of all of the subunits contact the aqueous phase, only one or possibly two penetrate into or through the lipid bilayer (Silvius *et al.*, 1978; Lewis *et al.*, 1986). By reconstituting the ATPase in vesicles prepared from phosphatidylcholine of varying fatty acid compositions, the effect of membrane fluidity on enzyme activity could be assessed. As judged by Arrhenius plots of Na^+/Mg^{2+}-ATPase activity, the enzyme is fully active when most of the phospholipids are in the liquid-crystalline state. At temperatures above the phase transition, however, the enzyme is fairly nonspecific for its fatty acid requirement (McElhaney, 1989). The effect of cholesterol was investigated by reconstituting the ATPase into dimyristoyl-phosphatidylcholine vesicles containing increasing amounts of cholesterol (McElhaney, 1989). At cholesterol concentrations between 13 and 24 mole%, the ATPase activity is slightly lowered at temperatures above the transition point and slightly increased at temperatures below the transition point. These small amounts of cholesterol have no measurable effect on the transition temperature itself. On the other hand, higher concentrations of cholesterol (33 and 45 mole%) markedly inhibit the ATPase activity and lower the temperature at which a change in the slope of the Arrhenius plot occurs. Parallel calorimetric measurements reveal that increasing the concentration of cholesterol in the dimyristoyl-phosphatidylcholine vesicles quenches the thermotropic transition seen in cholesterol-free vesicles. McElhaney (1989) attributes the cholesterol-mediated effects on the Na^+/Mg^{2+}-ATPase of *A. laidlawii* B to the ability of cholesterol to decrease the fluidity of the lipid bilayer above the phase transition since the directions of the changes generally correlate with the dependency of this ATPase for liquid-crystalline state lipids. As to whether the lack of a specific stimulatory effect of the acholeplasma ATPase for cholesterol reflects the sterol independence of *A. laidlawii* for growth could be explored by doing comparable experiments with a purified ATPase from mycoplasma. This of course must await the purification of one of the several suitable ATPases described for *Mycoplasma* species (Benyoucef *et al.*, 1982; Linker and Wilson, 1985; Shirvan *et al.*, 1989).

4.2. Ion Transport

Related to the discussion of ATPases *per se* is the topic of ion transport and cell volume regulation. The semipermeable plasma membrane surrounding wall-less cells is continually traversed in an inward direction by ions and water molecules which must be extruded by ion pumps in order to prevent osmotic lysis (Wilson, 1954). The cholesterol content of mycoplasma membranes is known to affect the intracellular ion concentration (Waitzkin and Abraham, 1981). Indeed, the cell volume of *M. capricolum* is inversely related to the cholesterol content of the membrane (Romano *et al.*, 1986). In *My. mycoides* subsp. *capri,* cholesterol leads to an increase in the intracellular potassium level possibly by modulating the membrane potential (Le Grimellec and Leblanc, 1978). A secondary effect of cholesterol in these cells is to inhibit potassium efflux probably by decreasing the solubility of this ion in the lipid bilayer. Yeagle (1988) has proposed a model of permeability whereby cholesterol reduces the incidence of defects in the bilayer caused by the *trans–gauche* isomerizations of the fatty acyl chains through which small molecules can diffuse.

4.3. Phospholipid Synthesis

The idea that cholesterol may influence phospholipid metabolism was raised by Horwitz *et al.* (1978) when they noted that the phospholipid/cholesterol ratio in mammalian cells is constant despite wide variations in the medium lipid composition. One of the cholesterol-sensitive steps in phospholipid synthesis in mammalian cells appears to be the rate-limiting step for phosphatidylcholine synthesis catalyzed by ATP:phosphocholine cytidylyltransferase (Cornell and Goldfine, 1983). Several steps in phospholipid synthesis in yeast are also influenced by sterol. Among these are the methylations of phosphatidylethanolamine to phosphatidylcholine (Ramgopal *et al.*, 1990).

Finding that the inhibition of cholesterol synthesis gives rise to the sequential inhibition of phospholipid synthesis followed by DNA, RNA, and protein syntheses, Cornell and Horwitz (1980) made the further proposal that cholesterol synthesis *per se* or an exogenous source of cholesterol may be the primary factor limiting cell cycling. The notion that cholesterol synthesis is a driving force for cell proliferation gained strong support recently when it was discovered that intermediates of the isoprenoid pathway, e.g., farnesyl pyrophosphate, are involved in the covalent modifications required to convert members of the ras family to their active form (Casey *et al.*, 1989). Moreover, in yeast trace amounts of ergosterol are reported to stimulate a protein kinase possibly involved in cell cycle control (Dahl *et al.*, 1987).

Prompted by the observation of sterol synergism in *M. capricolum* described above, J. S. Dahl *et al.* (1980) went on to explore the biochemical basis for this phenomenon in mycoplasma. They found that the small amount of cholesterol required for the synergistic effect coordinately controls the rates of phospholipid, RNA, and protein syntheses in a sequential manner (Dahl and Dahl, 1983). The addition of a small amount of cholesterol to the medium of cells grown on lanosterol elicits a stimulation first of phospholipid synthesis followed by the syntheses of RNA and protein. The effect on phospholipid metabolism is specific for the incorporation of unsaturated fatty acid uptake. Studies of the kinetics of fatty acid uptake by resting cells showed that the apparent K_m of oleate uptake in lanosterol-grown cells was lowered from 17 μM to 3 μM by the inclusion of a synergistic amount of cholesterol in the growth medium. By contrast, the apparent K_m for palmitate uptake was unchanged (Dahl *et al.*, 1981). A survey of several enzymes involved in phosphatidylglycerol biosynthesis revealed that a possible candidate for sterol-mediated regulation of unsaturated phospholipid synthesis in *M. capricolum* is the fatty acid-activating enzyme (Dahl, 1988). A positive role for cholesterol in determining the ratio of saturated to unsaturated fatty acid in mycoplasma phospholipids had been noted previously when it was shown that the phospholipids of cholesterol-poor *M. mycoides* subsp. *capri* are preferentially acylated with palmitate rather than oleate (Rottem *et al.*, 1973b).

5. CONCLUDING REMARKS

A role for cholesterol as an architectural component of lipid bilayers which endows the membrane with characteristic physical properties including an intermediate fluid state, resistance to thermotropic gel-to-liquid-crystalline phase transitions and lateral phase separations, decreased surface area, and decreased permeability is supported by the studies with mycoplasmas. In eukaryotes, the remarkable sequestration of sterol to the plasma membrane as opposed to internal membranes implies that the benefits derived from cholesterol are most consequential to membranes which delineate the internal content of the cell from the external environment. This general conclusion is supported by the sterol dependence of mycoplasmas in view of their single membrane structure and lack of a cell wall.

The absence of a universal sterol distribution across the living world, arising principally from the dearth of prokaryotes exhibiting either a sterol growth requirement or *in vivo* sterol biosynthesis, is curious. Rather than dismissing the possibility that prokaryotes as a rule have no need for a membrane component analogous in function to cholesterol, several investigators have suggested molecules that might act as sterol surrogates. Among these are the bacterial hopanoids and carotenoids (Nes, 1974; Ourisson *et al.*, 1979; Bloch, 1983). Although the synthesis of hopanoids has never been reported to occur in mollicutes, the synthesis of carotenoids is a general characteristic of the non-sterol-requiring *Acholeplasma* species (Smith, 1979). Moreover, the inclusion of carotenoids in membranes of *A. laidlawii* correlates with a decrease in membrane fluidity (Rottem and Markowitz, 1979). Worth noting as well is the occurrence of another related molecular entity in the lipids of *Acholeplasma* and *Thermoplasma* species (also non-sterol-requiring mollicutes), i.e., the polyisoprenoid subunit. Whereas all of the polar lipids of *Thermoplasma* species are comprised of C_{40} isoprenoids (acyclic, monocyclic, or bicyclic) in tetraether linkage to two glycerol molecules (Langworthy, 1979), *Acholeplasma* species can synthesize in addition to the fatty acylated lipids, a C_{20} polyprenyl glucoside (McElhaney, 1989). Remarkably, a feature which distinguishes cholesterol, carotenoids, and isoprenyl alcohols is their commonality of origin. All are comprised of basic isoprenoid subunits arising from intermediates in the sterol biosynthetic pathway. Whether or not their common origins can be extended to a similarity of function is at the moment an open question. Nevertheless, some of cholesterol's structural attributes responsible for its membrane behavior, i.e., the amphipathic structure and the bulky wedge shape of its tetracyclic nucleus and isooctyl side chain, are found at least in a general sense in the carotenoids and isoprenylated lipids.

Despite intensive investigations on sterol–phospholipid interactions in the last decade, the exact role of sterol in mycoplasma membranes remains elusive. The proposal that cholesterol performs roles in membranes apart from a bulk function is suggested by two lines of evidence. First, there is a threshold level of sterol below which mycoplasma will not grow. This level of sterol is thought to be too low to fulfill a bulk lipid requirement. Moreover, the phenomenon of sterol synergism implies more than one role for sterol distinguished by differences, both quantitative and qualitative, in sterol requirement. The alternate role for sterol is proposed to involve specific protein–sterol interactions (Dahl and Dahl, 1988). Experimental inquiry into an alternate role for cholesterol in mycoplasma has pointed to a possible involvement of cholesterol in phospholipid biosynthesis and potassium transport. Although there is a suggestion that these two processes may be interrelated (Dahl, 1988), the details of this relationship as well as the exact mechanism of the sterol-mediated control require further investigation.

Other potential targets for sterol-mediated regulation are suggested by experiments showing the coordinate regulation of phospholipid, RNA, and protein syntheses by cholesterol in *M. capricolum* (Dahl and Dahl, 1983). A similar phenomenon is known to occur in eukaryotes where it has been shown that inhibition of cholesterol synthesis depresses macromolecular synthesis as well (Cornell and Horwitz, 1980). In the latter case, roles for both cholesterol and nonsterol isoprenoids have been proposed (Quesney-Huneeus *et al.*, 1983; Goldstein and Brown, 1990). The recent discovery that ras is covalently modified by a farnesyl group brings to prominence the role for the cholesterol biosynthetic pathway in the control of cell proliferation (Casey *et al.*, 1989). Identification of prenylated proteins in eukaryotes begs the question whether similar covalent modifications of proteins also occur in mycoplasmas. This becomes especially relevant in view of the finding that many mycoplasma proteins are covalently modified with long-chain fatty acids (Dahl *et al.*, 1983). Classes of acylated proteins have been described in both prokaryotes and eukaryotes among which are growth regulatory proteins such as src, ras, and protein kinases in eukaryotes (James and Olson, 1990). The presence of hydroxymethylglutaryl-CoA reductase, a key enzyme in the isoprenoid pathway, is firmly established in *Acholeplasma* species (Smith, 1979; Glasfeld *et al.*, 1990). This enzyme and the ensuing pathway proximal to farnesyl pyrophosphate is, however, probably absent in mycoplasmas (Smith, 1979). Perhaps a counterpart to the role of sterol and nonsterol isoprenoids in eukaryotic cell cycle control is the role assumed by phospholipid in prokaryotic DNA synthesis (Lowey *et al.*, 1990). At the molecular level the two processes must be quite different, but the overall outcome, i.e., the coordination of membrane biogenesis and cell proliferation, is the same. Ultimately the answers to questions raised about unifying roles for sterols in mycoplasmas and other sterol-dependent cells, including the requisite sterol–protein interactions, might arise from future studies on the comparative molecular biology of the particular organisms and proteins involved.

6. REFERENCES

Asano, K., and Asano, A., 1988, Binding of cholesterol and inhibitory peptide derivatives with the fusogenic hydrophobic sequence of F-glycoprotein of HVJ (Sendai virus): Possible implication in the fusion reaction, *Biochemistry* **27**:1321–1329.

Backer, J. M., and Dawidowicz, E. A., 1981a, Mechanism of cholesterol exchange between phospholipid vesicles, *Biochemistry* **20**:3805–3810.

Backer, J. M., and Dawidowicz, E. A., 1981b, Transmembrane movement of cholesterol in small unilamellar vesicles detected by cholesterol oxidase, *J. Biol. Chem.* **256**:586–588.

Benyoucef, M., Rigaud, J. L., and Leblanc, G., 1982, Cation transport mechanisms in *Mycoplasma mycoides* var. *capri* cells. Na+-dependent K+ accumulation, *Biochem. J.* **208**:539–547.

Bernheimer, A. W., and Davidson, M., 1965, Lysis of pleuropneumonia-like organisms by staphylococcal and streptococcal toxins, *Science* **148**:1229–1231.

Bhakoo, M., Lewis, R. N. A. H., and McElhaney, R. N., 1987, Isolation and characterization of a novel monoacylated glucopyranosyl neutral lipid from the plasma membrane of *Acholeplasma laidlawii* B, *Biochim. Biophys. Acta* **922**:34–45.

Bird, C. W., Lynch, J. M., Pirt, F. J., Reid, W. W., Brooks, C. J. W., and Middleditch, B. S., 1971, Steroids and squalene in *Methylococcus capsulatus* grown on methane, *Nature* **230**:473–474.

Bittman, R., 1988, Sterol exchange between mycoplasma membranes and vesicles, in *Biology of Cholesterol* (P. L. Yeagle, ed.), pp. 173–195, CRC Press, Boca Raton, Fla.

Bittman, R., and Rottem, S., 1976, Distribution of cholesterol between the outer and inner halves of the lipid bilayer of mycoplasma cell membranes, *Biochim. Biophys. Acta* **71**:318–324.

Black, P. N., Said, B., Ghosn, C. R., Beach, J. V., and Nunn, W. D., 1987, Purification and characterization of an outer membrane-bound protein involved in long-chain fatty acid transport in *Escherichia coli*, *Biochemistry* **26**:1412–1419.

Blau, L., and Bittman, R., 1978, Cholesterol distribution between the two halves of the lipid bilayer of human erythrocyte ghost membranes, *J. Biol. Chem.* **253**:8366–8368.

Bloch, K., 1965, The biological synthesis of cholesterol, *Science* **150**:19–28.

Bloch, K., 1976, On the evolution of a biosynthetic pathway, in: *Reflections on Biochemistry* (A. Kornberg, B. L. Horecker, L. Cornudella, and J. Oro, eds.), Pergamon Press, Elmsford, N.Y., pp. 143–150.

Bloch, K., 1983, Sterol structure and membrane function, *CRC Crit. Rev. Biochem.* **14**:47–92.

Bloj, B., and Zilversmit, D. B., 1976, Asymmetry and transposition rates of phosphatidylcholine in rat erythrocyte ghosts, *Biochemistry* **15**:1277–1283.

Brown, M. S., and Goldstein, J. L., 1986, A receptor-mediated pathway for cholesterol homeostasis, *Science* **232**:34–47.

Carruthers, A., and Melchoir, D. L., 1984, Human erythrocyte hexose transporter activity is governed by bilayer lipid composition in reconstituted vesicles, *Biochemistry* **23**:6901–6911.

Casey, P. J., Solski, P. A., Der, C. S., and Buss, J., 1989, P21[ras] is modified by a farnesyl isoprenoid, *Proc. Natl. Acad. Sci. USA* **86**:1167–1177.

Clark, A. J., and Bloch, K., 1959, Function of sterols in *Dermestes vulpinus*, *J. Biol. Chem.* **234**:2583–2588.

Clejan, S., and Bittman, R., 1984a, Kinetics of cholesterol and phospholipid exchange between *Mycoplasma gallisepticum* cells and lipid vesicles, *J. Biol. Chem.* **259**:441–448.

Clejan, S., and Bittman, R., 1984b, Distribution and movement of sterols with different side chain structures between the two leaflets of the membrane bilayer of mycoplasma cells, *J. Biol. Chem.* **259**:449–455.

Clejan, S., Bittman, R., and Rottem, S., 1978, Uptake, transbilayer distribution, and movement of cholesterol in growing *Mycoplasma capricolum* cells, *Biochemistry* **17**:4579–4583.

Clejan, S., Bittman, R., and Rottem, S., 1981, Effect of sterol structure and exogenous lipids on the transbilayer distribution of sterols in the membranes of *Mycoplasma capricolum*, *Biochemistry* **20**:2200–2203.

Cluss, R. G., Johnson, J. K., and Somerson, N. L., 1983, Liposomes replace serum for cultivation of fermenting mycoplasmas, *Appl. Environ. Microbiol.* **46**:370–374.

Colbeau, A., Nachbaur, J., and Vignais, P. M., 1971, Enzymic characterization and lipid composition of rat liver subcellular membranes, *Biochim. Biophys. Acta* **249**:462–492.

Coleman, R., and Finean, J. B., 1966, Preparation and properties of isolated plasma membranes from guinea pig tissue, *Biochim. Biophys. Acta* **125**:197–206.

Cornell, R. B., and Goldfine, H., 1983, The coordination of sterol and phospholipid synthesis in cultured myogenic cells. Effect of cholesterol synthesis inhibition on the synthesis of phosphatidylcholine, *Biochim. Biophys. Acta* **750**:504–520.

Cornell, R. B., and Horwitz, A. F., 1980, Apparent coordination of the biosynthesis of lipids in cultured cells: Its relationship to the regulation of the membrane sterol:phospholipid ratio and cell cycling, *J. Cell Biol.* **86**:810–819.

Dahl, C., and Dahl, J., 1988, Cholesterol and cell function, in: *Biology of Cholesterol* (P. L. Yeagle, ed.), pp. 147–171, CRC Press, Boca Raton, Fla.

Dahl, C. E., Dahl, J. S., and Bloch, K., 1980a, Effect of alkyl-substituted precursors of cholesterol on artificial and natural membranes and on the viability of *Mycoplasma capricolum*, *Biochemistry* **19**:1462–1467.

Dahl, C. E., Dahl, J. S., and Bloch, K., 1980b, Effects of cycloartenol and lanosterol on artificial and natural membranes, *Biochem. Biophys. Res. Commun.* **92**:221–228.

Dahl, C. E., Dahl, J. S., and Bloch, K., 1983, Proteolipid formation in *Mycoplasma capricolum*, *J. Biol. Chem.* **258**:11814–11818.

Dahl, C., Biemann, H. P., and Dahl, J., 1987, A protein kinase antigenically related to pp60^{v-src} possibly involved in yeast cell cycle control: Positive in vivo regulation by sterol, *Proc. Natl. Acad. Sci. USA* **84**:4012–4016.

Dahl, J., 1988, Uptake of fatty acid by *Mycoplasma capricolum*, *J. Bacteriol.* **170**:2022–2026.

Dahl, J. S., and Dahl, C. E., 1983, Coordinate regulation of unsaturated phospholipid, RNA, and protein synthesis in *Mycoplasma capricolum* by cholesterol, *Proc. Natl. Acad. Sci. USA* **80**:692–696.

Dahl, J. S., Dahl, C. E., and Bloch, K., 1980, Sterols in membranes: Growth characteristics and membrane properties of *Mycoplasma capricolum* cultured on cholesterol and lanosterol, *Biochemistry* **19**:1467–1472.

Dahl, J. S., Dahl, C. E., and Bloch, K., 1981, Effect of cholesterol on macromolecular synthesis and fatty acid uptake by *Mycoplasma capricolum*, *J. Biol. Chem.* **256**:87–91.

de Kruyff, B., Demel, R. A., and vanDeenen, L. L. M., 1972, The effect of cholesterol and epicholesterol incorporation on the permeability and the phase transition of intact *Acholeplasma laidlawii*, *Biochim. Biophys. Acta* **255**:331–347.

de Kruyff, B., deGreef, W. J., vanEyk, R. V. W., Demel, R. A., and vanDeenen, L. L. M., 1973, The effect of different fatty acid and sterol composition on the erythritol flux through the cell membrane of *Acholeplasma laidlawii*, *Biochim. Biophys. Acta* **298**:479–499.

Demel, R. A., and de Kruyff, B., 1976, The function of sterols in membranes, *Biochim. Biophys. Acta* **457**:109–132.

Edward, D. G., and Fitzgerald, W. A., 1951, Cholesterol in the growth of organisms of the pleuropneumonia group, *J. Gen. Microbiol.* **5**:576–586.

Efrati, H., Shinitzky, M., and Razin, S., 1980, Effects of charged cholesteryl esters on mycoplasma growth, *FEBS Lett.* **122**:59–63.

Efrati, H., Rottem, S., and Razin, S., 1981, Lipid and protein membrane components associated with cholesterol uptake by mycoplasmas, *Biochim. Biophys. Acta* **641**:386–394.

Efrati, H., Oschry, Y., Eisenberg, S., and Razin, S., 1982, Preferential uptake of lipids by mycoplasma membranes from human plasma low-density lipoproteins, *Biochemistry* **21**:6477–6482.

Efrati, H., Wax, Y., and Rottem, S., 1986, Cholesterol uptake capacity of *Acholeplasma laidlawii* is affected by the composition and content of membrane glycolipids, *Arch. Biochem. Biophys.* **248**:282–288.

Glasfeld, A., Leanz, G. F., and Benner, S. A., 1990, The stereospecificities of seven dehydrogenases from *Acholeplasma laidlawii*, *J. Biol. Chem.* **265**:11692–11699.

Goldstein, J. L., and Brown, M. S., 1990, Regulation of the mevalonate pathway, *Nature* **343**:425–430.

Gross, Z., and Rottem, S., 1984, The preservation of *Mycoplasma capricolum* cell intactness after phospholipase A2 treatment, *Biochim. Biophys. Acta* **778**:372–378.

Gross, Z., Rottem, S., and Bittman, R., 1982, Phospholipid interconversions in *Mycoplasma capricolum*, *Eur. J. Biochem.* **122**:169–174.

Horwitz, A. F., Wright, A., Ludwig, P., and Cornell, R., 1978, Interrelated lipid alterations and their influence on the proliferation and fusion of cultured myogenic cells, *J. Cell Biol.* **77**:334–357.

Hyslop, P. A., Morel, B., and Sauerheber, R. D., 1990, Organization and interaction of cholesterol and phosphatidylcholine in model bilayer membranes, *Biochemistry* **29**:1025–1038.

Ipsen, J. H., Mouritsen, O. G., and Bloom, M., 1990, Relationships between lipid membrane area, hydrophobic thickness, and acyl-chain orientational order. The effects of cholesterol, *Biophys. J.* **57**:405–412.

James, G., and Olson, E. N., 1990, Fatty acylated proteins as components of intracellular signalling pathways, *Biochemistry* **29**:2623–2633.

Jones, O. T., and McNamee, M. G., 1988, Annular and nonannular binding sites for cholesterol associated with the nicotinic acetylcholine receptor, *Biochemistry* **27**:2364–2374.

Kahane, I., and Razin, S., 1977, Cholesterol-phosphatidylcholine dispersions as donors of cholesterol to mycoplasma membranes, *Biochim. Biophys. Acta* **471**:32–38.

Kannenberg, E., and Poralla, K., 1982, The influence of hopanoids on growth of *Mycoplasma mycoides*, *Arch. Microbiol.* **133**:100–102.

Kannenberg, E., Blume, A., McElhaney, R. N., and Poralla, K., 1983, Monolayer and calorimetric studies of phosphatidylcholines containing branched-chain fatty acids and of their interactions with cholesterol and with a bacterial hopanoid in model membranes, *Biochim. Biophys. Acta* **733**:111–116.

Lange, Y., Dolde, J., and Steck, T., 1981, The rate of transmembrane movement of cholesterol in the human erythrocyte, *J. Biol. Chem.* **256**:5321–5323.

Langworthy, T. A., 1979, Special features of thermoplasmas, in: *The Mycoplasmas I: Cell Biology* (M. F. Barile and S. Razin, eds.), Academic Press, New York, pp. 495–513.

Le Grimellec, C., and Leblanc, G., 1978, Effect of membrane cholesterol on potassium transport in *Mycoplasma mycoides* var. *capri* (PG3), *Biochim. Biophys. Acta* **514**:152–163.

Lelong, I., Luu, B., Mersel, M., and Rottem, S., 1988, Effect of 7β-hydroxycholesterol on growth and membrane composition of *Mycoplasma capricolum*, *FEBS Lett.* **232**:354–358.

Lewis, R. N. A. H., George, R., and McElhaney, R. N., 1986, Structure–function investigations of the membrane sodium–magnesium-ATPase from *Acholeplasma laidlawii* B: Studies of reactive amino acid residues using group-specific reagents, *Arch. Biochem. Biophys.* **247**:201–210.

Liebel, W. J., Firestone, L. L., Legler, D. C., Braswell, L. M., and Miller, K. W., 1987, Two pools of cholesterol in acetylcholine receptor-rich membranes from Torpedo, *Biochim. Biophys. Acta* **897**:249–260.

Linker, C., and Wilson, T. H., 1985, Characterization and solubilization of the membrane-bound ATPase from *Mycoplasma gallisepticum*, *J. Bacteriol.* **163**:1258–1262.

Lowey, B., Marczynski, G. T., Dingwall, A., and Shapiro, L., 1990, Regulatory interactions between phospholipid synthesis and DNA replication in *Caulobacter crescentus*, *J. Bacteriol.* **172**:5523–5530.

McElhaney, R. N., 1982, Effect of membrane lipids on transport and enzymatic activities, *Curr. Top. Membr. Transp.* **17**:317–380.

McElhaney, R. N., 1989, The influence of membrane lipid composition and physical properties on membrane structure and function in *Acholeplasma laidlawii*, *CRC Crit. Rev. Microbiol.* **17**:1–32.

McIntosh, T. J., 1978, The effect of cholesterol on the structure of phosphatidylcholine bilayers, *Biochim. Biophys. Acta* **513**:43–58.

McLean, L. R., and Phillips, M. C., 1981, Mechanisms of cholesterol and phosphatidylcholine exchange or transfer between unilamellar vesicles, *Biochemistry* **20**:2893–2900.

McNamee, M. G., Ellena, J. F., and Dalziel, A. W., 1982, Lipid–protein interactions in membranes containing the acetylcholine receptor, *Biophys. J.* **37**:103–104.

Melchior, D. L., and Rottem, S., 1981, The organization of cholesterol esters in membranes of *Mycoplasma capricolum, Eur. J. Biochem.* **117:**147–153.

Michelangeli, F., East, J. M., and Lee, A. G., 1990, Structural effects on the interaction of sterols with the (Ca^{2+} + Mg^{2+})-ATPase, *Biochim. Biophys. Acta* **1025:**99–108.

Montesano, R., Perrelet, A., Vassalli, P., and Orci, L., 1979, Absence of filipin–sterol complexes from large coated pits on the surface of cultured cells, *Proc. Natl. Acad. Sci. USA* **76:**6391–6395.

Nes, W. R., 1974, Role of sterols in membranes, *Lipids* **9:**596–612.

Odriozola, J. M., Waitzkin, E., Smith, T. L., and Bloch, K., 1978, Sterol requirement of *Mycoplasma capricolum, Proc. Natl. Acad. Sci. USA* **75:**4107–4109.

Ourisson, G., Albrecht, P., and Rohmer, M., 1979, The hopanoids: Paleochemistry and biochemistry of a group of natural products, *Pure Appl. Chem.* **51:**709–729.

Pinto, W. J., Lozano, R., Sekula, B. C., and Nes, W. R., 1983, Stereochemically distinct roles for sterol in *Saccharomyces cerevisiae, Biochem. Biophys. Res. Commun.* **112:**47–54.

Quesney-Huneeus, V., Galick, H. A., Siperstein, M. D., Erickson, S. K., Spencer, T. A., and Nelson, J. A., 1983, The dual role of mevalonate in the cell cycle, *J. Biol. Chem.* **258:**378–385.

Ramgopal, M., and Bloch, K., 1983, Sterol synergism in yeast, *Proc. Natl. Acad. Sci. USA* **80:**712–715.

Ramgopal, M., Zundel, M., and Bloch, K., 1990, Sterol effects on phospholipid biosynthesis in the yeast strain GL7, *J. Lipid Res.* **31:**653–658.

Razin, S., 1974, Correlation of cholesterol to phospholipid content in membranes of growing mycoplasmas, *FEBS Lett.* **47:**81–85.

Razin, S., 1982, Sterols in mycoplasma membranes, *Curr. Top. Membr. Transp.* **17:**183–205.

Razin, S., and Argaman, M., 1963, Lysis of mycoplasma, bacterial protoplasts, spheroplasts, and L-forms by various agents, *J. Gen. Microbiol.* **30:**155–172.

Rilfors, L., Lindblom, G., Wieslander, A., and Christiansson, A., 1984, Lipid bilayer stability in biological membranes, in: *Membrane Fluidity* (M. Kates and L. A. Manson, eds.), Plenum Press, New York, pp. 205–245.

Rodriguez, R. J., Taylor, F. R., and Parks, L. W., 1982, A requirement for ergosterol to permit growth of yeast sterol auxotrophs on cholestanol, *Biochem. Biophys. Res. Commun.* **106:**435–441.

Rodwell, A. W., 1969, The supply of cholesterol and fatty acids for the growth of mycoplasmas, *J. Gen. Microbiol.* **58:**29–37.

Rodwell, A. W., 1983a, *Mycoplasma gallisepticum* requires exogenous phospholipid for growth, *FEMS Microbiol. Lett.* **17:**265–268.

Rodwell, A. W., 1983b, Defined and partly defined media, in: *Methods in Mycoplasmology,* Volume 1 (S. Razin and J. G. Tully, eds.), Academic Press, New York, pp. 163–172.

Romano, N., Shirvan, M. H., and Rottem, S., 1986, Changes in membrane lipid composition of *Mycoplasma capricolum* affect the cell volume, *J. Bacteriol.* **167:**1089–1091.

Rottem, S., 1980, Membrane lipids of mycoplasmas, *Biochim. Biophys. Acta* **604:**65–90.

Rottem, S., 1981, Cholesterol is required to prevent crystallization of *Mycoplasma arginini* phospholipids at physiological temperature, *FEBS Lett.* **133:**161–164.

Rottem, S., and Markowitz, O., 1979, Carotenoids act as reinforcers of the *Acholeplasma laidlawii* lipid bilayer, *J. Bacteriol.* **140:**944–948.

Rottem, S., and Verkleij, A. J., 1982, Possible association of segregated lipid domains of *Mycoplasma gallisepticum* membranes with cell resistance to osmotic lysis, *J. Bacteriol.* **149:**338–345.

Rottem, S., Yashouv, J., Ne'eman, Z., and Razin, S., 1973a, Cholesterol in mycoplasma membranes. Composition, ultrastructure, and biological properties of membranes from *Mycoplasma mycoides* var. *capri* cells adapted to grow with low cholesterol concentrations, *Biochim. Biophys. Acta* **323**:495–508.

Rottem, S., Cirillo, V. P., de Kruyff, B., Shinitzky, M., and Razin, S., 1973b, Cholesterol in mycoplasma membranes. Correlation of enzymic and transport activities with physical state of lipids in membranes of *Mycoplasma mycoides* var. *capri* adapted to grow with low cholesterol concentrations, *Biochim. Biophys. Acta* **323**:509–519.

Rottem, S., Slutzky, G. M., and Bittman, R., 1978, Cholesterol distribution and movement in the *Mycoplasma gallisepticum* cell membrane, *Biochemistry* **17**:2723–2726.

Schroeder, F., 1981, Use of a fluorescent sterol to probe the transbilayer distribution of sterols in biological membranes, *FEBS Lett.* **135**:127–130.

Schubert, D., and Boss, K., 1982, Band 3 protein–cholesterol interactions in erythrocyte membranes, *FEBS Lett.* **150**:4–8.

Shirvan, M. H., Schuldiner, S., and Rottem, S., 1989, Volume regulation in *Mycoplasma gallisepticum:* Evidence that Na^+ is extruded via a primary Na^+ pump, *J. Bacteriol.* **171**:4417–4424.

Silvius, J. R., Read, B. D., and McElhaney, R. N., 1978, Membrane enzymes: Artifacts in Arrhenius plots due to temperature dependence of substrate-binding affinity, *Science* **199**:902–904.

Simmonds, A. C., East, J. M., Jones, O. T., Rooney, E. K., McWhirter, J., and Lee, A. G., 1982, Annular and non-annular binding sites on the $(Ca^{2+} + Mg^{2+})$-ATPase, *Biochim. Biophys. Acta* **693**:398–406.

Slutzky, G. M., Razin, S., Kahane, I., and Eisenberg, S., 1977, Cholesterol transfer from serum lipoproteins to mycoplasma, *Biochemistry* **16**:5158–5163.

Smith, P. F., 1964, Comparative physiology of pleuropneumonia-like and L-type organisms, *Bacteriol. Rev.* **28**:97–125.

Smith, P. F., 1979, The composition of membrane lipids and lipopolysaccharides, in: *The Mycoplasmas I: Cell Biology* (M. F. Barile and S. Razin, eds.), Academic Press, New York, pp. 231–257.

Smith, P. F., and Lynn, R. J., 1958, Lipid requirements for the growth of pleuropneumonialike organisms, *J. Bacteriol.* **76**:264–269.

Smith, P. F., and Rothblat, G. H., 1962, Comparison of lipid composition of pleuropneumonia-like and L-type organisms, *J. Bacteriol.* **83**:500–506.

Stremmel, W., Strohmeyer, G., Borchard, F., Kochwa, S., and Berk, P., 1985, Isolation and partial characterization of a fatty acid binding protein in rat liver plasma membrane, *Proc. Natl. Acad. Sci. USA* **82**:4–8.

Waitzkin, E. D., and Abraham, E. H., 1981, Effect of sterol structure on intracellular sodium and potassium in *Mycoplasma capricolum, Proc. 16th Ann. Conf. Microbeam Anal. Soc.*, pp. 226–228.

White, J., Kielan, M., and Helenius, A., 1983, Membrane fusion proteins of enveloped animal viruses, *Q. Rev. Biophys.* **16**:151–195.

Wilson, T. H., 1954, Ionic permeability and osmotic swelling, *Science* **120**:104–105.

Yeagle, P. L., 1985, Cholesterol and the cell membrane, *Biochim. Biophys. Acta* **822**:267–287.

Yeagle, P. L., 1988, Cholesterol and the cell membrane, in: *Biology of Cholesterol* (P. L. Yeagle, ed.), CRC Press, Boca Raton, Fla., pp. 121–145.

Yeagle, P., Martin, R. B., Lala, A. K., Lin, H. K., and Bloch, K., 1977, Differential effects of cholesterol and lanosterol on artificial membranes, *Proc. Natl. Acad. Sci. USA* **74**:4924–4926.

Chapter 6

Extramembranous Structure in Mycoplasmas

F. Chris Minion and Ricardo F. Rosenbusch

1. INTRODUCTION

The murein sacculus surrounds the inner membrane of most prokaryotes giving shape and stability to the organism (Costerton and Irvin, 1981). It is composed of peptidoglycan, a polymer of carbohydrates and amino acids. In chlamydia, the outer surface is composed of a rigid, disulfide-interlocked protein layer which replaces the function of the peptidoglycan layer. In these organisms, the degree of cross-linking coincides with differentiation from a metabolically inactive, extracellular elementary body to a metabolically active, pleomorphic, intracellular reticulate body (Hackstadt *et al.*, 1985). Mollicutes, however, lack cell walls, peptidoglycan and protein matrices (Plackett, 1959). Their single membrane must serve to separate cytoplasmic components from the environment, provide shape, and prevent lysis under changing environmental conditions. In this respect their membrane more closely resembles eukaryotic membranes than other eubacteria. It is a single cholesterol-containing phospholipid bilayer to which is attached or embedded a variety of proteins and carbohydrates. The exact nature of the outer surface of mycoplasmas, except for the protein component which has

F. Chris Minion and Ricardo F. Rosenbusch   Veterinary Medical Research Institute, Iowa State University, Ames, Iowa 50011.

Subcellular Biochemistry, Volume 20: Mycoplasma Cell Membranes, edited by Shlomo Rottem and Itzhak Kahane. Plenum Press, New York, 1993.

been studied extensively in many species, is not known, and variations occur as a consequence of environmental or genetic changes (Rosengarten and Wise, 1990; Watson *et al.*, 1988).

Many bacteria also contain a mucilaginous coating, the capsule, that is not essential for viability. Being highly hydrated, capsules have been viewed as being barriers to water diffusion. They also serve to protect invading microbes against host immune responses. Capsules have been described in several mollicute species, but their function and composition remain obscure. Morphologically, they appear to more closely resemble the glycocalyx of eukaryotes than the classical capsules of the more traditional eubacteria. This chapter focuses on the available information concerning the mycoplasma outer surface as regards extramembranous structures. Of primary emphasis is the composition and morphology of capsules and capsular materials. Other aspects of membrane structure and function are given elsewhere in this volume.

2. MORPHOLOGY AND ELECTRON MICROSCOPY

Our knowledge of the surface of mollicutes has been obtained primarily through the use of electron microscopy. The small size of the organism and the limited thickness of the capsule prevent a clear delineation of encapsulated organisms by light microscopy, even when combined with the India ink contrast technique (Domermuth *et al.*, 1964). The complexity of the growth media or tissue environment imposes severe limitations on the ability to resolve capsule under these conditions. Interference (Normarsky) microscopy has allowed the visualization of capsular material in clumps of mycoplasma cells (Furness *et al.*, 1976), but the majority of morphological information has arisen from electron microscopic observations.

The lack of a rigid cell wall has complicated morphological analysis of mycoplasmas because of artifacts arising during fixation and drying protocols. Boatman (1979) reviews the basic morphology and problems inherent with examination of mycoplasmas by electron microscopy. Fixation in glutaraldehyde does not necessarily prevent distortions due to changes in osmotic pressure or drying from arising, and hypertonic solutions proved of more value than other osmolarities. No single solution solved all of the technical difficulties, however, suggesting that a comparative study with different solutions will be necessary with any morphological analysis of mycoplasmas.

Early suggestions of a glycocalyxlike structure came from observations of an asymmetry in the limiting membrane of several different mycoplasma species (Domermuth *et al.*, 1964). This was believed to have been due to an additional surface layer. Additional evidence came from observations of an electron-translucent 7 to 11-nm space that separated it from eukaryotic cell membranes during active infections or in contaminated cell cultures (Boatman, 1979; Boatman *et al.*, 1976; Salih and Rosenbusch, 1988; Tajima *et al.*, 1982). The lack of electron density at the mycoplasma–host cell interface in these studies may have occurred when fixation with osmium tetroxide prior to dehydration resulted in the collapse of the capsule or the refractory nature of capsule to normal electron microscopic staining procedures. It is generally believed that, like other bacteria, mollicute capsules are highly hydrated and require stabilization prior to fixation to maintain their integrity. Visualization of mycoplasma capsules has been accomplished by staining with polycationic compounds such as ruthenium red (Figure 1). Ruthenium red complexes with osmium tetroxide to stain polyanionic compounds (Luft, 1964). Tannic acid (Wilson and Collier, 1976) and potassium tellurite (Green and Hanson, 1973) have also been used on occasion to visualize extracellular layers. Another technique that has been gaining acceptance is the use of antibodies to stabilize the layer and increase contrast (Tajima *et al.*, 1985) (Figure 2). For these studies, sera from infected animals have been used, and in some instances, mucosal secretions have been used as well (Almeida and Rosen-

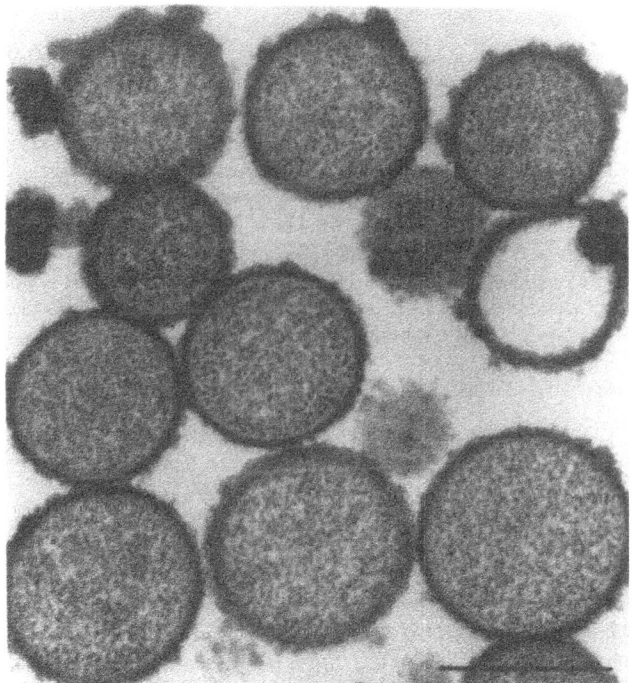

FIGURE 1. *Ureaplasma urealyticum* stained with ruthenium red. *In vitro*-grown organisms were prepared and stained according to Robertson and Smook (1976). Bar represents 0.5 μm. (Reproduced courtesy of J. Robertson, University of Alberta, Alberta, Canada.)

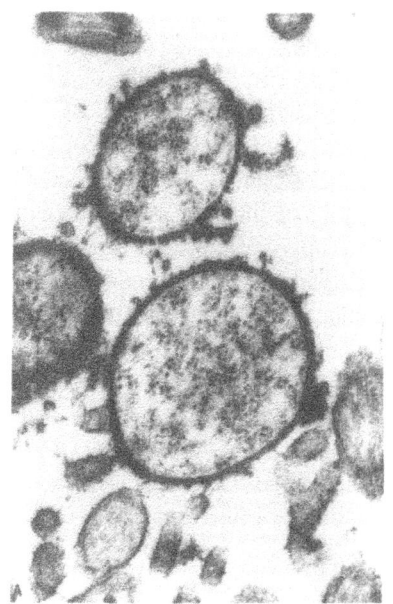
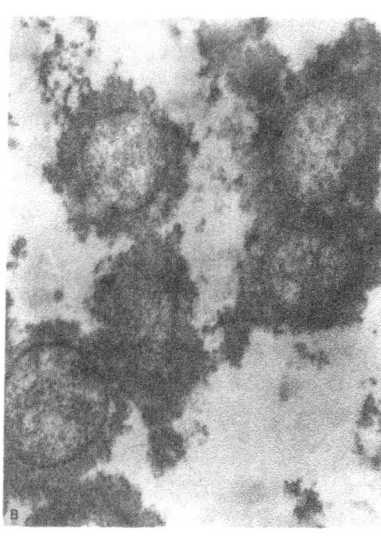

FIGURE 2. (A) Effect of antibody stabilization on capsule morphology. Tissue was taken from the bronchiolar epithelium of a pig inoculated intranasally with *in vivo*-passaged virulent strain MI-3 of *M. hyopneumoniae* and killed 5 weeks after inoculation. The tissue was fixed in a mixture of glutaraldehyde and ruthenium red and sections were stained with uranyl acetate and lead citrate. Capsular material stained black with ruthenium red is present on the outer surface of the limiting membrane. × 22,000. (B) Tissue was taken from the bronchiolar epithelium of a pig inoculated intranasally with *in vivo*-passaged virulent strain MI-3 of *M. hyopneumoniae* and killed 3 weeks after inoculation. The tissue was treated with homologous convalescent antiserum, fixed and stained as described above. Mycoplasmas lying free in the bronchiolar lumen are surrounded by a thick layer of antibody-stabilized capsule. × 19,200. (Reproduced courtesy of M. Tajima and Y. Yagihashi, Nippon Institute for Biological Science, Tokyo, Japan.)

busch, 1991). Although none have been reported, similar studies should be applicable using antibodies directed against purified capsular material. Concanavalin A–iron dextran has also been used to add contrast to capsule material (Rurangirwa *et al.*, 1987). This technique has the added potential of deriving some information regarding capsule composition because of the specificity of lectin–carbohydrate interactions.

Mycoplasma capsules have most often been described as being 20–40 nm thick (Howard and Gourlay, 1974; Tajima and Yagihashi, 1982). Capsules up to 40 nm thick were seen surrounding *Mycoplasma hyopneumoniae* in infected porcine lung tissue (Tajima and Yagihashi, 1982). Capsules of *My. mycoides* subsp. *mycoides* were 30 nm thick when the organism was grown *in vitro* (Howard and Gourlay, 1974). Little is known concerning the effects of environmental factors on capsule production. In some species, capsule seems to be produced constitutively, while in others capsule is produced in measurable quantities only during *in vivo* growth. Also, growth *in vivo* may tend to produce thicker capsules, but a direct comparison between *in vitro* and *in vivo* growth on capsule size and thickness has not been done, except for *M. dispar*. This species produces little capsule *in vitro* as determined by reactivity to capsule-specific antibodies after several *in vitro* passages, but is able to produce significant amounts of capsule either *in vivo* or when subsequently grown on tissue culture cells (Almeida and Rosenbusch, 1991).

Several mycoplasma species produce a measurable capsule including *M. mycoides* subsp. *mycoides* (Buttery and Plackett, 1960), *M. dispar* (Howard and Gourlay, 1974), *M. gallisepticum* (Tajima *et al.*, 1979), *M. hominis* (Furness *et al.*, 1976), *M. hyopneumoniae* (Horn, 1970; Tajima and Yagihashi, 1982), *M. meleagridis* (Green and Hanson, 1973), *M. pneumoniae* (Wilson and Collier, 1976), *M. pulmonis* (Taylor-Robinson *et al.*, 1981), *M. synoviae* (Ajufo and Whithear, 1978), *M. mobile* (Rosengarten *et al.*, 1988), and *Spiroplasma citri* (Cole *et al.*, 1973). *Ureaplasma urealyticum* expresses capsule (Robertson and Smook, 1976), but there was great variation among strains. In contrast, *U. diversum* has been described to have only a very thin exopolymer (Boatman *et al.*, 1976). An extracellular polysaccharide described for *M. capricolum* (Rurangirwa *et al.*, 1987) may also be considered as capsular in nature in view of its reported size by size-exclusion chromatography (200 kDa). Capsules have not been described among the acholeplasmas, anaeroplasmas, or asteroleplasmas and therefore may be unique to mammalian pathogenic species.

In addition to capsular structures, fibrils are often found associated with mycoplasmas. Black *et al.* (1972) reported hairlike projections on the surface of human T mycoplasmas (now named *U. urealyticum*). Similar structures have been reported with *M. pulmonis* and *M. gallisepticum* (Razin, 1973). Tajima and Yagihashi (1982) reported fibrillar material 5 nm in diameter and up to 200 nm in length associated with *M. hyopneumoniae* membranes. This fibrillar material seems to be associated with heavily encapsulated strains. Capsule polymers are generally viscous and fibrillar structures are often seen with capsulated bacteria (Cagle, 1975). Therefore, fibrils may be composed of capsule components and may simply represent a second morphological form. This is supported by the findings that *M. mycoides* subsp. *mycoides* produces macroscopically visible "threads" during growth in broth, while noncapsulated strains do not (Gourlay and Thrower, 1968). Fibrils have also been observed between mycoplasmas and eukaryotic cell surfaces in the open space between the two cell membranes, suggesting a role for these structures in attachment. These structures could also have arisen as preparation artifacts, and further examination is needed to assess their function.

3. CAPSULE COMPOSITION AND CHEMICAL STRUCTURE

Our current knowledge concerning composition of capsules is insufficient. Preparation of purified capsule material has been problematic. The difficulty of separating *de novo*-produced material from that adsorbed to the mycoplasma surface during growth (Yaguzhinskaya, 1976) has not often been rectified, and therefore it may be important to combine biochemical and immunological analysis to capsule fractions to ensure purity.

The purification of capsular material has been accomplished by using hot phenol extraction procedures followed by ethanol precipitation or ion-exchange chromatography (Buttery and Plackett, 1960; Rurangirwa *et al.*, 1987). Treatment of organisms with proteases, i.e., Proteinase K, prior to extraction has improved the quality of the preparations and eliminated steps in the purification procedure. It should be pointed out, however, that membrane lipoglycans will be removed by hot phenol extraction and can be a source of contamination (Smith, 1984). These materials are integral components of the membrane and are not considered as capsule materials. Prolonged exposure to buffered saline can also be used for extraction of capsule (Almeida *et al.*, 1992). The ability to extract capsule in saline may relate to the observations of patchy ruthenium red staining material on mycoplasma surfaces. Capsule may be loosely attached to the mycoplasma membranes and there are reports of releasing carbohydrates from mycoplasma cells by sonication (Terry and Zupnik, 1973) although this has not been confirmed (Razin, 1978). Minion *et al.* (1984) reported releasing a hemagglutination-active material from *M. pulmonis* during mild sonication and, therefore, the release of materials from mycoplasmas by mild sonication may occur only under certain growth or treatment conditions that have yet to be accurately defined. In addition to the procedures mentioned above, other methods used for isolation of capsules from gram-positive bacteria may by useful in mycoplasma studies (Chomarat *et al.*, 1989).

Further purification of capsular material may be accomplished by the binding of mycoplasma polysaccharide fractions to specific lectins (Kahane and Tully, 1976; Schiefer *et al.*, 1978a,b). The studies by Kahane and Tully (1976) and Schiefer *et al.* (1974) on the binding of lectins to mycoplasma whole cells and membranes served to stimulate thinking in this area. The capsule of *U. urealyticum*-bound concanavalin A lectin (Robertson and Smook, 1976; Whitescarver *et al.*, 1975), and the capsule of *M. dispar*-bound *Ricinus communis* I lectin (Almeida and Rosenbusch, 1991) have been studied in this fashion. When lectin affinity is used as part of a purification protocol, elution of the bound polysaccharide must be effected by competition (Kahane and Schiefer, 1983) with a disaccharide or monosaccharide at fairly high concentrations (usually 0.2 M), and a gel filtration step will be necessary to separate the polysaccharide from the mono- or disaccharide.

Evidence for the chemical composition of mycoplasma capsules can be derived from two independent sources. Indirect evidence for its carbohydrate nature can be derived from electron microscopic studies of preparations stained with a polycationic electron-dense reagent, i.e., ruthenium red (Luft, 1964) or cationized ferritin (Schiefer *et al.*, 1976), together with the osmium tetroxide fixative. Also, the ability of specific lectins to interact with different mycoplasmas (Kahane and Tully, 1976; Schiefer *et al.*, 1974) suggests a mycoplasma-specified surface carbohydrate structure. This information is of limited value in this regard, since lectins could interact with other carbohydrate-containing macromolecules such as glycolipids and lipoglycans. The lectin studies can also give some information as to the type(s) of carbohydrates incorporated into the capsule because of the specific nature of their interactions. By coupling lectin-specific interactions with electron microscopy, qualitative information can be obtained on capsule composition (Robertson and Smook, 1976). Taken in total, these lines of evidence indicate that mycoplasma capsules are polyanionic and thus may be composed of acidic carbohydrates or lipids.

Chemical analysis of only a few mycoplasma capsules has been reported. The first capsule to be studied was that from *M. mycoides* subsp. *mycoides*. Buttery and Plackett (1960) reported a high-molecular-weight galactan isolated by warm phenol. Further studies revealed it was linked in a $\beta(1-6)$ configuration (Buttery, 1970; Plackett and Buttery, 1964). Using similar techniques, Rurangirwa *et al.* (1987) isolated a polysaccharide from the F-38 strain. This polysaccharide was composed of equal molar quantities of glucose, galactose, mannose, fucose, galactosamine, and glucosamine. In contrast, a polysaccharide from *M. mycoides* subsp. *capri* (Jones *et al.*, 1965) and an unspeciated bovine mycoplasma (Plackett *et al.*, 1963) contained glucose as the sole carbohydrate. In more recent studies, the capsule of *M. dispar* has been shown to be composed of a polymer of galacturonic acid (Rosenbusch *et al.*, unpublished). In *A. laidlawii*, a polyhexosamine has been isolated that contained galactosamine and glucosamine (Terry and Zupnik, 1973). It represents about 4% of the membrane dry weight (Engelman and Morowitz, 1968) and is apparently loosely associated with the membrane (Terry and Zupnik, 1973). This weak association has not been confirmed (Razin, 1978), but there is general agreement that capsule production is sensitive to environmental stimuli and variation between laboratories may not be uncommon.

4. BIOLOGICAL AND PATHOGENIC CHARACTERISTICS OF CAPSULE

The most obvious role of capsule is in adherence of the organism to host tissues. This has been suggested by numerous studies (Green and Hanson, 1973;

Howard *et al.*, 1974; Wilson and Collier, 1976). This could result from a latch effect by the association of multiple repeating units with cell receptors resulting in irreversible binding (Robb, 1984). Once bound, mycoplasmas can be detached only with great difficulty (Razin, 1978). Binding of mycoplasmas to host cells also has been described as a multifactorial process (Minion *et al.*, 1984) that may involve capsular materials. Attachment of *M. dispar* to erythrocytes appeared to be mediated by ruthenium red-stainable capsular material and fine extracellular threads bridging gaps between membranes (Howard *et al.*, 1974). Similar observations were made with *M. hyopneumoniae* attaching to porcine respiratory epithelium (Tajima and Yagihashi, 1982). Strains of *M. hyopneumoniae* that were extensively passaged in vitro did not exhibit these fibrillar structures and were less pathogenic to pigs.

In general, capsular polysaccharides have antiphagocytic properties associated with their electronegative charges and the formation of microcolonies surrounded by a glycocalyx of exopolysaccharides which are resistant to enzymatic degradation (Isenberg, 1988). In mycoplasmas, production of capsule does not appear to be required to avoid phagocytosis since in general, mycoplasmas are difficult to phagocytose. *M. pulmonis* was taken up and killed by mouse alveolar macrophages only in the presence of specific antisera (Davis *et al.*, 1980). The role of capsule in the antiphagocytic process was not addressed, but capsule may be released from the organism upon trypsin treatment, and trypsin-treated mycoplasmas are readily taken up and killed by macrophages (Davis *et al.*, 1980). The ability to attach and multiply on the surface of cultured macrophages and neutrophils without stimulation of ingestion was described for mycoplasmas pathogenic for the respiratory tract of cattle (Howard *et al.*, 1980). Both the capsulated *M. dispar* and the noncapsulated *M. bovis* were capable of inhibiting the ability of bovine neutrophils to phagocytose *E. coli*. Thus, capsular polysaccharides *per se* do not appear to enhance or diminish the antiphagocytic properties exhibited by mycoplasmas, but in some species they may be an important virulence factor.

Capsules nonspecifically bind factor H, a normal inhibitor of the alternative pathway of complement activation (Kasper, 1986), and can prevent the nonspecific deposition of IgG on bacterial surfaces (Absolom, 1988). All but one of 15 mycoplasma strains tested for serum resistance were resistant to killing by gnotobiotic calf serum and this was attributed to lack of activation of the alternative complement pathway (Howard, 1980). This suggests a role for capsule in protection against nonspecific host immune defenses.

Although purified capsular material is poorly immunogenic (R. F. Rosenbusch, unpublished observations), convalescent animals can have anticapsule serum antibodies (Rurangirwa *et al.*, 1987; Buttery, 1970). Alternatively, in some species, capsule may be immunosuppressive, preventing an appropriate immune response. The galactan of *M. mycoides* subsp. *mycoides* does not induce antibody responses in immunized animals unless combined with Freund's adjuvant (Hudson *et al.*, 1967). It has been suggested that the mycoplasma galactan possesses serological similarity to pneumogalactan, a product of normal lung epithelial cells (Gourlay and Shifrine, 1966; Shifrine and Gourlay, 1965). A similar immunological anergy may be associated with the capsule of *M. dispar* since cattle may have low serum antibody titers against this species (Howard, 1983). In addition, immune responses to capsular polysaccharides are normally T-independent and absent in the very young (Stein, 1985). These features are also seen in immune responses against *M. dispar* (Howard and Gourlay, 1983).

Mycoplasma capsules may have toxic effects on specific eukaryotic cells in the body. This is expressed as capillary thrombosis and pulmonary edema in calves given purified galactan intravenously (Buttery *et al.*, 1976). Lethal effects for chick embryos inoculated in the chorioalantoic cavity included hemorrhagic lesions, but this toxic effect was significantly reduced with more purified galactan preparations (Villemot *et al.*, 1962). Impurities of concern include nucleic acids and lipoglycans since these latter compounds from acholeplasmas have been shown to elicit pyrogenic responses in rabbits and clotting of *Limulus* amoebocyte lysates (Seid *et al.*, 1980). Another effect ascribed to the galactan of *M. mycoides* subsp. *mycoides* is the deposition of fibrin around chronic lung lesions in cattle (Buttery *et al.*, 1980), and is mimicked by placing the organism within diffusion chambers in the peritoneal cavity or tissues of calves and rabbits (Buttery *et al.*, 1980; Lloyd, 1966). Cattle given galactan developed persistent mycoplasmemia when concurrently infected by subcutaneous injection, and polyarthritis was seen in calves similarly infected (Lloyd *et al.*, 1971).

5. CONCLUDING REMARKS

The capsules of mycoplasmas, their structural features, chemical analysis, genetics, and role (if any) in pathogenesis have been incompletely studied. The complete chemical characterization of capsules from several different species may reveal that capsule polysaccharides are generally less complex than lipoglycans, where a large variety of monosaccharides are present. If this is true, genetic analysis of capsule genes may be relatively straightforward. Recent advances in mycoplasma genetics with transposons (Dybvig and Cassell, 1987; Mahairas and Minion, 1989b) and integrative vectors (Mahairas and Minion, 1989a) promise to provide the tools necessary to address this and many other problems in mycoplasmology.

A second important area needing further study is the toxic and immunomodulatory effects of mycoplasma capsules. Although it is clear that the capsule of *M. mycoides* subsp. *mycoides* is biologically active, little is known concerning capsules from other mycoplasmal species. A related question is the potential of mycoplasma capsules to serve as immunogens. Little is known about the host response to capsular material during active infections. Almost all studies of this type have measured antibody responses against protein antigens only.

Finally, development of knowledge about the molecular biology of capsule biosynthesis and regulation of expression may provide important background for further understanding of the biology and pathogenesis of mycoplasma infections. The infectious process is complex and the contribution of extramembranous structures to pathogenesis may prove substantial.

REFERENCES

Absolom, D. R., 1988, The role of bacterial hydrophobicity in infection: bacterial adhesion and phagocytic ingestion, *Can. J. Microbiol.* **34:**287–298.

Ajufo, J. C., and Whithear, K. G., 1978, Evidence for a ruthenium red-staining extracellular layer as the haemagglutinin of the WVU 1853 strain of *Mycoplasma synoviae, Aust. Vet. J.* **54:**502–504.

Almeida, R. A., and Rosenbusch, R. F., 1991, Capsulelike material of *Mycoplasma dispar* induced by in vitro culture with bovine cells is antigenically related to similar structures expressed in vivo, *Infect. Immun.* **59:**3119–3125.

Almeida, R. A., Wannemuehler, M. J., and Rosenbusch, R. F., 1992, Interaction of *Mycoplasma dispar* with bovine alveolar macrophages, *Infect. Immun.* **60:**2914–2919.

Black, F. T., Birch-Andersen, A., and Freundt, E. A., 1972, Morphology and ultrastructure of human T-mycoplasmas, *J. Bacteriol.* **111:**254–259.

Boatman, E. S., 1979, Morphology and ultrastructure of the mycoplasmatales, in: *The Mycoplasmas,* Volume I (M. F. Barile and S. Razin, eds.), Academic Press, New York, pp. 63–101.

Boatman, E. S., Cartwright, F., and Kenny, G., 1976, Morphology, morphometry and electron microscopy of HeLa cells infected with bovine mycoplasma, *Cell Tissue Res.* **170:**1–16.

Buttery, S. H., 1970, Hapten inhibition of the reaction between *Mycoplasma mycoides* polysaccharide and bovine antisera, *Immunochemistry* **7:**305–310.

Buttery, S. H., and Plackett, P., 1960, A specific polysaccharide from *Mycoplasma mycoides, J. Gen. Microbiol.* **23:**357–368.

Buttery, S. H., Lloyd, L. C., and Titchen, D. A.,, 1976, Acute respiratory, circulatory and pathological changes in the calf after intravenous injections of the galactan from *Mycoplasma mycoides* subsp. *mycoides, J. Med. Microbiol.* **9:**379–391.

Buttery, S. H., Cottew, G. S., and Lloyd, L. C., 1980, Effect of soluble factors from *Mycoplasma mycoides* subsp. *mycoides* on the collagen content of bovine connective tissues, *J. Comp. Pathol.* **90:**303–314.

Cagle, G. D., 1975, Fine structure and distribution of extracellular polymer surrounding selected aerobic bacteria, *Can. J. Microbiol.* **21:**395–408.

Chomarat, M., Ichiman, Y., and Yoshida, K., 1989, Protection of mice by a pseudodiffuse strain of *Staphylococcus aureus* possessing polyvalent capsular type antigen, *J. Med. Microbiol.* **28:**129–136.

Cole, R. M., Tully, J. G., and Popkin, T. J., 1973, Ultrastructure of the agent of citrus "stubborn" disease, *Ann. N.Y. Acad. Sci.* **225:**471–493.

Costerton, J. W., and Irvin, R. T., 1981, The bacterial glycocalyx in nature and disease, *Annu. Rev. Microbiol.* **35:**299–324.

Davis, J. K., Delozier, K. M., Asa, K., Minion, F. C., and Cassell, G. H., 1980, Interactions between murine alveolar macrophages and *Mycoplasma pulmonis* in vitro, *Infect. Immun.* **29:**590–599.

Domermuth, C. H., Nielsen, M. H., Freundt, E. A., and Birch-Andersen, A., 1964, Ultrastructure of *Mycoplasma* species, *J. Bacteriol.* **88:**727–744.

Dybvig, K., and Cassell, G. H., 1987, Transposition of gram-positive transposon Tn*916* in *Acholeplasma laidlawii* and *Mycoplasma pulmonis, Science* **235:**1392–1394.

Engelman, D. M., and Morowitz, H. J., 1968, Characterization of the plasma membrane of *Mycoplasma laidlawii:* IV. Structure and composition of membrane and aggregated components, *Biochim. Biophys. Acta* **150:**385–396.

Furness, G., Whitescarver, J., Trocola, M., and DeMaggio, M., 1976, Morphology, ultrastructure, and mode of division of *Mycoplasma fermentans, Mycoplasma hominis, Mycoplasma orale,* and *Mycoplasma salivarium, J. Infect. Dis.* **134:**224–229.

Gourlay, R. N., and Shifrine, M., 1966, Antigenic cross-reactions between the galactan from *Mycoplasma mycoides* and polysaccharides from other sources, *J. Comp. Pathol.* **76:**417–425.

Gourlay, R. N., and Thrower, K. J., 1968, Morphology of *Mycoplasma mycoides* thread-phase growth, *J. Gen. Microbiol.* **55:**155–159.

Green, F., III, and Hanson, R. P., 1973, Ultrastructure and capsule of *Mycoplasma meleagridis, J. Bacteriol.* **116:**1011–1018.

Hackstadt, T., Todd, W. J., and Caldwell, H. D., 1985, Disulfide-mediated interactions of the chlamydial major outer membrane protein: Role in the differentiation of chlamydia, *J. Bacteriol.* **161:**25–31.

Horn, R. W., 1970, The ultrastructure of mycoplasma and mycoplasma-like organisms, *Micron* **2:**19–38.

Howard, C. J., 1980, Variation in susceptibility of bovine mycoplasmas to killing by the alternative complement pathway in bovine serum, *Immunology* **41:**561–568.

Howard, C. J., 1983, Mycoplasmas and bovine respiratory disease: Studies related to pathogenicity and immune response—A selective review, *Yale J. Biol. Med.* **56:**789–797.

Howard, C. J., and Gourlay, R. N., 1974, An electron-microscopic examination of certain bovine mycoplasmas stained with ruthenium red and the demonstration of a capsule on *Mycoplasma dispar, J. Gen. Microbiol.* **83:**393–398.

Howard, C. J., and Gourlay, R. N., 1983, Immune response of calves following the inoculation of *Mycoplasma dispar* and *Mycoplasma bovis, Vet. Microbiol.* **8:**45–56.

Howard, C. J., Gourlay, R. N., and Collins, J., 1974, Serological comparison and hemagglutinating activity of *Mycoplasma dispar, J. Hyg.* **73:**457–466.

Howard, C. J., Gourlay, R. N., and Taylor, G., 1980, Immunity to mycoplasma infections of the calf respiratory tract, *Adv. Exp. Med. Biol.* **137:**711–726.

Hudson, J. R., Buttery, S., and Cottew, G. C., 1967, Investigations into the influence of the galactan of *Mycoplasma mycoides* on experimental infection with that organism, *J. Pathol. Bacteriol.* **94:**257–273.

Isenberg, H. D., 1988, Pathogenicity and virulence: Another view, *Clin. Microbiol. Rev.* **1:**40–53.

Jones, A. S., Tittensor, J. R., and Walker, R. T., 1965, The chemical composition of nucleic acids and other macromolecular constituents of *Mycoplasma mycoides* var. *capri, J. Gen. Microbiol.* **40:**405–411.

Kahane, I., and Schiefer, H. G., 1983, Characterization of carbohydrate components of mycoplasma membranes, in: *Methods of Mycoplasmology,* Volume I (S. Razin and J. G. Tully, eds.), Academic Press, New York.

Kahane, I., and Tully, J. G., 1976, Binding of plant lectins to mycoplasma cells and membranes, *J. Bacteriol.* **128:**1–7.

Kasper, D. L., 1986, Bacterial capsules. Old dogmas and new tricks, *J. Infect. Dis.* **153:**407–415.

Lloyd, L. C., 1966, Tissue necrosis produced by *Mycoplasma mycoides* in intraperitoneal diffusion chambers, *J. Pathol. Bacteriol.* **92**:225–229.

Lloyd, L. C., Buttery, S. H., and Hudson, J. R., 1971, The effect of the galactan and other antigens of *Mycoplasma mycoides* var. *mycoides* on experimental infection with that organism in cattle, *J. Med. Microbiol.* **4**:425–439.

Luft, J. H., 1964, Electron microscopy of cell extraneous coats as revealed by ruthenium red staining, *J. Cell Biol.* **23**:54a.

Mahairas, G. G., and Minion, F. C., 1989a, Transformation of *Mycoplasma pulmonis:* Demonstration of homologous recombination, introduction of cloned genes, and the preliminary description of an integrating shuttle system, *J. Bacteriol.* **171**:1775–1780.

Mahairas, G. G., and Minion, F. C., 1989b, Random insertion of the gentamicin resistance transposon Tn*4001* in *Mycoplasma pulmonis, Plasmid* **21**:43–47.

Minion, F. C., Cassell, G. H., Pnini, S., and Kahane, I., 1984, Multiphasic interactions of *Mycoplasma pulmonis* with erythrocytes defined by adherence and hemagglutination, *Infect. Immun.* **44**:394–400.

Plackett, P., 1959, On the probable absence of "mucocomplex" from *Mycoplasma mycoides, Biochim. Biophys. Acta* **35**:260–262.

Plackett, P., and Buttery, S. H., 1964, A galactofuranose disaccharide from the galactan of *Mycoplasma mycoides, Biochem. J.* **90**:201–205.

Plackett, P., Buttery, S. H., and Cottew, G. S., 1963, Carbohydrates of some mycoplasma strains, in: *Recent Progress in Microbiology VIII* (N. E. Gibbons, ed.), University of Toronto Press, Toronto, p. 533.

Razin, S., 1973, Physiology of mycoplasmas, *Adv. Microb. Physiol.* **10**:1–80.

Razin, S., 1978, The mycoplasmas, *Microbiol. Rev.* **42**:414–470.

Robb, I. D., 1984, Stereo-biochemistry and functions of polymers in microbial adhesion and aggregation, in: *Microbial Adhesion and Aggregation* (K. C. Marshall, ed.), Springer-Verlag, Berlin, pp. 39–49.

Robertson, J., and Smook, E., 1976, Cytochemical evidence of extramembranous carbohydrates on *Ureaplasma urealyticum* (T-strain mycoplasma), *J. Bacteriol.* **128**:658–660.

Rosengarten, R., and Wise, K. S., 1990, Phenotypic switching in mycoplasmas: Phase variation of diverse surface lipoproteins, *Science* **247**:315–318.

Rosengarten, R., Kirchhoff, H., Kerlen, G., and Seack, K.-H., 1988, The surface layer of *Mycoplasma mobile* 163K and its possible relevance to cell cohesion and group motility, *J. Gen. Microbiol.* **134**:275–281.

Rurangirwa, F. R., McGuire, T. C., Magnuson, N. S., Kibor, A., and Chema, S., 1987, Composition of a polysaccharide from mycoplasma (F-38) recognized by antibodies from goats with contagious pleuropneumonia, *Res. Vet. Sci.* **42**:175–178.

Salih, B. A., and Rosenbusch, R., 1988, Attachment of *Mycoplasma bovoculi* to bovine conjunctival epithelium and lung fibroblasts, *Am. J. Vet. Res.* **49**:1661–1664.

Schiefer, H. G., Gerhardt, U., Brunner, H., and Krupe, M., 1974, Studies with lectins on the surface carbohydrate structures of mycoplasma membranes, *J. Bacteriol.* **120**:81–88.

Schiefer, H. G., Krauss, H., Brunner, H., and Gerhardt, U., 1976, Ultrastructural visualization of anionic sites on mycoplasma membranes by polycationic ferritin, *J. Bacteriol.* **127**:461–468.

Schiefer, H. G., Krauss, H., Schummer, U., Brunner, H., and Gerhardt, U., 1978a, Cytochemical localization of surface carbohydrates on mycoplasma membranes, *Experientia* **34**:1011–1012.

Schiefer, H. G., Krauss, H., Schummer, U., Brunner, H., and Gerhardt, U., 1978b, Studies with ferritin-conjugated concanavalin A on carbohydrate structures of mycoplasma membranes, *FEMS Microbiol. Lett.* **3**:183–185.

Seid, R. C., Jr., Smith, P. F., Guevarra, G., Hochstein, H. D., and Barile, M. F., 1980, Endotoxin-like activities of mycoplasma lipopolysaccharides (lipoglycans), *Infect. Immun.* **29**:990–994.

Shifrine, M., and Gourlay, R. N., 1965, Serologic relationship between galactans from normal bovine lung and from *Mycoplasma mycoides*, *Nature* **208**:498–499.

Smith, P. F., 1984, Lipoglycans from mycoplasmas, *CRC Crit. Rev. Microbiol.* **11**:157–185.

Stein, K. E., 1985, Network regulation of the immune response to bacterial polysaccharide antigens, *Curr. Top. Microbiol. Immunol.* **119**:57–74.

Tajima, M., and Yagihashi, T., 1982, Interaction of *Mycoplasma hyopneumoniae* with the porcine respiratory epithelium as observed by electron microscopy, *Infect. Immun.* **37**:1162–1169.

Tajima, M., Nunoya, T., and Yagihashi, T., 1979, An ultrastructural study on the interaction of *Mycoplasma gallisepticum* with the chicken tracheal epithelium, *Am. J. Vet. Res.* **40**:1009–1014.

Tajima, M., Yagihashi, T., and Miki, Y., 1982, Capsular material of *Mycoplasma gallisepticum* and its possible relevance to the pathogenic process, *Infect. Immun.* **36**:830–833.

Tajima, M., Yagihashi, T., and Nunoya, T., 1985, Ultrastructure of mycoplasmal capsules as revealed by stabilization with antiserum and staining with ruthenium red, *Jpn. J. Vet. Sci.* **47**:217–223.

Taylor-Robinson, D., Furr, P. M., Davies, H. A., Manchee, R. J., and Bove, J. M., 1981, Mycoplasmal adherence with particular reference to the pathogenicity of *Mycoplasma pulmonis*, *Isr. J. Med. Sci.* **17**:599–603.

Terry, T. M., and Zupnik, J. S., 1973, Weak association of glucosamine-containing polymer with the *Acholeplasma laidlawii* membrane, *Biochim. Biophys. Acta* **291**:218–224.

Villemot, J. M., Provost, A., and Queval, R., 1962, Endotoxin from *Mycoplasma mycoides*, *Nature* **193**:906–907.

Watson, H. L., McDaniel, L. S., Blalock, D. K., Fallon, M. T., and Cassell, G. H., 1988, Heterogeneity among strains and a high rate of variation within strains of a major surface antigen of *Mycoplasma pulmonis*, *Infect. Immun.* **56**:1358–1363.

Whitescarver, J., Castillo, F., and Furness, G., 1975, The preparation of membranes of some human T-mycoplasmas and the analysis of their carbohydrate content, *Proc. Soc. Exp. Biol. Med.* **150**:20–22.

Wilson, M. H., and Collier, A. M., 1976, Ultrastructural study of *Mycoplasma pneumoniae* in organ culture, *J. Bacteriol.* **125**:332–339.

Yaguzhinskaya, O. E., 1976, Detection of serum proteins in the electrophoretic patterns of total proteins of mycoplasma cells, *J. Hyg.* **77**:189–198.

Chapter 7

Spiralins

J. M. Bové, X. Foissac, and Colette Saillard

Spiroplasmas are unique among the mollicutes in that they are helical and motile, and yet they have no flagella, periplasmic filaments, or other organelles of locomotion. As true mollicutes they lack a cell wall; their cell envelope is reduced to a single cytoplasmic membrane. The molecular bases of spiroplasma helicity and motility are not better understood today than 20 years ago when the spiroplasmas were first discovered. What are the proteins responsible for these properties? Two proteins have drawn attention in this respect: spiralin, because it is the most abundant protein in the spiroplasma membrane, and the fibril protein, because it can associate into long fibrils. Both proteins have been extensively studied, their genes have been sequenced and mapped on the spiroplasma genome (Chevalier *et al.*, 1990a,b; Williamson *et al.*, 1991). Yet their role in spiroplasmas is still not understood. This chapter is essentially devoted to recent studies of spiralin and updates a recent review (Bové *et al.*, 1989).

1. DEFINITION

When membrane proteins of *S. citri* strain C189 were separated by SDS-PAGE and stained with amido black, one zone of the gel was most heavily

J. M. Bové, X. Foissac, and Colette Saillard   Laboratory of Cellular and Molecular Biology, INRA and the University of Bordeaux II, 33883 Villenave d'Ornon Cedex, France.

Subcellular Biochemistry, Volume 20: Mycoplasma Cell Membranes, edited by Shlomo Rottem and Itzhak Kahane. Plenum Press, New York, 1993.

Table I
Amino Acid Composition of Spiralin

| Amino acid | No./molecule[a] | % of total (mole%) | |
		Sequence[a]	Protein[b]
Ala	32	13.28	12.84
Val	32	13.28	11.81
Lys	27	11.20	10.84
Thr	23	9.54	9.64
Asp	13	5.39	
Asn	17	7.05	13.62
Glu	13	5.39	
Gln	9	3.73	10.02
Ile	16	6.64	6.35
Leu	13	5.39	4.79
Ser	12	4.98	6.68
Gly	12	4.98	5.00
Pro	9	3.73	3.50
Tyr	5	2.07	2.45
Phe	5	2.07	1.83
Cys	2	0.83	0.55
Met	1	0.41	0
Arg	0	0	0
His	0	0	0
Trp	0	0	0

[a]Determined from sequence of *S. citri* (R8A2HP) spiralin gene (Chevalier *et al.*, 1990a).
[b]Determined from spiralin purified from *S. citri* (R8A2HP) (Wroblewski *et al.*, 1984).

stained; the protein in this zone was called spiralin (Wroblewski *et al.*, 1977). Spiralin was thus defined as the most abundant protein in the cell membrane of *S. citri;* the name "spiralin" does not infer any special function. Spiralin represents more than 20% of the total membrane protein. It has been purified to homogeneity and shown to have a molecular weight of about 26,000. The amino acid composition of spiralin has been determined (Wroblewski *et al.*, 1977, 1984). The protein is unusual in that it lacks methionine, histidine, tryptophan, and arginine (Table I).

Spiralin as originally described came from *S. citri* strain C189. Proteins with compositions quasi-identical to spiralin-C189 were also purified from three other helical strains of *S. citri:* strains R8A2 and Scaph (Wroblewski *et al.*, 1984) as well as strain SP-A (Archer and Townsend, 1981). Crossed immunoelectrophoresis showed that the spiralins purified from these three strains and from strain C189 were antigenically similar. One- and two-dimensional polyacrylamide gel electrophoresis has shown that all *S. citri* strains tested, including the nonhelical strain ASP 1, contain spiralin, but according to the strain considered, spiralin can have slightly different electrophoretic mobilities and immunological properties (Mouchès *et al.*, 1979, 1983). Possible reasons for this will be examined below (see Section 7).

Spiralins from *S. melliferum* strains B88 and BC3 have also been purified (Wroblewski *et al.*, 1984; Archer and Townsend, 1981). The *S. melliferum* spiralin (spiralin-Sm) and the *S. citri* spiralin (spiralin-Sc) had the same solubility properties, electrophoretic mobilities, and molecular weights. Amino acid compositions were also similar; in particular, the two proteins lacked methionine and tryptophan. In spite of these similarities, the two spiralins were found by crossed immunoelectrophoresis to be antigenically dissimilar and to have no common epitopes (Archer and Townsend, 1981; Wroblewski *et al.*, 1984). In ELISA, monospecific antibodies against spiralin from *S. citri* strain R8A2 detected only *S. citri* strains and not *S. melliferum* strain BC3 (Mouchès and Bové, 1983). However, in growth inhibition, metabolic inhibition, and deformation tests, minor cross-reactions were observed between antispiralin serum and *S. melliferum* strain BC3 (Whitcomb *et al.*, 1983). More recently, however, it was shown by rocket immunoelectrophoresis, quantitative immunoblotting, and spiroplasma deformation test that spiralin-Sc and spiralin-Sm are antigenically related, but that probably no more than two epitopes simultaneously saturable with antibodies are shared by the two proteins (Fontenelle *et al.*, 1987; Zaaria *et al.*, 1990). At least one of these epitopes is accessible to antibodies on the spiroplasma cell surface.

2. LOCALIZATION OF SPIRALIN IN SPIROPLASMA CELLS

Townsend and Plaskitt (1985) used immunoferritin and immunogold labeling to localize protein p25, the spiralin of *S. melliferum* strain BC3, in spiroplasma cells. Immunoferritin labeling with anti-p25 antibody easily detected the p25 antigen on the spiroplasma cell surface, and the staining was uniformly distributed over the cell surface. Immunogold labeling of ultrathin sections of spiroplasma cells showed that p25 was associated only with the plasma membrane. There was heavy labeling with gold particles around the circumference of the cell, but a significant proportion of the gold particles appeared to be located also on the cytoplasmic side of the plasma membrane. The spiralin-like protein p25 was thus detected on both sides of the spiroplasma membrane, suggesting a transmembrane distribution of p25. Wroblewski (1978, 1981) has also proposed that spiralin forms oligomers that are able to span the *S. citri* membrane. Cross-linking experiments have shown that transmembrane spiralin was involved in homooligomers and that, among these, the dimer was the most abundant form.

The dimers appeared to be stabilized by intermolecular disulfide bonds (Wroblewski, 1981).

3. PROPERTIES OF SPIRALIN AS DEDUCED FROM STUDIES OF THE PROTEIN

3.1. Spiralin, an Integral, Amphiphilic Membrane Protein

Several lines of evidence indicate that spiralin is an integral (intrinsic) membrane protein, as opposed to peripheral (extrinsic) membrane proteins. Integral proteins are embedded in and interact extensively with the hydrocarbon chains of membrane lipids; they can be released only by agents that compete for these nonpolar interactions. Organic solvents and detergents are such agents. In line with these characteristics, spiralin could be quantitatively extracted from the spiroplasma membrane with sodium deoxycholate (DOC) and sodium dodecyl sulfate (SDS) (Wroblewski *et al.*, 1977). It was not possible to release spiralin with detergent-free buffers, even in the presence of 2-mercaptoethanol and EDTA. The efficiency of mild (ionic and neutral) detergents for solubilizing spiralin was as follows: deoxycholate > sodium lauroyl sarcosinate, cholate, taurocholate, taurodeoxycholate > Triton X-100 > Brij 58 > Tween 20, indicating that mild ionic detergents were more effective than neutral ones (Brij 58, Tween 20), Triton X-100 occupying an intermediate position (Wroblewski, 1979).

Furthermore, spiralin was able to bind Triton X-100, a mild, nonionic and nondenaturing detergent (Wroblewski *et al.*, 1977). The zwitterionic, nondenaturing detergent SB12 was also bound (Wroblewski *et al.*, 1987). The ability to bind such detergents characterizes proteins that have hydrophobic as well as hydrophilic properties (amphiphilic, amphipathic proteins), as opposed to ordinary soluble hydrophilic proteins, which bind little or no detergent. Spiralin is thus an amphiphilic protein. The amphiphilic nature of spiralin can be further demonstrated by charge-shift electrophoresis alone (Helenius and Simons, 1977) or combined with immunoelectrophoresis (Bhakdi *et al.*, 1977; Wroblewski, 1979). In the presence of both Triton X-100, a neutral detergent, and DOC, an anionic, negatively charged detergent, an amphiphilic protein will bind both detergents to form a ternary protein–Triton X-100–DOC complex. This complex will be more negatively charged than the binary complex containing only protein and Triton X-100, and, in comparison with the latter, it displays a more anodal migration (anodal shift). Similarly, when the second detergent is cationic, such as cetyltrimethylammonium bromide (CTAB), the ternary complex will show a more cathodal migration (cathodal shift). In agreement with its amphiphilic nature, the complex of spiralin with Triton X-100 (Wroblewski, 1979) or SB12 (Wroblewski *et al.*, 1987) exhibited a cathodal shift in the presence of CTAB and an anodal shift in the presence of DOC.

An additional method for demonstrating the amphiphilic nature of a protein has been developed by Simons *et al.* (1978). First, a soluble complex between the protein and a detergent such as Triton X-100 or DOC is formed (see above). Next, the detergent is progressively removed by dialysis. During this process the protein molecules associate to form water-soluble aggregates. The protein aggregates obtained resemble detergent micelles in their structure. The protein monomers forming the micelle are probably arranged so that the surface is polar and the interior apolar. Hydrophobic interactions provide the driving force for their formation. Such micelles have been obtained from the spiralin–DOC complex by extensive removal of DOC by dialysis in the presence of Bio-Beads SM-2. The micelles, when analyzed by ultracentrifugation in a sucrose gradient, gave only one band at a buoyant density of 1.2. Electron microscopy showed the spiralin micelles to be globular aggregates with diameters in the range of 15–30 nm (Wroblewski *et al.*, 1987).

Finally, when the spiralin–DOC complex was dialyzed as above to remove the detergent but when, furthermore, removal of DOC was done in the presence of (egg yolk) phospholipids, the phospholipid molecules associated to form liposomes (lipid bilayer membrane) and spiralin was found to be associated with the liposomes. Spiralin could not be released from the liposomes by treatment with 0.5% NaCl, indicating that spiralin was not adsorbed to the liposomes by ionic bonds but was actually integrated in the bilayer (Wroblewski *et al.*, 1987).

In summary, spiralin binds detergents under nondenaturing conditions, can be inserted into the liposome bilayer, and is capable of forming micelles. The capacity to form water-soluble micelles suggests that spiralin belongs to a category of integral membrane proteins characterized by large hydrophilic domain(s) protruding on the membrane surface and small hydrophobic portion(s) within the lipid bilayer (Wroblewski *et al.*, 1987).

3.2. Spiralin, an Acyl Protein

Over the past few years, very interesting observations have been made concerning the integral proteins of the cell membrane of *M. capricolum* (Dahl *et al.*, 1983; Dahl and Dahl, 1984) and *A. laidlawii* (Dahl *et al.*, 1984; Nystrom *et al.*, 1986). Several of these proteins (20 to 30 of more than 50) were found to be covalently modified by acyl (fatty acid) chains. Acylation with palmitate dominated over that with oleate. Phospholipids appeared to be the direct acyl donors to membrane proteins. There were one to two acyl chains per protein molecule. The majority of the acyl chains could be released with mild alkaline hydrolysis (0.1 M KOH or 1 M hydroxylamine), indicating that the bonds between the acyl chains and the proteins were ester bonds involving the carboxyl group of the fatty acid and a hydroxyl group, probably of serine or threonine residues. The number of membrane acyl proteins in the mollicutes was at least twice that in bacteria. Calculation of the mean hydrophobicities for the membrane acyl proteins revealed that these proteins were surprisingly hydrophilic or, in other words, that they had significantly lower hydrophobicities than nonacylated integral membrane proteins. It has been suggested that acylation aids in both the insertion and membrane anchoring of the acyl proteins. The introduction of saturated fatty acyl groups into membrane proteins potentially facilitates both protein–protein and protein–lipid interactions by optimizing van der Waals contacts, enhancing membrane integrity in organisms (mollicutes) that lack a peptidoglycan cell wall.

Acylation of the membrane protein of *S. melliferum* strain B88 has been investigated recently (Wroblewski et al., 1989). Among 51 membrane peptides, 17 including spiralin but not the fibril protein, proved to be acylated with palmitate through ester bonds. The enrichment in serine residues of the water-soluble fraction from spiralin micelles treated with trypsin suggests that acylation involves serine residues. Threonine residues might also be involved (see below).

It will be discussed below (Section 6) that the primary translational product of the spiralin gene might well be a prespiralin with a signal (leader) sequence at the hydrophobic N-terminus (Mouchès *et al.*, 1985; Le Hénaff *et al.*, 1991). Cleavage of prespiralin to mature spiralin would be close to cysteine-24. Since spiralin is acylated, it has been speculated that the N-terminus of the mature protein is a diglyceride-cysteine or an *N*-acyl diglyceride-cysteine, by analogy with bacterial lipoproteins (Wu and Tokunaga, 1986). Such modifications of the N-terminal cysteine would also explain the blocking of the N-terminus of the mature protein.

4. PROPERTIES OF SPIRALIN AS DEDUCED FROM ITS GENE

The first mollicute gene to have been cloned and expressed in a bacterial host, *Escherichia coli,* was the spiralin gene (Mouchès *et al.*, 1985). This work was carried out in 1983–1984 at a time when it was not yet known that in *Mycoplasma* spp. (Yamao *et al.*, 1985) and *Spiroplasma* spp. (Renaudin *et al.*, 1986), the triplet UGA is not a stop codon but codes for tryptophan. Hence, mollicute genes containing UGA tryptophan codons cannot be fully expressed in transformed bacterial clones, since the bacterial ribosome stops at the first UGA codon encountered. Only genes without UGA tryptophan codons can be fully translated in the bacterial host. This is precisely the case of spiralin: the protein contains no tryptophan and its gene has no UGA codon!

Cloning and Sequencing Spiralin Genes

Spiralin Gene of *S. citri* and Spiralin-Sc

A library of cloned genome sequence of *S. citri* R8A2 was constructed by incorporation of *Hind*III restriction fragments into plasmid pBR328 and cloning in *E. coli* (Mouchès *et al.*, 1985). The bacterial clone harboring recombinant plasmid pES1 was selected by its ability to express spiralin as detected by ELISA. Plasmid pES1 was subcloned into plasmid pES3'. The 5-kbp spiroplasma insert of pES3' was entirely sequenced (Chevalier *et al.*, 1990a). Six ORFs were found (Figure 1). The translational product of one of these (ORF IV) had a size (241 amino acids totaling 25,282 Da) and an amino acid composition very similar to those of spiralin purified from *S. citri* (Table I). Four amino acids—alanine, valine, lysine, and threonine—account for more than 47.8% of all amino acids in the protein. Spiralin contains a low number of tyrosine, phenylalanine, and cysteine residues and lacks arginine, histidine, tryptophan, and internal methionine.

The spiralin gene represents a monocistronic transcription unit with its own promoter and terminator. The -10 region of the promoter has the sequence TGTAAT, only one nucleotide short of the consensus sequence of the -10 region (TATAAT) characteristic of promoters recognized by *E. coli* RNA polymerase carrying sigma factor σ^{70}, or the *B. subtilis* enzyme functioning with σ^{43}. The -35 region (TGTTATTT) has the first four nucleotides of the consensus sequence TGTTGACA. The spiralin promoter when inserted into the promoter selection vector pKK 232.8 permitted expression of chloramphenicol acetyltransferase, confirming its promoter function. The spiralin terminator hairpin has a stem totaling 13 base pairs of which only 2 are $G \equiv C$; the stem was followed by a stretch of 6 U bases, characteristic of terminators independent of termination factors such as rho.

Ten nucleotides upstream of the ATG initiation codon, the spiralin transcription unit has a Shine–Dalgarno (SD) ribosome binding site complementary to the 3'OH end of the 16 S ribosomal RNA of *S. citri*. This 16 S rRNA has recently been sequenced (Grau *et al.*, 1991) and when its sequence at the 3'OH end is aligned with those of *E. coli* and *B. subtilis*, the spiroplasma 16 S rRNA is respectively 7 and 4 bases longer than in *E. coli* and *B. subtilis*. This accounts for the fact that base pairing between the SD sequence and the 16 S rRNA occurs over as much as 8 bases in a row, a situation typical of gram-positive bacteria where SD sequences are longer than those found in gram-negative bacteria.

The codon usage for spiralin was determined from the nucleotide sequence of its gene. A preferential use of A- and T-rich codons was observed, especially when A or T occurred at the 3' end of codons specifying the same amino acid; less than 8.0% of the codons (18 of 241) had a C or a G at the third (3') position. This codon usage reflects the high A+T content (74%) of *S. citri* DNA.

As shown in Figure 1 besides the spiralin gene, the spiroplasma DNA fragment was found to contain five additional ORFs. Sequence analysis software programs have unambiguously identified the products of four ORFs as ribosomal protein S2 (ORF I), elongation factor Ts (ORF II), ATP-dependent 6-phosphofructokinase (ORF V), and pyruvate kinase. ORF III remains unidentified; its product (protein X) could be a regulatory protein acting at an inverted repeat sequence located between the promoter and the ATG initiation codon of ORF I (rps B). As shown in Figure 1, rps B and tsf (ORF II) are divergent genes with

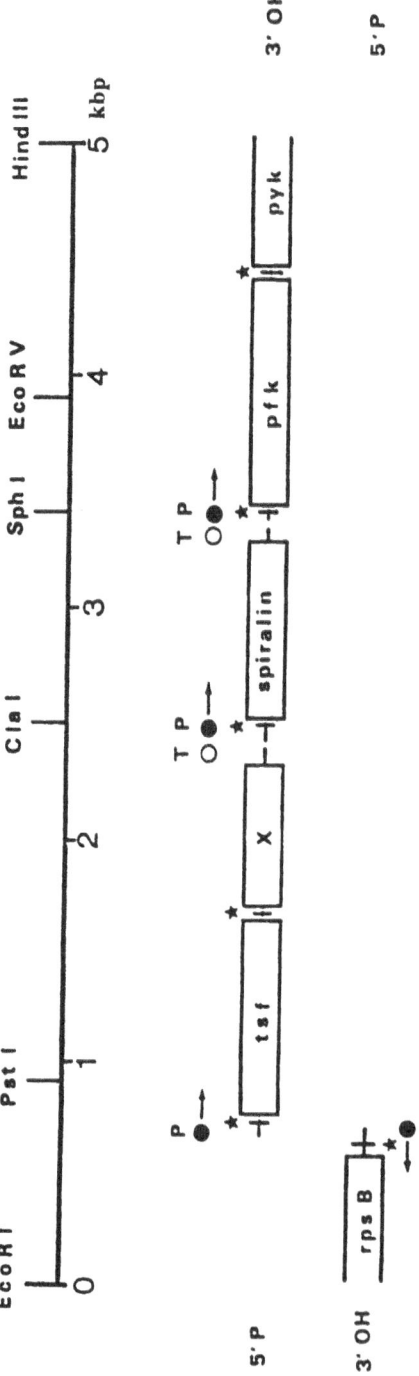

FIGURE 1. Localization of ORFS and organization of corresponding genes on the *S. citri* DNA insert of plasmid PES3′. For each ORF, the direction of transcription is indicated by an arrow. Promoters (P) and terminators (T) are represented by solid and open circles, respectively. Ribosome-binding sites are indicated by asterisks. (From Chevalier *et al.*, 1990a.)

overlapping promoters; this is the first description of divergent genes in mol-licutes. In *E. coli*, rps B and tsf are not divergent as they are in *S. citri*. Finally, the spiralin gene and the flanking genes have been mapped on the *S. citri* genome (F. Ye, F. Laigret, and J. M. Bové, unpublished).

Spiralin Gene of *S. melliferum* and Spiralin-Sm

To clone the gene encoding *S. melliferum* spiralin (spiralin-Sm), we as-sumed that there was sequence homology between the spiralin gene of *S. citri* and that of *S. melliferum*. This assumption was based on two arguments: there is 68% homology between the DNA of *S. citri* and that of *S. melliferum* (Bové *et al.*, 1989), and the amino acid compositions of spiralin-Sc and spiralin-Sm are quite similar (Wroblewski *et al.*, 1984). We, therefore, used the gene of spiralin-Sc as a probe in Southern hybridization to detect the spiralin-Sm gene in re-stricted *S. melliferum* DNA (Chevalier *et al.*, 1990b). A 4.6-kbp *Cla*I DNA fragment from *S. melliferum* strongly hybridized with the probe. This fragment was inserted in pBR322 and cloned in *E. coli*. It was further subcloned in the replicative forms of M13mp18 and M13mp19, and the nucleotide sequence of the insert (*Hind*III–*Sph*I fragment) was determined.

The nucleotide sequence of the *Hind*III–*Sph*I fragment was aligned with the sequence of the *S. citri* spiralin gene. The alignment shows a homology of 88.6%. The *S. melliferum* sequence had all of the restriction sites of the *S. citri* sequence, except for an *Rsa*I site. In the *S. melliferum* DNA, a 12-bp insertion occurred at nucleotide 398, and two small deletions of 3 and 6 bp were observed at nucleotides 702 and 710, respectively. A total of 75 base substitutions oc-curred in the *S. melliferum* DNA. The amino acid sequence of the *S. melliferum* spiralin was deduced from the nucleotide sequences. Spiralin-Sm is composed of 242 amino acids (spiralin-Sc: 241), giving a molecular mass of 25,430 Da (spiralin-Sc: 25,282 Da). Spiralin-Sm shares with spiralin-Sc the following char-acteristics. Four residues–Ala, Val, Lys, and Thr—represent more than 51% of all amino acids in the proteins. The two proteins contain a low number of Tyr, Phe, and Lys residues and lack Trp and internal Met residues. However, spiralin-Sm contains two Arg residues and one His residue, while spiralin-Sc does not. In all, 187 of 242 amino acids of spiralin-Sm (77%) are shared by spiralin-Sc. The 31 N-terminal amino acid residues, including the putative signal sequence (first 23 amino acids), are totally conserved between the two spiralins.

5. SECONDARY STRUCTURE AND TOPOLOGY OF SPIRALIN IN THE SPIROPLASMA MEMBRANE

According to computer analyses of secondary structures, spiralin contains three regions with α-helix potential. Only the putative α-helix involving residues

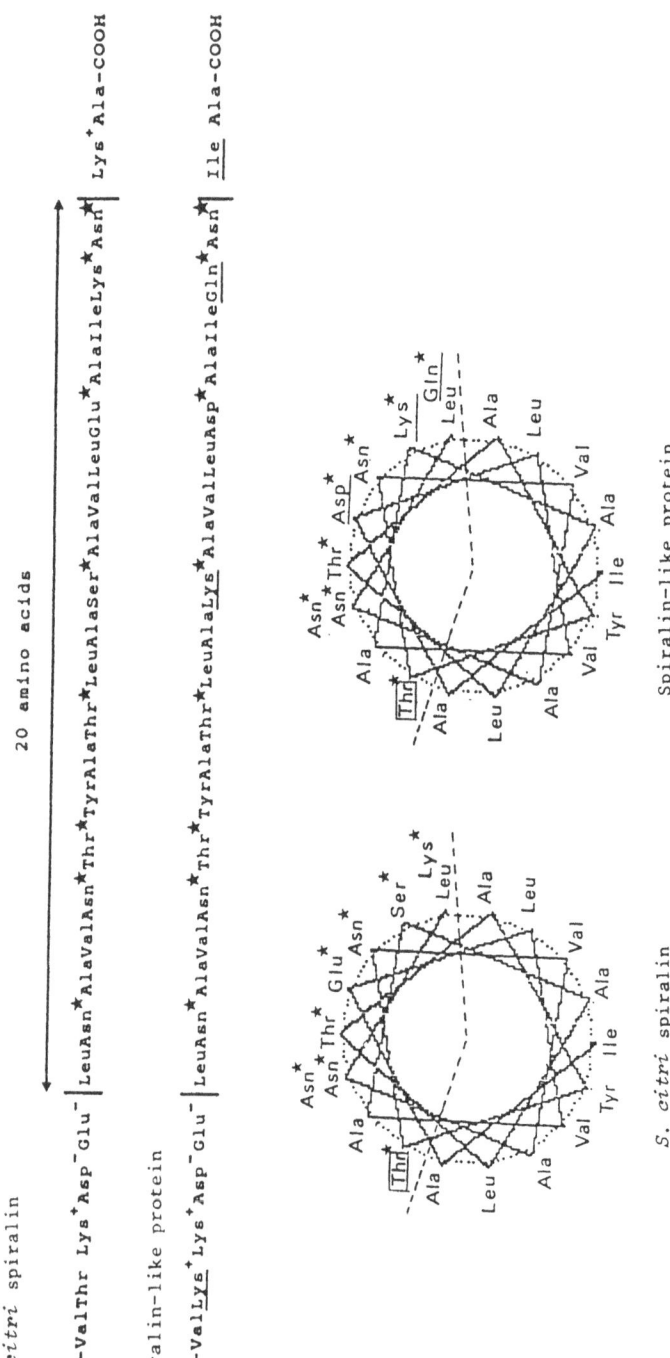

FIGURE 2. Comparison of the putative transmembrane α-helices of *S. citri* spiralin and *S. melliferum* spiralin. (A) Amino acid sequences. The α-helix region is indicated by the arrows. (B) Axial projections of the α-helices. Hydrophilic (polar or charged) amino acids in the helices are indicated by asterisks. Amino acids that are different in spiralin from the two species are underlined. (From Chevalier *et al.*, 1990b.)

162 to 181 has the size required for membrane spanning and could represent the transmembrane domain of spiralin, which is well known to be an integral membrane protein. Axial projection of this α-helix (Figure 2) shows that the distribution of the amino acid residues determines a hydrophobic face and a hydrophilic face, involving two-thirds and one-third of the α-helix surface, respectively. The amphiphilic nature of the α-helix suggests that several spiralin molecules assemble in the membrane to form a homooligomer in such a way that the hydrophobic faces of their α-helices are on the outside of the oligomer, turned toward the lipid bilayer, while the hydrophilic domains are turned toward the inside of the oligomer (Figure 3), away from the hydrophobic environment of the lipid bilayer. This interpretation agrees with biochemical data on spiralin. Indeed, spiralin seems to occur as dimers, tetramers, and large oligomers in the spiroplasma membranes. Disulfide bonds involving cysteine have been implicated in the formation of oligomers. While this might be so in the case of spiralin which contains two Cys residues, it does not seem to apply to the spiralin of *S. melliferum* BC3[T], which contains only one Cys, or to that of strain B88, which lacks Cys. It is also known that the spiralin of *S. melliferum* B88 can be acylated with palmitic acid through an ester linkage involving the hydroxyl group of Ser or Thr residues. The precise position of the involved residue in the molecule is, however, unknown. (See also Section 7.) The boxed Thr residue shown in the α-helix projection in Figure 2 could be the acylation target. The covalently bound acyl chain would help to anchor the α-helix, and hence the whole spiralin molecule, in the lipid bilayer. Strengthening of spiralin anchoring in the spiroplasma membrane through acylation has been proposed.

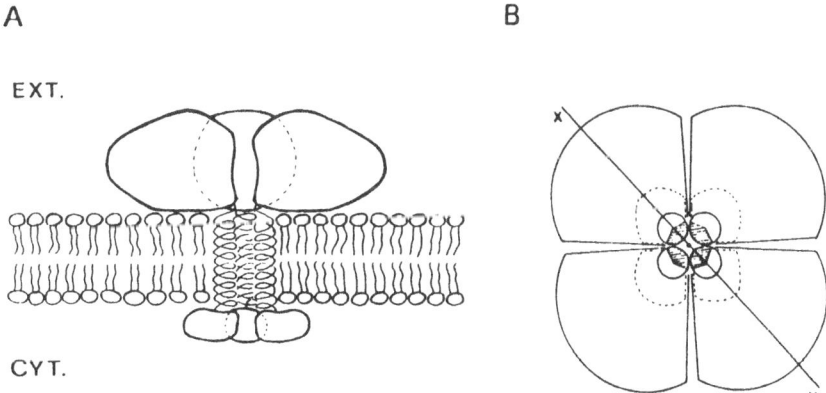

A   B

FIGURE 3. Topological model of spiralin in the cell membrane of spiroplasma. The C-terminus of spiralin is part of the small internal domain. A lipid component on the N-terminus of spiralin might be inserted into the external side of the spiroplasma membrane (not shown). Panel A is a section of panel B along the XY line. EXT., exterior of the cell; CYT., cytoplasm. (From Chevalier *et al.*, 1990b.)

6. SPIRALIN AND PRESPIRALIN

As shown by the hydropathy profile (Chevalier *et al.*, 1990b), the putative transmembrane α-helix is located between two hydrophilic domains, a large N-terminal one (residues 21 to 161) and a small C-terminal one (residues 182 to 241). The large domain is thought to be external for the following reasons. It is linked to the N-terminal hydrophobic end (residues 1 to 20). This hydrophobic end, strongly conserved between spiralin-Sc and spiralin-Sm, is probably required to initiate translocation of the nascent polypeptide chain through the membrane, followed by translocation of the large hyrophilic domain until the putative transmembrane α-helix anchors the protein in the membrane. In this way, the small C-terminal domain remains on the cytoplasmic face of the membrane. Biochemical and immunobiological data suggest a model in which two hydrophilic domains protrude on the two faces of the membrane.

Does the N-terminal hydrophobic end of spiralin represent a cleavable signal (leader) sequence? If so, the translational product of the spiralin gene would be a prespiralin and the actual membrane-bound spiralin in spiroplasma would result from maturation of the prespiralin by removal of the signal sequence. The most direct approach to examine this possibility would be to sequence the N-terminal end of spiralin as purified from spiroplasma membranes. Unfortunately, this N-terminal end is blocked and sequencing is not feasible. Therefore, the Marchalonis and Weltman (1971) procedure was used (Le Hénaff *et al.*, 1991). In this technique, the amino acid sequence of a protein as deduced from the nucleotide sequence of its gene is fictitiously shortened by 1 to 50 residues from each terminus and the compositions thus obtained are compared with that of the purified protein. When this method was applied to spiralin, the best fit between the compositions of purified spiralin and the fictitiously shortened spiralin was obtained when the first 24 amino acids were subtracted from the N-terminal side of the protein. These N-terminal amino acids would thus represent a signal sequence with the cleavage site close to cysteine-24. The structure of this signal sequence fits the basic design of signal peptides as it contains an essential apolar core of 17 residues (leucine-4 to serine-20) flanked by a short basic sequence (Met-Lys-Lys) on the N-terminal site. The stretch Val-Val-Ala-Cys on the C-terminal end of the apolar core is to some extent evocative of the consensus sequence Leu-(Ala or Ser)-(Gly or Ala)$^{-1}$-Cys^{+1} of bacterial lipoprotein modification/processing site (Hayashi and Wu, 1990) including the cleavage site sequence Ala-Ile-Ser^{-1}-Cys^{+1} of the *Mycoplasma hyorhinis* p37 protein.

Several forms of spiralin are produced in *E. coli* with molecular weights ranging from 28,000 to 30,500, as determined by PAGE (Mouchès *et al.*, 1985). It was suggested as early as 1985 that these forms might represent a variety of posttranslational modifications, the 30,500 protein being a prespiralin with a signal (leader) sequence and the 28,000 protein lacking this sequence. In light of the preceding discussion, this suggestion gains considerable weight.

7. POLYMORPHISM OF SPIRALIN

When the total proteins of several *S. citri* strains are submitted to SDS-PAGE, spiralin does not show the same electrophoretic mobility (EM) for all strains analyzed (Figure 4). The strains can be grouped according to the EM of spiralin. The spiralins of strains Israel, Asp 1, Alc 254, 78, M4, MH2, and Corse (group 1) have the highest EM (Figure 4). Next comes group 2 with strains R8A2 HP, C189, and Hinckley, followed by strain R8A2B (group 3) and finally strain Palmyre (group 4) whose spiralin has the lowest EM. These results show that *S. citri* spiralins of different EM occur.

Furthermore, in Western blots, monoclonal antibodies against spiralin of *S. citri* Israel, a strain of EM group 1, detect not only spiralin of the homologous Israel strain but also spiralins of the other strains of EM groups 1, 2, and 3. However, there is no reaction with the spiralin of the Palmyre strain of EM group 4. Antiserum against total *S. citri* proteins reacts with all spiralins tested. Thus, when analyzed by SDS-PAGE and Western blotting, spiralin shows polymorphism.

To discover the reasons for these differences at the molecular level, we have sequenced the spiralin gene of the following *S. citri* strains: R8A2B (group 3), Corse, Asp 1, Alc 254, 78 (group 1) and Palmyre (group 4). The spiralin

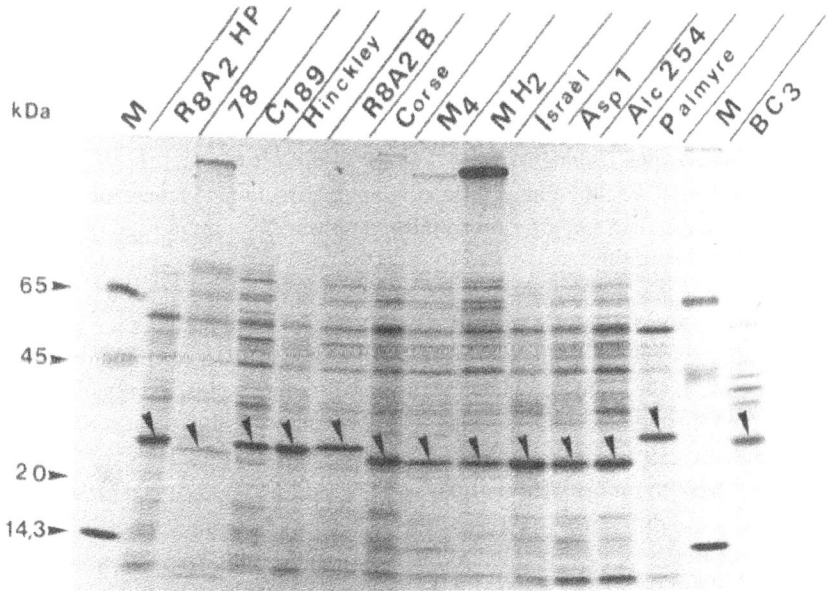

FIGURE 4. PAGE Pattern of SDS-solubilized cellular proteins from 12 *S. citri* strains and strain BC3 of *S. melliferum*. Spiralin is indicated by arrowheads. M, molecular mass marker.

sequences of *S. citri* strain R8A2 HP (group 2) taken as the reference strain and of *S. melliferum* strain BC3 were known from previous work (Chevalier *et al.*, 1990a,b). Sequencing was carried out directly on the PCR amplified DNA of the spiralin gene with primers D and D′ encompassing the whole gene (Figure 5).

Comparison of these sequences has shown that the spiralins (or rather pre-spiralins) of all *S. citri* strains tested have the same number of amino acids: 241, except strain Palmyre with 242 amino acids. The first 35 amino acids at the N-terminus are fully conserved except for Phe-11 which is replaced by another hydrophobic amino acid, leucine, in strain Palmyre (Table II). In particular, all of the proteins have cysteine in position 24. Hence, the signal sequence discussed above (see Section 6) in the case of *S. citri* strain R8A2 HP is highly conserved; it occurs in all prespiralins examined, and is probably cleaved at the same site in all prespiralins, so as to leave the same amino acid, cysteine-24, as the N-termi-nus of the mature spiralins. From here on, we will use the term *prespiralin* for the primary translational product of the spiralin gene, and *spiralin* for the putative mature product.

All *S. citri* prespiralins studied including those of R8A2 and C189 have two cysteine residues, one at position 24 and the second at position 197. Yet, only one cysteine residue is detected chromatographically in spiralin purified from *S. citri*. *N*-acylation of the N-terminal cysteine would render this cysteine residue undetectable. Only cysteine 197 would be detected.

The sequence of the spiralin gene of *S. melliferum* strain BC3 is fully identical to that of *S. citri* over the first 32 N-terminal amino acids. In particular, there is a cysteine at position 24 which could become, as in *S. citri* spiralin, the N-terminus of the mature protein. Blanchard (1986) did not detect cysteine in spiralin purified from *S. melliferum* BC3. This is probably because the putative N-terminal cysteine residue in *S. melliferum* spiralin is blocked, by *N*-acylation for instance, and would escape chromatographic detection. As there are no other cysteine residues in *S. melliferum* spiralin, it appears as if there is no cysteine in *S. melliferum* spiralin.

Since, at the N-terminal end, the amino acid sequence of all *S. citri* pre-spiralins examined is identical well beyond cysteine-24, the putative cleavage region, it is reasonable to assume that the same signal sequence is removed from all spiralins. If the signal sequence measures 23 amino acids, its removal from the prespiralins would yield spiralins with 218 amino acids except for Palmyre spiralin with 219 amino acids. Thus, the spiralins (except Palmyre spiralin) would all have the same size, and the same *N*-modified N-terminal amino acid. Therefore, the differences in EM shown by these spiralins cannot be due to differences in size or modification of the N-terminus. Palmyre spiralin, however, with one additional amino acid residue (proline-240, Table II) would be expected to have a lower EM, as is indeed the case (Figure 4).

Table II shows that in comparison with the amino acid sequence of pre-

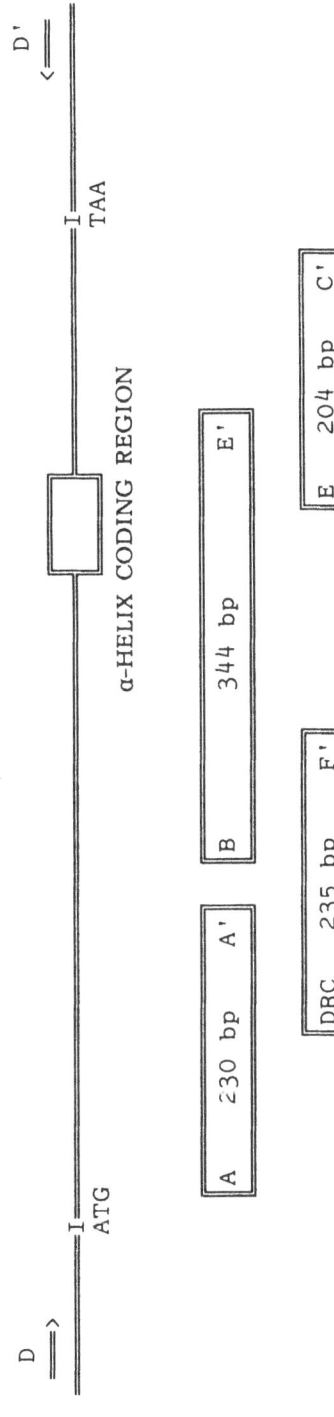

FIGURE 5. Position of the amplified probes AA′, BE′, DRCF′, and EC′ on the *S. citri* spiralin gene. Primers D and D′ are indicated by arrows. D is located upstream of the spiralin gene promoter and D′ is at the beginning of the phosphofructokinase ORF.

Table II
Amino Acid Changes in Prespiralin of Different *S. citri* Strains

Position in prespiralin	11	36	39	40	54	67	79	90	92	100	101	122	139	143	196
R8A2 HP	Phe	Lys	Ala	Val	Asn	Lys	Ala	Lys	*Thr*	Glu	Glu	Glu	Gly	Asp	Asn
R8A2 B	—	—	—	Ala	—	—	—	—	—	—	Gly	—	—	—	—
Corse	—	—	—	Ala	—	—	—	—	—	—	Gly	—	—	—	—
Asp 1/Alc 254	—	—	—	Ala	—	—	—	Glu	—	Gln	Gly	Lys	*Ser*	Asn	—
78	—	—	Val	Ala	—	—	—	Glu	—	Gln	Gly	Lys	*Ser*	Asn	—
Palmyre	Leu	Asn	—	Ala	Lys	Gln	Asp	—	Glu	—	—	—	—	—	Asp

Position in prespiralin	198	199	200	201	204	207	217	227	234	235	236	237	238	239	240[a]
R8A2 HP	Asp	Ala	Gly	Asp	Ala	Asp	Glu	Val	*Thr*	Lue	Ala	Pro	Pro	Lys	
R8A2 B	—	—	Glu	—	—	—	—	—	—	—	—	—	—	—	
Corse	—	—	—	—	—	—	—	—	—	—	—	—	—	Asn	
Asp 1/Alc 254	—	—	—	Asn	Glu	—	—	—	Ile	—	—	—	Ala	Asn	
78	—	—	—	Asn	Glu	—	—	—	Ile	—	—	—	Ala	Asn	
Palmyre	Asn	Glu	—	—	*Thr*	Asn	Lys	Phe	Ile	Asn	Lys	—	Val	*Thr*	Pro

[a] Amino acid present only in Palmyre spiralin.
Italics indicate an amino acid that can be acylated.

spiralin R8A2 HP, strain R8A2 B (group 3) has a prespiralin with only 1 amino acid change, namely at position 200 (Gly → Glu). There are 3 changes for strain Corse (group 1), 12 for strains Asp 1 and Alc 254 (group 1) (two strains found to have identical prespiralin sequences), 13 for strain 78 (group 1), and 22 for strain Palmyre (group 4). However, the number of these changes does not seem to be the only reason for spiralin polymorphism since the EM of strain Corse spiralin with 3 changes is very similar to that of strains Asp 1/Alc 254 with 12 changes and strain 78 with 13 changes. What seems more important is the nature of the change and especially the changes that affect amino acids that can be acylated: serine, threonine, and cysteine. Regarding cysteine, there are no differences between the prespiralins: all have two Cys residues, one at position 24 and one at position 197, i.e., position 1 and 218 in the putative spiralin. As for serine, all prespiralins have the same serine residues as the reference prespiralin R8A2 HP. However the spiralins of strains Asp 1/Alc 254 and 78 have an additional Ser residue instead of a Gly at position 139 (Table II). Regarding threonine, the prespiralins of strains R8A2 B and Corse are identical to strain R8A2 HP spiralin. However, strains Asp 1/Alc 254 and 78 have lost a Thr residue at position 234 and have an Ile instead. Finally, spiralin of strain Palmyre has the same number of Thr residues as R8A2 HP spiralin; it has lost Thr-92 and -234 for Glu and Ile, respectively, but it has gained two Thr residues, one at position 204 in replacement of an Ala residue, and one at position 239 in replacement of a Lys residue.

It has recently been shown that M2 protein of influenza A virus is acylated posttranslationally with palmitate (Veit *et al.*, 1991). When analyzed by SDS-PAGE, the acylated protein had a lower EM than the unacylated protein. This result indicates that acylation can have a profound effect on protein migration in SDS-PAGE, and could be involved in spiralin polymorphism.

At this time we do not know where internal acylation of the various spiralins occurs. The acylation could also be affected by modifications of the spatial arrangement of spiralin as a result of amino acid changes other than those affecting serine and threonine (see Table II). These changes could also affect exposure of epitopes to monoclonal antibodies. It is thus urgent to determine the precise extent of spiralin acylations.

Finally, lipid modification of the N-terminus of spiralin, possibly a cysteine residue, could be involved in attaching the N-terminus of spiralin on the external site of the spiroplasma membrane.

8. SPIRALINS IN SPIROPLASMA SPECIES OTHER THAN *S. CITRI* AND *S. MELLIFERUM*

Evidence for the presence of a spiralin gene in the spiroplasma serovars of Table III was obtained in the following way. Primers D and D' (Figure 5)

Table III
Intensity Results Obtained by Hybridization between DD' PCR Products and the Four Amplified Probes

	A/A'	DRC/F'	B/E'	E/C'
S. citri R8A2HP group I-1	+++[a]	+++	+++	+++
S. melliferum BC3 group I-2	+++	+++	+++	ND
S. kunkelii E275 group I-3	+	−	+++	−
Spiroplasma species 277F group I-4	++	+	+++	+++
Spiroplasma species LB12 group I-5	++	+	+++	++
Spiroplasma species M55 group I-6	++	+++	+++	+++
Spiroplasma species N525 group I-7	++	+	+++	++
S. phoeniceum P40 group I-8	+	−	++	−
S. mirum SMCA group V	+++	++	+++	+
Spiroplasma species EA1 group VIII	++	+	+++	+++
Spiroplasma species DF1 group XVII	+++	++	+++	+++

[a] +++ indicates a hybridization intensity equal to that obtained with *S. citri* strain R8A2HP. Lower intensities are represented by ++ and +. No hybridization is indicated by −. ND, not determined.

encompassing the spiralin gene of *S. citri* R8A2 HP were used in a PCR reaction. For all spiroplasmas tested, the amplified DNA had a size close to 1030 bp, the size expected and observed for *S. citri* R8A2 HP amplified DNA. To examine the sequence homology of the amplified DNAs with the spiralin gene, the four probes indicated in Figure 5 were used in Southern hybridizations with the amplified DD' DNAs. The four probes covered the whole spiralin gene and were obtained by PCR using specific primers (A–A', B–E', DRC–F', E–C') with *S. citri* R8A2 HP DNA. All amplified DD' DNAs hybridized with all four probes except the DNAs of *S. kunkelii* and *S. phoeniceum* which were recognized by only two probes. The highest homologies to *S. citri* R8A2 HP DNA were obtained with the DNAs of *S. melliferum* (group I-2), *Spiroplasma* sp. strain M55 (group I-6), and *Spiroplasma* sp. strain DF1 (group XVII). Good hybridizations were also obtained in the case of *Spiroplasma* sp. strain 277F (group I-4), *S. mirum* (group V), *Spiroplasma* sp. strain EA1 (group VIII), *Spiroplasma* sp. strain LB12 (group I-5), and *Spiroplasma* sp. strain N525 (group I-7). Unexpectedly, *S. kunkelii* (group I-3) and *S. phoeniceum* (group I-8) gave the lowest hybridizations. These two spiroplasmas are in serogroup I to which also *S. citri* (group I-1) belongs, and like *S. citri*, they are plant pathogens.

In all, these results show that all of the spiroplasmas studied have a gene with more or less homology to the spiralin gene of *S. citri*. We conclude that these spiroplasmas, like *S. citri* and *S. melliferum*, have spiralin genes.

9. CONCLUSION

Spiralin is the most abundant membrane protein of *S. citri*. A spiralin protein is also present in *S. melliferum*. The spiralin genes of these two spiroplasmas have been identified and sequenced; they show considerable sequence homology. Several other spiroplasmas have been examined and shown to have sequence homology with the spiralin gene of *S. citri*. Spiralin is probably characteristic of all members of the genus *Spiroplasma*. The function of spiralin is still unknown, even though study of the protein itself has progressed. Spiralin as seen in *S. citri* and *S. melliferum* is an amphiphilic, integral membrane protein. The primary translational product is probably a prespiralin with a hydrophobic N-terminal end of which the first 23 amino acids up to Cys 24 represent a signal (leader) sequence required to guide and insert the protein into the spiroplasma membrane. The leader sequence is highly conserved in all strains of *S. citri* examined and is the same in *S. citri* and *S. melliferum*. Cleavage of the signal sequence is seen to occur close to Cys 24. The spiralin gene is expressed in *E. coli*. The protein product from *E. coli* has a lower EM than the protein from *S. citri*. This suggests that the protein seen in *E. coli* is prespiralin (241 amino acid residues) and that found in *S. citri* is mature spiralin ($\approx$ 218 residues).

Mature spiralin is anchored in the spiroplasma membrane by an α-helix and has a small C-terminal hydrophilic domain on the internal side of the membrane and a large, hydrophilic N-terminal domain on the external side. The putative lipid component of the N-terminus could be inserted in the external side of the membrane and involved in maintaining the large hydrophilic domain in position. Axial projection of the α helix reveals a hydrophilic face involving one-third of the α-helix surface and a hydrophobic face over two-thirds of the surface. The amphiphilic nature of the α-helix suggests that several spiralin molecules assemble in the membrane to form a homooligomer in such a way that the hydrophobic faces of their α-helix are on the outside of the oligomer, turned toward the lipid bilayer, while the hydrophilic domains are turned toward the inside of the oligomer, away from the hydrophobic environment of the lipid bilayer. The N-terminal amino acid of mature spiralin after removal of the leader sequence could well be Cys-24 of prespiralin. It is known that the N-terminus of spiralin is blocked. Since spiralin is known to be acylated, *N*-acylation of the terminal cysteine could be responsible for blocking. The spiralin genes from several strains of *S. citri* have been sequenced. Up to 15 amino acid changes have been detected between the spiralin of reference strain R8A2 HP and the spiralin of the various strains. How these changes are involved in spiralin polymorphism and affect properties such as electrophoretic mobility and reactivity to antibodies is under study.

10. REFERENCES

Archer, D. B., and Townsend, R., 1981, *J. Gen. Microbiol.* **123:**61–68.

Bhakdi, S., Bhakdi-Lehnen, B., and Bjerrum, O. J., 1977, *Biochim. Biophys. Acta* **470:**35–44.

Blanchard, A., 1986, Thèse de l'Université de Rennes.

Bové, J. M., Carle, P., Garnier, M., Renaudin, R., and Saillard, C., 1989, in: *The Mycoplasmas,* Volume V (J. G. Tully and R. F. Whitcomb, eds.), Academic Press, New York, pp. 243–365.

Chevalier, C., Saillard, C., and Bové, J. M., 1990a, *J. Bacteriol.* **172:**2693–2703.

Chevalier, C., Saillard, C., and Bové, J. M., 1990b, *J. Bacteriol.* **172:**6090–6097.

Dahl, C. E., and Dahl, J. S., 1984, *J. Biol. Chem.* **259:**10771–10776.

Dahl, C. E., Dahl, J. S., and Block, K., 1983, *J. Biol. Chem.* **258:**11814–11818.

Dahl, C. E., Sacktor, N. C., and Dahl, J. S., 1985, *J. Bacteriol.* **162:**445–447.

Fontenelle, C., Zaaria, A., Blot, M. F., and Wroblewski, H., 1987, *Rev. Inst. Pasteur Lyon* **20:**127–132.

Grau, O., Laigret, F., Carle, P., Tully, J. G., Rose, D. L., and Bové, J. M., 1991, *Int. J. Syst. Bacteriol.* **41**(4):473–478.

Hayashi, S., and Wu, H. C., 1990, *J. Bioenerg. Biomembr.* **22:**451–471.

Helenius, A., and Simons, K., 1977, *Proc. Natl. Acad. Sci. USA* **74:**529–532.

Le Hénaff, M., Brenner, C., Fontenelle, C., Delamarche, C., and Wroblewski, H., 1991, *C. R. Acad. Sci. Ser. III* **312**(5):189–195.

Marchalonis, J. J., and Weltman, N., 1971, *Comp. Biochem. Physiol.* **38B:**609–625.

Mouchès, C., and Bové, J. M., 1983, *Yale J. Biol. Med.* **56:**723–727.

Mouchès, C., Vignault, J. C., Tully, J. G., Whitcomb, R. F., and Bové, J. M., 1979, *Curr. Microbiol.* **2:**69–74.

Mouchès, C., Candresse, T., McGarrity, G. J., and Bové, J. M., 1983, *Yale J. Biol. Med.* **56:**431–437.

Mouchès, C., Candresse, T., Barroso, G., Saillard, C., Wroblewski, H., and Bové, J. M., 1985, *J. Bacteriol.* **164:**1094–1099.

Nystrom, S., Johansson, K. E., and Wieslander, A., 1986, *Eur. J. Biochem.* **156:**85–94.

Renaudin, J., Pascarel, M. C., Saillard, C., Chevalier, C., and Bové, J. M., 1986, *C. R. Acad. Sci. Ser. III* **303**(13):539–540.

Simons, K., Helenium, A., Leonard, K., Sarvas, M., and Gething, M. J., 1978, *Proc. Natl. Acad. Sci. USA* **75:**5306–5310.

Townsend, R., and Plaskitt, K. A., 1985, *J. Gen. Microbiol.* **131:**983–992.

Veit, M., Klenk, H. D., Kendal, A., and Rott, R., 1991, *J. Gen. Virol.* **72:**1461–1465.

Whitcomb, R. F., Tully, J. G., and Wroblewski, H., 1983, *Curr. Microbiol.* **9:**7–12.

Williamson, D. L., Renaudin, J., and Bové, J. M., 1991, *J. Bacteriol.* **173:**4353–4362.

Wroblewski, H., 1978, *Zentralbl. Bakteriol. A* **241:**179–180.

Wroblewski, H., 1979, *J. Bacteriol.* **140:**738–741.

Wroblewski, H., 1981, *J. Bacteriol.* **145:**61–67.

Wroblewski, H., Johansson, K. E., and Hjerten, S., 1977, *Biochim. Biophys. Acta* **465**:275–289.

Wroblewski, H., Robic, D., Thomas, D., and Blanchard, A., 1984, *Ann. Microbiol. (Inst. Pasteur)* **135A**:73–82.

Wroblewski, H., Blanchard, A., Nystrom, S., Wieslander, A., and Thomas, D., 1987, *Isr. J. Med. Sci.* **23**:439–441.

Wroblewski, H., Nystrom, S., Blanchard, A., and Wieslander, A., 1989, *J. Bacteriol.* **171**:5039–5047.

Wu, H. C., and Tokunaga, M., 1986, *Curr. Top. Microbiol. Immunol.* **125**:127–156.

Yamao, F., Muto, A., Kawauchi, Y., Iwani, M., Iwagami, S., Azumi, Y., and Osawa, S., 1985, *Proc. Natl. Acad. Sci. USA* **82**:2306–2309.

Zaaria, A., Fontenelle, C., Le Hénaff, M., and Wroblewski, H., 1990, *J. Bacteriol.* **172**:5494–5496.

Chapter 8

Adherence of Mycoplasma to Cell Surfaces

Itzhak Kahane and Shulamith Horowitz

1. INTRODUCTION

Many of the 114 mycoplasma species known to date (Tully, 1993) have been found to adhere to host cells. The majority of the species are parasites and pathogens.

Since evidence has been provided for at least several pathogenic mycoplasmas (e.g., *M. pneumoniae, M. pulmonis, M. gallisepticum*) that their adherence is a prerequisite for their pathogenicity, it is not surprising that adherence has been quite extensively studied in the past three decades. The thrust was fueled by the rationale that by revealing the nature of the adherence, better ways would be found for protection from infection by inhibition of adherence and also by development of effective vaccine.

Mycoplasmas are the smallest and simplest free-living organisms (Razin, 1987). It could, therefore, be predicted that their adherence will, at the most, resemble that of the more specialized eubacteria such as *Escherichia coli* or *Pseudomonas aeruginosa*. In these, the adherence is a direct recognition of the

Itzhak Kahane   Department of Membrane and Ultrastructure Research, The Hebrew University–Hadassah Medical School, Jerusalem 91010, Israel.   **Shulamith Horowitz**   Mycoplasma Laboratory, Department of Microbiology and Immunology, Ben Gurion University of the Negev, and Soroka Medical Center, Beersheva 84105, Israel.

Subcellular Biochemistry, Volume 20: Mycoplasma Cell Membranes, edited by Shlomo Rottem and Itzhak Kahane. Plenum Press, New York, 1993.

bacterial adhesins to carbohydrate moieties on the host membrane receptors. With quite a surprise, we can summarize the many studies of mycoplasma adherence by stating that this is not the case. The findings indicate that: (1) mycoplasmas do not use a single type of mechanism for adherence; (2) many times, more than one type of mycoplasma membrane component is involved in the adherence; and, moreover, (3) it was demonstrated most recently that at least some mycoplasmas cannot be considered only surface parasites, adhering to the outer surface of the host cell, but some adhere and later penetrate through the membrane into the host cell and probably survive there.

2. MORPHOLOGICAL STUDIES

Morphological studies of the adherence of mycoplasmas were conducted on a large variety of mycoplasmas interacting with cells and tissues, both *in vivo* and *in vitro*. The majority of these studies have been reviewed (Barile, 1979; Bredt *et al.*, 1981; Razin *et al.*, 1981; Razin, 1985), and therefore only the conclusions will be stated and the more recent findings added. The studies were conducted both by regular or fluorescence microscopy and by electron microscopy (both SEM and TEM) and most of them revealed that the mycoplasmas are membrane parasites avidly adhering to the outer surface of the host cell membrane. Ultra-structure studies using TEM mostly revealed that despite the very intimate association of the mycoplasmas and the host cells, their membranes were easily distinguished. Studies in which SEM was used added another touch, as if etching was caused by the mycoplasmas adhering to the host cells. A complementary aspect was recently reported by Robertson and Sherburne in their study of hemadsorption by colonies of *Ureaplasma urealyticum*. They showed guinea pig erythrocytes embedded in ureaplasma colonies and craters left when erythrocytes were dislodged (Robertson and Sherburne, 1991). In only a few cases were there some indications for fusion of the mycoplasmas with the host cell. One interesting example for this is given in the studies of Lindsey and Cassell (1973) on *M. pulmonis,* where the organisms are in close proximity to areas resembling coated pits. We have studied adhering *M. pneumoniae* and, in several hundred serial sections, have not encountered such a phenomenon (Kahane, Saada, and Kessel, unpublished results). In studies of the well-defined asymmetrical, flask-shaped mycoplasmas (e.g., *M. pneumoniae*) or the fish mycoplasma (*M. mobile*), mostly the tip area, which is the leading head when these microorganisms glide, was found to be the area by which the mycoplasmas adhere to the host cell. However, some mycoplasmas were adhering via other areas of their membranes. The almost classical figure of mycoplasma (*M. pneumoniae*) adhering to the membrane of the host cell squeezed between two cilia, as reported by Collier

(1979) and others, suggested that only the membrane can be the site of interaction. This view should be slightly modified, and cilia should also be considered to have receptors for mycoplasmas. The data for this derive from two sources. First are studies conducted by Loveless and Feizi (1989) on the human bronchial epithelium, which is the primary site of infection of *M. pneumoniae*. They have used MAb and lectins specific to the sialo-oligosaccharides of the receptor to which *M. pneumoniae* adhere. These probes mostly labeled the apical and microvillar border and the cilia. On the other hand, the secretory and mucus lacked them (Loveless and Feizi, 1989). The second body of evidence for this point is that two mycoplasmas—*M. equigenitalium* when studied with the equine uterine tube (Bermudez *et al.*, 1980) and *U. urealyticum* in adherence studies using bovine Fallopian tube mucosal cells in culture (Saada *et al.*, 1993)—were found to adhere both to the cilia and to the membrane. Since these studies were conducted on cells of the female genital organs, it may be that such adherence is a property of these cells, and this and the nature of the receptors have to be further studied. The recent studies of the pathogenic *M. fermentans* (incognitus strain) produced a considerable amount of information on the morphological and ultrastructural aspects of this organism's adherence and infection. Lo and his colleagues employed immunohistochemical methods in thymus, liver, spleen, or brain from 22 patients with AIDS as well as two placentas delivered by patients with AIDS (Lo *et al.*, 1989) and revealed characteristic structures of mycoplasma organisms. More recently, in patients with AIDS-associated nephropathy, mycoplasma-like structures were detected by electron microscopy and monoclonal antibodies in the patients' kidneys. There was no evidence of the incognitus strain mycoplasma infection in renal parenchymal cells in AIDS patients with normal renal histology (Bauer *et al.*, 1991).

Penetration of the cells by *U. urealyticum* was also seen when we studied their adherence and pathogenicity in an *in vitro* model using bovine Fallopian tube epithelial cells (Saada *et al.*, 1993).

The morphological studies highlighted some major questions:

1. In light of the observation that all flask-shaped, pathogenic mycoplasmas possess gliding motility, are they attracted to the host cells by chemotaxis?
2. What are the components on the membranes of the mycoplasmas and the host cell membranes which are involved in the adherence process?
3. Does membrane fusion occur during adherence?
4. How do the mycoplasmas that are intracellular penetrate the cells?
5. How do mycoplasmas reach the host cell membrane, especially since it is covered most of the time by a layer of mucus which is moved by the cilia?

3. ADHERENCE: THE MOLECULAR ASPECTS

The Host Cell Receptors

Several general approaches were used for the study of the host cell receptors. Obviously, many studies were conducted using combinations of the approaches:

1. Enzymatic treatment of the host cell membranes with proteases, lipases, or glycosidases in order to abolish or alter the receptor. A complementary approach was to first eliminate the receptor and later to let it resynthesize.
2. Competition experiments. In these, by using an analogue of the receptor, the binding of the mycoplasmas to the host cell is blocked. The analogues used were ligands or specific antibodies or, much preferred, their Fab fragments.
3. Binding of isolated receptors to mycoplasma membranes or vice versa. Binding of the membranes or soluble membrane components to isolated receptor that is bound to solid surface.

Sialoglycoconjugate Receptors

The initial studies were conducted by Thomas's group, who treated erythrocytes with neuraminidase (at that time called receptor destroying enzyme). Such treatment reduced *M. gallisepticum* and *M. pneumoniae* adherence to the RBC, indicating the involvement of sialic acid of the RBC in the adherence (Gesner and Thomas, 1966). Similar studies were conducted by several groups (for review, see Kahane and Marchesi, 1973).

So far, the most detailed experiments on the nature of the sialoreceptor were conducted using *M. pneumoniae*. Based on restoration of binding to neuraminidase-treated erythrocytes using CMP-sialic acid and purified sialyltransferases, adhesion of the erythrocytes on surface-grown sheet cultures of *M. pneumoniae* specifically recognized sialic acid linked $\alpha(2-3)$ to *N*-acetyllactosamine sequences (Loomes *et al.*, 1984). This was supported by Roberts *et al.* (1989), using several purified glycoproteins, including laminin, fetuin, and human chorionic gonadotropin and saturable adhesion of *M. pneumoniae* when adsorbed to plastic. The findings of Roberts *et al.* (1989) that adhesion to the proteins is energy linked supported earlier findings of Kahane's group (Kahane *et al.*, 1982).

Several other mycoplasmas adhere to sialoglycoconjugates of the host cell receptors. These include *M. genitalium*, *M. synoviae*, *M. gallisepticum* (as reviewed by Razin, 1985), and *U. urealyticum*, as indicated by our recent studies (Saada *et al.*, 1991). The sialoglycoproteins α-acid glycoprotein and fetuin interfered less than glycophorin in the adherence of *U. urealyticum* to human RBC.

The linkage of sialic acid in the glycoconjugate is important, as indicated in the above-mentioned and other experiments. The end result is that the linkage is not the same even for the mycoplasmas which recognize sialoreceptors. Moreover, even this group, with its heterogeneity for sialoglycoconjugates on the host cell receptors, may also recognize other sites. Among them are sulfatides (Krivan *et al.*, 1989; Saada *et al.*, 1991).

Sulfatide Receptors

Using a virulent strain of *M. pneumoniae*, Krivan *et al.* (1989) found that these organisms adhered strongly to sulfatide and other sulfated glycolipids, but did not bind to various sialoglycolipids. In a recent study, we found that *U. urealyticum* shares this property with *M. pneumoniae* (Saada *et al.*, 1991). One wonders about other adhering mycoplasmas.

Other Receptor Sites

The studies also indicate that other types of recognition may occur between mycoplasmas and host cells. These include proteins, e.g., in the adherence of *M. salivarium* or *M. hominis* and glycoproteins, e.g., in *M. hyorhinis,* and sites of hydrophobic nature, e.g., in *M. pulmonis*. With the latter, an interesting model was suggested by Minion *et al.* (1984) in which, during the initial stages, host membrane components are moved aside to expose the hydrophobic surface to which mycoplasmas adhere. This mechanism may also take place in adherence of other mycoplasmas, even those which initially adhere to sialoglycoconjugates, since with both *M. gallisepticum* and *M. pneumoniae* the adherence to such sites is reversible in the initial stages and later is irreversible (Kahane *et al.*, 1982).

The finding that mycoplasmas may adhere via a variety of host cell receptors was also clearly demonstrated in the experiments of Yayoshi *et al.* (1984) in which mutants of *M. pneumoniae* were formed by *N*-methyl-*N*-nitro-nitrosoguanidine. One of these lost its ability to adhere to RBC, but retained its adherence to lung cells.

Geary *et al.* (1990) have indicated that human lung fibroblasts possess a 100-kDa receptor which is an asialoglycoprotein and is common for *M. pneumoniae, M. genitalium,* and *M. gallisepticum,* but not for *M. pulmonis*. Its counterparts in the mycoplasmas will be discussed next.

4. THE MOLLICUTES COUNTERPARTS: ADHESINS

Being membrane proteins, characterization of the adhesins was a slow process and encountered many difficulties when only "classical" methods, including membrane solubilization and affinity chromatography, were available.

The rapid progress of molecular genetics (Klinkert *et al.*, 1985; Inamine *et al.*, 1988; for reviews, see Baseman, this volume; Razin and Jacobs, 1992) revolutionized this research despite the problems encountered by the fact that mycoplasmas employ the UGA codon for tryptophan which is usually a stop codon in *E. coli*.

Most studies have focused on P1, the major adhesin of *M. pneumoniae*, and on P140, the P1 analogue adhesin of *M. genitalium*. They are the focus of Baseman's chapter in this volume. Therefore, only their major features which are important for our discussions will be considered here.

M. pneumoniae **P1.**   This adhesin has an affinity for sialoglycoconjugates. It is a product of the P1 operon (Figure 1), and is composed of the *P1* gene (*ORF-5*) flanked by the predicted genes designated *ORF-4* and *ORF-6* (Inamine *et al.*, 1988). Its deduced amino acid sequence was reported by several groups. As for many membrane proteins, P1 has a leading sequence of 59 amino acids and the mature protein has 1560 amino acids. Its estimated molecular weight is 169,750. The major features of its primary structure are that it has no cysteines, and at the carboxy-terminus, 13 of the 26 amino acids are prolines. The extensive studies of Jacobs's group shed light on the possible topology of P1 and on its immunogenic properties and were recently reviewed in depth by Razin and Jacobs (1992). The other open reading frames of this operon have been studied by gene fusion between the NH_2-terminus of the RNA replicase of the *E. coli* bacteriophage MS2 and selected regions of ORF-4 and ORF-6. The corresponding fusion proteins synthesized in *E. coli* were used to immunize mice. Antisera directed against ORF-4-related sequences did not recognize *M. pneumoniae* in Western blot analysis, whereas antisera directed against ORF-6-derived fusion protein reacted with two *M. pneumoniae* proteins of 40 and 90 kDa. In addition, some of the antisera also recognized proteins that formed in SDS-PAGE a protein ladder between 115 and 145 kDa (Sperker *et al.*, 1991). The role of these proteins is not clear, but as we know, P1, which is the major adhesin of *M. pneumoniae*, seems to be associated with some auxiliary proteins and these will be discussed later.

M. genitalium **Adhesin (P140).**   This adhesin, except for having a mass of 140 kDa, much resembles P1 of *M. pneumoniae* and is discussed in great detail by Baseman (this volume).

Adhesins with Affinity for Asialoglycoconjugates.   The studies of Geary *et al.* (1990) using the 100-kDa lung fibroblast receptor, which is an asialoglycoprotein, indicated that this receptor was common for *M. pneumoniae*, *M. genitalium*, and *M. gallisepticum*. The receptor and antireceptor sera were used to probe these organisms for the corresponding binding proteins. A 32-kDa protein of *M. pneumoniae*, a 90-kDa protein of *M. genitalium*, and a 139-kDa protein of *M. gallisepticum* were recognized (Geary *et al.*, 1990).

M. gallisepticum **Adhesins.**   This avian mycoplasma is a respiratory tract

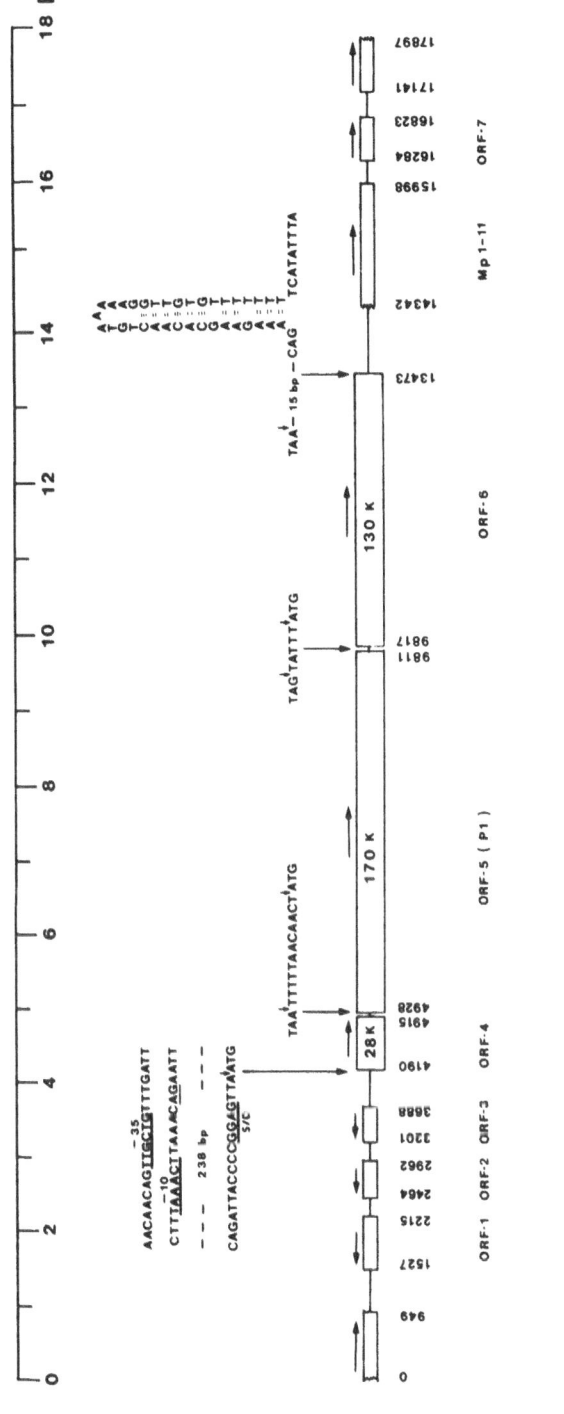

FIGURE 1. Functional map derived from the analysis of the nucleotide sequence of the 18-kb region containing the P1 operon. ORFs are indicated by the open boxes with their limits given by nt positions and the 5'-to-3' orientation shown by arrows. The P1 operon is composed of the *P1* gene (*ORF-5*) flanked by the predicted genes designated *ORF-4* and *ORF-6*; the estimated sizes of the encoded proteins are given within the respective boxes. The limits of the P1 operon are indicated by the horizontal arrows below the map. Relevant flanking and intervening nucleotide sequences are shown above their map locations. The −35, −10, transcription start point, and Shine–Dalgarno (S/D; RBS) sequences are underlined. Vertical arrows mark the beginning and end of the ATG codons and TAA or TAG stop codons, respectively, which delineate the ORFs of the P1 operon. The sequences indicative of a transcription terminator are depicted as a stem-and-loop structure ($\Delta G = -8.6$ kcal/mole); K = kDa. (Reproduced from Inamine *et al.*, 1988, by permission of Elsevier Science Publishers.)

pathogen. Several attempts were made to isolate its adhesin. Kahane *et al.* (1984) obtained, by affinity chromatography, a fraction with a major polypeptide of 75 kDa.

It was mentioned above that *M. gallisepticum* recognized the same asialoglycoprotein receptor as *M. pneumoniae* and *M. genitalium* (Geary *et al.*, 1990). The resemblance of these organisms was further studied by Dallo *et al.* (1990) who reported that P1 antibody reacted in immunoblots with a 155-kDa polypeptide of *M. gallisepticum* and that the probes of the genes of P1 and P140 hybridized by Southern blots with genomic *M. gallisepticum* DNA. On the other hand, the cytadherence-inhibiting MAbs to P1 and P140 did not react with *M. gallisepticum* 155-kDa putative adhesin and the probe of the P1 gene that encodes the 13-amino-acid sequence involved in cytadherence (Dallo *et al.*, 1988) did not hybridize with *M. gallisepticum* (Dallo *et al.*, 1990).

The similarities in some of the properties of the adhesins of the three mycoplasmas led Dallo and Baseman to propose that they share a family of adhesin-related genes (Dallo *et al.*, 1990).

M. pulmonis. *M. pulmonis* causes chronic murine respiratory mycoplasmosis and genital disease. Therefore, the presence of this pathogen is of major concern in studies involving laboratory mice and rats. Attempts to characterize the adhesin led to identification of an extracellular surface membrane protein which, when electrophoresed in SDS-PAGE, formed a "ladder" pattern and is a predominantly recognized antigen in naturally occurring infection (Horowitz *et al.*, 1987). This antigen, now designated variable V-1, was shown to have a subunit structure of varying size—the largest (upper band) with an apparent molecular mass of about 131 kDa, and a successive series of bands with a uniform spacing of approximately 3 kDa.

Monoclonal and monospecific polyclonal antibodies specific for V-1 show various forms of this antigen among *M. pulmonis* strains (Watson *et al.*, 1988), differing in their structure (electrophoretic pattern), in their charge (pI), and in some of the epitopes. This heterogeneous antigen might be associated with virulence, as suggested by Cassell *et al.* (1988), since it was present only in the virulent strains of *M. pulmonis* (Davidson *et al.*, 1990).

Studies of Lai *et al.* (1991) suggest an adhesin of *M. pulmonis*. This polypeptide with an estimated mass of 66 kDa was purified by affinity chromatography, using as a ligand an adherence-blocking monoclonal antibody (SE12F4) that inhibited *M. pulmonis*'s growth and, *in vitro*, prevented the attachment of the organisms to fibroblasts or to RBC.

Mice vaccinated with this adhesin molecule and challenged with a virulent strain of *M. pulmonis* were shown to be efficiently protected. Moreover, this protective capacity was passively transferred to other mice by injection of sera from the vaccinated mice. It seems, therefore, that in the case of *M. pulmonis* (Lai *et al.*, 1991), in contrast to that of *M. pneumoniae* (Hirschberg *et al.*, 1991;

Razin and Jacobs, 1992), vaccination with the adhesin does confer full protection (Lai *et al.*, 1991).

U. urealyticum. Immunoblotting of *U. urealyticum* serovars with rabbit antisera revealed an intensely straining complex with a "ladder" pattern when some serovars were reacted with homologous antisera (Horowitz *et al.*, 1986). This complex, similar to V-1 antigen of *M. pulmonis*, consisted of a series of proteins with a molecular weight and pI heterogeneity. This antigen was detected by immunoblotting with human sera from patients with chronic pelvic inflammatory disease and from patients who had experienced a premature rupture of membranes associated with intraamniotic infection with *U. urealyticum* (Horowitz *et al.*, 1989). Findings supported the possible relevance of such an antigen to pathogenicity for *U. urealyticum* were recently obtained. It was demonstrated that after successful treatment of ureaplasma infection, when patients became clinically cured and culturally negative, the "ladder" antigen of *U. urealyticum* was not detected using the convalescent sera of these patients (Horowitz *et al.*, 1990). Further investigations are needed to analyze the expression of this antigen in disease in comparison with asymptomatic human carriers.

The variability of this antigen (Watson *et al.*, 1990), as well as that of *M. pulmonis*, resemble the observations of Wise's group, reported for *M. hyorhinis*, with regard to its role in phase variation and possible role in mycoplasma–host cell interaction and pathogenicity (e.g., Rosengarten and Wise, 1990).

M. hominis. Most of the antigens of *M. hominis* described so far are proteins that were defined by means of functional assays (Brown *et al.*, 1983). Some membrane proteins have been reported to be responsible for the antigenic differences among strains of this organism, and some were accounted for differences in the attachment of *M. hominis* strains to eukaryotic cells (Izumikawa *et al.*, 1987). Twelve antigens of strain PG-21 detected by immunoblotting were identified as surface antigens (Cassell *et al.*, 1988) as they were iodolabeled by the Bolton–Hunter method, previously shown to label primarily mycoplasma surface proteins (Horowitz *et al.*, 1987). The molecular masses of these antigens ranged from 36.4 to 102.7 kDa. The latter, 102.7 kDa, was the most immunogenic, as judged by the intense reactivity of almost all patients' sera. It was also detected in all seven reference strains of *M. hominis*, while the other surface proteins demonstrated antigenic heterogeneity among the strains. Others have reported 11 surface antigens within the same molecular mass ranges in a different strain of *M. hominis* (Alexander, 1987) and, in another report, it was shown that the major antigen detected in strain PG-21 is a 58-kDa protein, but its surface nature is not discussed (Schalla and Harrison, 1987).

M. iowae. Few reports are available concerning *M. iowae* (Jordan *et al.*, 1982; Gallagher and Rhoades, 1983; Yoder and Hofstad, 1964; Shareef *et al.*, 1990). The possession by *M. iowae* of capsular material or an attachment organelle as a "fuzzy" layer of material outside the organism's cell membrane was detected by electron microscopy (by TEM, but not by SEM). The outer surface of *M. iowae* cells seems to possess fine fibrils 50–150 nm in length. Interlocking sets of fibrils between adjacent cells were observed. Mycoplasma cells were found in the cryptic lumen and close to the epithelial microvilli of the cloaca in infected tissues, but no intimate contact was noticed between mycoplasma cells and the epithelial cells; however, fibrils were observed between mycoplasma and adjacent microvilli. Since a specialized attachment organelle, such as the terminal structure of *M. pneumoniae* or the bleb of *M. gallisepticum,* was not found, and neither was an intimate or generalized attachment observed such as seen for *M. pulmonis,* it was assumed that *M. iowae* fibrils are similar to those of *M. hyopneumoniae* (Tajima and Yagahashi, 1982), and that the attachment of *M. iowae* to epithelial cells might be mediated by these fibrils.

Other Mycoplasmas.   Earlier studies on adherence, putative adhesins, surface proteins, and also on carbohydrate moieties on the surface of mycoplasma which may also be involved in their adherence have been reviewed. Because little was added by these studies, they will not be reviewed here, especially because some of these aspects are referred to and are relevant to the material presented in another chapter in this volume, in which mycoplasma capsular materials are discussed.

However, another point that is worth mentioning is that much of the detail is still missing; e.g., that spiroplasmas, or at least one species, *Spiroplasma sabaudiense,* can cytadsorb. It was reported recently (Humphrey-Smith *et al.,* 1990) that during the infective cycle in mosquito cells *in vitro,* there is a morphogenesis from an extracellular helical form to a coccoid form concomitant with cytadsorption and subsequent phagocytosis, This is considered as evidence that cytadsorption is directly associated with cytopathology in pathogenic strains and that morphogenesis depends upon a receptor mechanism.

5. MOLECULAR ARRANGEMENT OF ADHESINS IN THE MEMBRANE AND THE POSSIBLE ROLE OF THE CYTOSKELETON AND PHOSPHORYLATED PROTEINS

The molecular architecture of the adhesins of *M. pneumoniae* and *M. genitalium* were discussed earlier in this chapter and in greater detail by Baseman (this volume). P1, the adhesin of *M. pneumoniae,* probably traverses the membrane several times (according to our current model, most probably seven) and its COOH-terminal segment is exposed in the cytoplasm. The properties of this segment are unique. Having 13 proline residues out of 26 amino acids, it should have a relatively rigid conformation which is important with regard to its physiological role. It was documented that in order for *M. pneumoniae* to adhere, most P1 adhesins were oriented in the tip area.

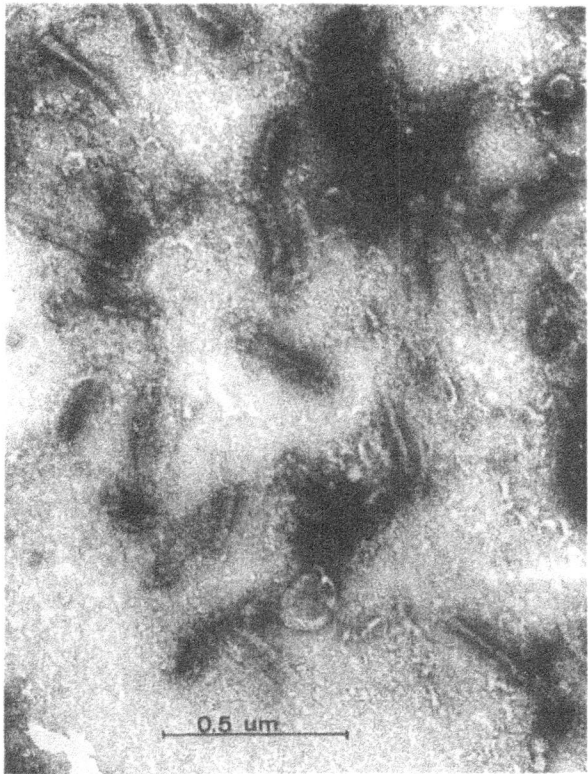

FIGURE 2. Micrograph of negatively stained *M. pneumoniae* Triton shell-enriched fraction. Note the enrichment of tip structures in the preparation. (I. Kahane, A.-B. Saada, S. Chitov, and M. Kessel, unpublished results.)

We have demonstrated that about 10% of P1 reside in the Triton shells of *M. pneumoniae* (Kahane *et al.*, 1985b). In earlier studies (Meng and Pfister, 1980), these Triton shells were shown to consist of filamentous structures, probably linked to the asymmetrical shape of *M. pneumoniae*, and it may be that similar structures occur in all flask-shaped mycoplasmas (Figures 2, 3). It is tempting to speculate that the proline-rich segment of the adhesin P1, which is exposed on the cytoplasmic side of the membrane, is linked with some components of the cytoskeleton.

What determines which P1 will be linked to the Triton shell's cytoskeleton warrants further investigation. The cytoskeleton was shown to be composed of a variety of polypeptides; at least one may be phosphorylated. We suggest that the cytoskeleton may also be involved in the gliding motility of the mycoplasmas. At this point, it should be recalled that the high-molecular-weight proteins

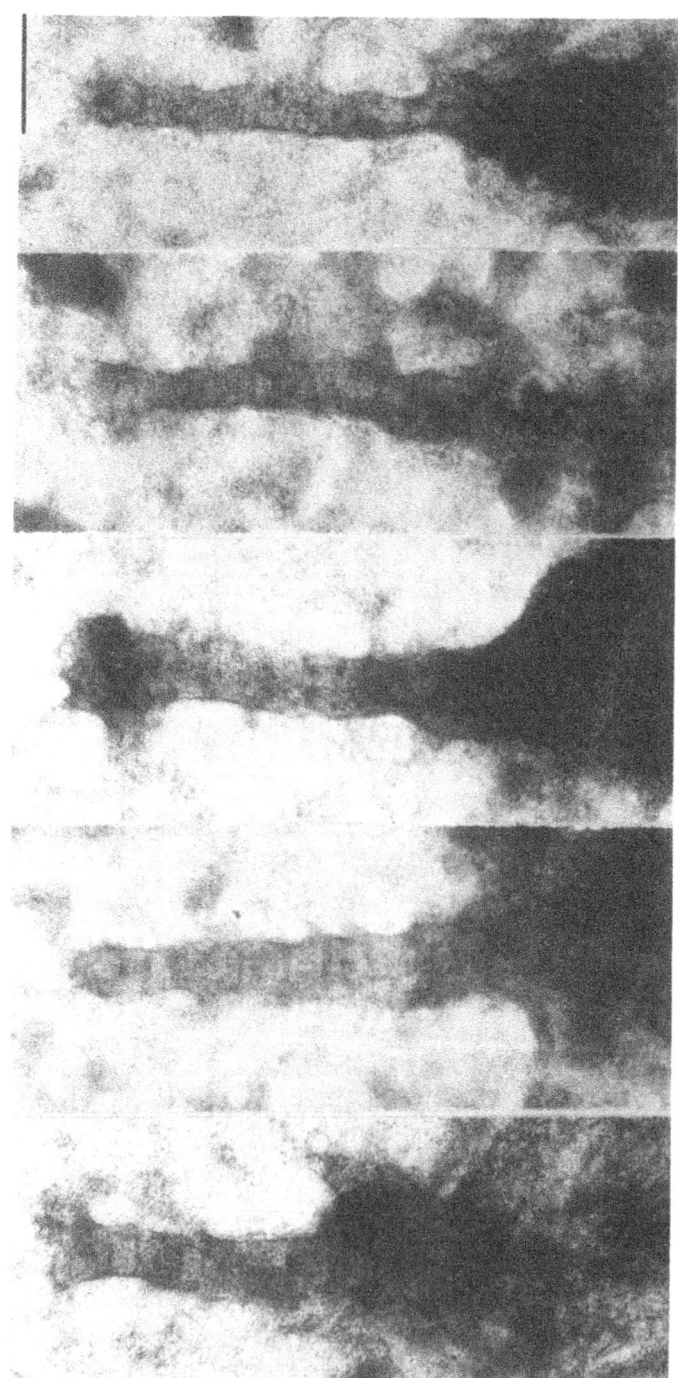

FIGURE 3. Micrographs of a gallery of tip structures obtained from the Triton shell-enriched fraction. Bar = 0.1 μm. (I. Kahane, A.-B. Saada, S. Chitov, and M. Kessel, unpublished results.)

(HMWI/A) of *M. pneumoniae* were documented to be involved in cytadherence. These accessory proteins are associated in the Triton shell and have been studied in detail by Krause's group. These studies indicate the regulation of synthesis of several proteins of the Triton shells which are somehow involved in the cytadherence of *M. pneumoniae* (Stevens and Krause, 1990, 1991), and refer to it as a phase variation in adherence. Another aspect which was earlier called phase variations was the finding that P1 is not always functional. It was shown that in any given population of wild-type *M. pneumoniae*, although all of the cells have P1, many of them do not adhere (Kahane *et al.*, 1985a). In *M. pneumoniae*, at least the major adhesin is linked to a substructure. The observation that not all mycoplasmas of a given culture adhere is a common finding with many cytadhering mycoplasmas.

6. THE ADVANTAGE OF ADHERENCE TO MOLLICUTES

In the previous sections, we discussed the various modes by which mycoplasmas adhere to host cells. The fact that several modes developed most probably indicates the major ecological advantage that the capacity to cytadhere provides the mycoplasma. Indeed, the studies by Kahane's group of the initial stages of the cytopathic process which mycoplasmas induce in the host cell indicate that the intimate association caused by the adherence allows the oxidative stress to develop in the host cell (Almagor *et al.*, 1985, 1986; Krause and Chen, 1990).

The cycle of events (Kahane, 1984; Almagor *et al.*, 1985) that is summarized in Table I results in the oxidation of lipids as well as proteins of the host cell membrane. The membrane becomes leaky and nutrients ooze from the host

Table I
Initial Events Caused by Mycoplasma Cytadherence

1. Adherence to host cell receptors (sialoglycoconjugates, protein, sulfatides)
2. Production of superoxide anions by mycoplasmas or induction of host cells for enhanced production of superoxide anions
3. Partial inhibition of host cell catalase
4. Partial inhibition of host cell Cu-Zn SOD
5. Increased oxidative stress in cytoplasm and for membrane components
6. Secondary oxidation and breakdown of fatty acids and production of the cross-linker malonyl dialdehyde
7. Cross-linkage of cytoplasmic and membrane components
8. Membrane becomes leaky
9. Cytoplasmic components ooze through host cell membrane
10. Cytadhering mycoplasmas benefit from better nutrition

cytoplasm through the membrane, thus providing the adhering mycoplasma with enriched nutrition. From these as well as other studies, it seems that the cytadherence of mycoplasmas, at least in the initial stages, is a separate process from the cytotoxic effects which they inflict and indeed, nonadhering mutants cannot produce the cytopathic effects. In chronic infections, adverse immunologic reaction to the adhesins may occur, as recently reviewed by Naot (1993).

7. CONCLUDING REMARKS

As has been indicated in this review, since no general rule can be drawn regarding mycoplasma adherence, the characteristics of the process have to be studied individually for each adhering mycoplasma. These studies should include characterization of the adhesin(s) involved and possible presence of auxiliary proteins. Elucidation of their gene(s) should provide the answer for the possible adhesin family. Much work lies ahead of us in understanding the phase variation of the adhesins.

Other points of interest are how the mycoplasmas approach the site of adherence and the properties related to the adherence. What are the processes that lead to the fusion and, more so, the penetration of the host cell membrane, a property that some mycoplasmas seem to have?

8. REFERENCES

Alexander, A. G., 1987, Analysis of protein antigen of *M. hominis:* Detection of polypeptide involved in the human immune response, *Isr. J. Med. Sci.* **23:**608–612.

Almagor, M., Kahane, I., Wiesel, J. M., and Yatziv, S., 1985, Human ciliated epithelial cells from nasal polyps as an experimental model for *Mycoplasma pneumoniae* infection, *Infect. Immun.* **48:**552–555.

Almagor, M., Kahane, I., Gilon, C., and Yatziv, S. 1986, Protective effects of the glutathione redox cycle and vitamin E on cultured fibroblasts infected by *Mycoplasma pneumoniae, Infect. Immun.* **52:**240–244.

Barile, B. F., 1979, Mycoplasma–tissue cell interactions, in: *The Mycoplasmas,* Volume II (J. G. Tully and R. F. Whitcomb, eds.), Academic Press, New York, pp. 425–474.

Bauer, F. A., Wear, D. J., Angritt, P., and Lo, S.-C., 1991, *Mycoplasma fermentans* (incognitus strain) infection in the kidneys of patients with acquired immunodeficiency syndrome and associated nephropathy. A light microscopic, immunhistochemical, and ultrastructural study, *Hum. Pathol.* **22:**63–69.

Bermudez, V., Miller, R., Rosendahl, S., and Johnson, W., 1980, *Proc. 7th Congr. IOM, Baden, Austria,* p. 1127.

Bredt, W., Feldner, J., and Kahane, I., 1981, Adherence of mycoplasmas to cells and inert surfaces: Phenomena, experimental models and possible mechanisms, *Isr. J. Med. Sci.* **17:**586–588.

Brown, M. B., Minion, C. F., Davis, J. K., Pritchard, D. G., and Cassell, G. H., 1983, Antigens of *Mycoplasma hominis, Sex. Trans. Dis.* **10:**247–254.

Cassell, G. H., Watson, H. J., Blalock, D. K., Horowitz, S. A., and Duffy, L. B., 1988, Protein antigens of genital mycoplasmas, *Rev. Infect. Dis.* **10:**S391–S398.

Collier, A. M., 1979, Mycoplasmas in organ culture, in: *The Mycoplasmas,* Volume II (J. G. Tully and R. F. Whitcomb, eds.), Academic Press, New York, pp. 475–493.

Dallo, S. F., Horton, J. R., Su, C.-J., and Baseman, J. B., 1988, Identification of P1 gene domain containing epitope(s) mediating *Mycoplasma pneumoniae* cytadherence, *J. Exp. Med.* **167:**718–723.

Dallo, S. F., Horton, J. R., Su, C.-J., and Baseman, J. B., 1990, Restriction fragment length polymorphism in the cytadhesin P1 gene of human clinical isolates of *Mycoplasma pneumoniae, Infect. Immun.* **58:**2017–2020.

Davidson, M. K., Lindsey, J. R., Davis, J. K., Parker, R. F., Ross, S. E., Watson, H. L., Tully, J. G., and Cassell, G. H., 1990, An alternative approach to identification of virulence mechanisms of *Mycoplasma pulmonis,* in: *Recent Advances in Mycoplasmology* (G. Stanek, G. H. Cassell, J. G. Tully, and R. F. Whitcomb, eds.), Fischer Verlag, Stuttgart, pp. 695–697.

Gallagher, J. E., and Rhoades, K. R., 1983, Scanning and light electron microscopy of selected avian strains of *Mycoplasma iowae, Avian Dis.* **27:**211–217.

Geary, S. J., Gabridge, M. G., and Gladd, M. F., 1990, Utilization of a 100 kDa human lung fibroblast receptor site to identify a 32 kDa *Mycoplasma pneumoniae* antigen involved in attachment, in: *Recent Advances in Mycoplasmology* (G. Stanek, G. H. Cassell, J. G. Tully, and R. F. Whitcomb, eds.), Fischer Verlag, Stuttgart, pp. 697–700.

Gesner, B., and Thomas, L. 1966, Sialic acid binding sites: Role in hemagglutination by *Mycoplasma gallisepticum, Science* **151:**590–591.

Hirschberg, L., Holme, T., and Krook, A., 1991, Human antibody response to the major adhesin of *Mycoplasma pneumoniae* increase in titers against synthetic peptides in patients with pneumonia, *Acta Pathol. Microbiol. Immun. Scand.* **99:**515–520.

Horowitz, S. A., Duffy, L., Garrett, B., Stephens, J., Davis, J. K., and Cassell, G. H., 1986, Can group and serovar-specific proteins be detected in *Ureaplasma urealyticum? Pediatr. Infect. Dis.* **5:**S325–S331.

Horowitz, S. A., Garrett, B., Davis, J. K., and Cassell, G. H., 1987, Isolation of *M. pulmonis* membranes and identification of surface antigens, *Infect. Immun.* **55:**1314–1320.

Horowitz, S. A., Bar-David, J., Mazor, M., and Gal, H., 1989, Detection of *Ureaplasma urealyticum* in pelvic inflammatory disease (PID) and premature rupture of membrane (PROM), *First World Congr. Infect. Dis. Obstet. Gynecol., Hawaii,* p. 60.

Horowitz, S. A., Bar-David, J., Mazor, M., Gal, H., and Saadon, R., 1990, Detection of *Ureaplasma urealyticum* specific antigens in PID and in PROM, *Lett. 8th Int. Congr. IOM, Istanbul,* p. 416.

Humphrey-Smith, I., Chastel, S., Grulet, O., and LeGoff, F., 1990, The infective cycle of *Spiroplasma sabaudiense in vitro* in *Aedes albopictus* (C636) cells and its potential for the biological control of Dengue virus, in: *Recent Advances in Mycoplasmology* (G. Stanek, G. H. Cassell, J. G. Tully, and R. F. Whitcomb, eds.), Fischer Verlag, Stuttgart, pp. 922–924.

Inamine, J. M., Loechel, S., and Hu, P.-C., 1988, Analysis of the nucleotide sequence of the P1 operon of *Mycoplasma pneumoniae, Gene* **73:**175–183.

Izumikawa, K., Chandler, D. K. F., Grabowski, M. W., and Barile, M. F., 1987, Attachment of *Mycoplasma hominis* to human cell cultures, *Isr. J. Med. Sci.* **23:**603–607.

Jordan, F. T. W., Erno, H., Cottew, G. S., Hinz, K. H., and Stipkovits, L., 1982, Characterization and taxonomic description of five mycoplasma serovars (serotypes) of avian origin and their elevation to species rank and further evaluation of the taxonomic status of *Mycoplasma synoviae, Int. Syst. Bacteriol.* **32:**108–115.

Kahane, I., 1984, *In vitro* studies on the mechanism of adherence and pathogenicity of mycoplasmas, *Isr. J. Med. Sci.* **20:**874–877.

Kahane, I., and Marchesi, V. T., 1973, Studies on the orientation of proteins in Mycoplasma and erythrocyte membranes, *Ann. N.Y. Acad. Sci.* **225:**38–45.

Kahane, I., Banai, M., Razin, S., and Feldner, J., 1982, Attachment of mycoplasmas to host cell membranes, *Rev. Infect. Dis.* **4**(Suppl.):S185–S192.

Kahane, I., Granek, J., and Reisch-Saada, A., 1984, The adhesins of *Mycoplasma gallisepticum* and *Mycoplasma pneumoniae*, *Ann. Microbiol. (Paris)* **135A:**25–32.

Kahane, I., Tucker, S., and Baseman, J. B., 1985a, Detection of *Mycoplasma pneumoniae* adhesin (P1) in the nonhemadsorbing population of virulent *Mycoplasma pneumoniae*, *Infect. Immun.* **49:**457–458.

Kahane, I., Tucker, S., Leith, D. K., Morrison-Plummer, J., and Baseman, J. B., 1985b, Detection of the major adhesin P1 in Triton shells of virulent *Mycoplasma pneumoniae*, *Infect. Immun.* **50:**944–946.

Klinkert, M. Q., Herrmann, R., and Schaller, H., 1985, Surface proteins of *M. hyopneumoniae* identified from an *Escherichia coli* expression plasmid library, *Infect. Immun.* **49:**329–335.

Krause, D. C., and Chen, Y. Y., 1990, Oxidative stress response and *Mycoplasma pneumoniae* toxicity for tracheal epithelium, in: *Recent Advances in Mycoplasmology* (G. Stanek, G. H. Cassell, J. G. Tully, and R. F. Whitcomb, eds.), Fischer Verlag, Stuttgart, pp. 703–708.

Krivan, H. C., Olson, L. D., Barile, M. F., Ginsburg, V., and Roberts, D., 1989, Adhesion of *Mycoplasma pneumoniae* to sulfated glycolipids and inhibition by dextran sulfate, *J. Biol. Chem.* **264:**9283–9288.

Lai, W. C., Bennett, M., Pakes, S. P., and Murphree, S. S., 1991, Potential subunit vaccine against *Mycoplasma pulmonis* purified by a protective monoclonal antibody, *Vaccine* **9:**177–184.

Lindsey, J. R., and Cassell, G. H. 1973, Experimental *Mycoplasma pulmonis* infection in pathogen-free mice, *Am. J. Pathol.* **72:**63–84.

Lo, S.-C., Dawson, M. S., Wong, D. M., Newton, P. B., III, Sonoda, M. A., Engler, W. F., Wang, R. V.-H., Shih, J. W.-K., Alter, H. J., and Wear, D. J., 1989, Identification of *Mycoplasma incognitus* infection in patients with AIDS: An immunochemical, *in situ* hybridization and ultrastructural study, *Am. J. Trop. Med. Hyg.* **41:**89–164.

Loomes, L. M., Uemura, K.-I., Childs, R. A., Paulson, J. C., Rogers, G. N., Scudder, P. R., Michalski, J.-C., Hounsell, E. F., Taylor-Robinson, D., and Feizi, T., 1984, Erythrocyte receptors for *Mycoplasma pneumoniae* are sialylated oligosaccharides of Ii antigen type, *Nature* **307:**560–563.

Loveless, R. W., and Feizi, T., 1989, Sialo-oligosaccharide receptors for *Mycoplasma pneumoniae* and related oligosaccharides of poly-N-acetyllactosamine series are polarized at the cilia and apical–microvillar domains of the ciliated cells in human bronchial epithelium, *Infect. Immun.* **57:**1285–1289.

Meng, K. E., and Pfister, R. M., 1980, Intracellular structures of *Mycoplasma pneumoniae* revealed after membrane removal, *J. Bacteriol.* **144:**390–399.

Minion, F. C., Cassell, G. H., Pnini, S., and Kahane, I., 1984, Multiphasic interactions of *Mycoplasma pulmonis* with erythrocytes defined by adhering and hemagglutination, *Infect. Immun.* **44:**394–400.

Naot, Y., 1993, Mycoplasmas as immunomodulators, in: *Rapid Diagnosis of Mycoplasmas* (I. Kahane and A. Adoni, eds.), Plenum Press, New York.

Razin, S., 1985, Mycoplasma adherence, in: *The Mycoplasmas*, Volume IV (S. Razin and M. F. Barile, eds.), Academic Press, New York, pp. 161–242.

Razin, S., 1987, Appealing attributes of mycoplasmas in cell biology research, *Isr. J. Med. Sci.* **23:**318–325.

Razin, S., and Jacobs, E., 1992, Mycoplasma adhesion, *J. Gen. Microbiol.* **138:**407–422.

Razin, S., Kahane, I., Banai, M., and Bredt, W., 1981, Adhesion of mycoplasmas to eukaryotic cells, *Ciba Found. Symp.* **80**:98–118.

Roberts, D. D., Olson, L. D., Barile, M. F., Ginsburg, V., and Krivan, H. C., 1989, Sialic acid-dependent adhesion of *Mycoplasma pneumoniae* to purified glycoproteins, *J. Biol. Chem.* **264**:9289–9293.

Robertson, J. A., and Sherbourne, R., 1991, Hemadsorption by colonies of *Ureaplasma urealyticum*, *Infect. Immun.* **59**:2203–2206.

Rosengarten, R., and Wise, K. S., 1990, Phenotypic switching in mycoplasmas: Phase variation of diverse surface lipoproteins, *Science* **247**:315–318.

Saada, A.-B., Terespolski, Y., Adoni, A., and Kahane, I., 1991, Adherence of *Ureaplasma urealyticum* to human erythrocytes, *Infect. Immun.* **59**:467–469.

Saada, A.-B., Rahamim, E., Kahane, I., and Beyth, Y., 1993, Detection of adherence of *Ureaplasma urealyticum* to Fallopian tube mucosa cells in culture, in: *Rapid Diagnosis of Mycoplasmas* (I. Kahane and A. Adoni, eds.), Plenum Press, New York.

Schalla, W. O., and Harrison, H. R. 1987, Western blot analysis of the human response to *Mycoplasma hominis*, *Isr. J. Med. Sci.* **23**:613–617.

Shareef, J. M., Willcox, J., and Kumar, P., 1990, Adherence of *Mycoplasma iowae* to epithelial mucosa of the cloaca, in: *Recent Advances in Mycoplasmology* (G. Stanek, G. H. Cassell, J. G. Tully, and R. F. Whitcomb, eds.), Fischer Verlag, Stuttgart, pp. 872–874.

Sperker, B., Hu, P.-C., and Herrmann, R., 1991, Identification of gene products of the P1 operon of *Mycoplasma pneumoniae*, *Mol. Microbiol.* **5**:299–306.

Stevens, M. K., and Krause, D. C., 1990, Disulfide-like protein associated with *Mycoplasma pneumoniae* cytadherence phase variation, *Infect. Immun.* **58**:3430–3433.

Stevens, M. K., and Krause, D. C. 1991, Localization of the *Mycoplasma pneumoniae* cytadherence-associated proteins HMW1 and HMW4 in cytoskeletonlike Triton shell, *J. Bacteriol.* **173**:1041–1050.

Tajima, M., and Yagahashi, T., 1982, Interaction of *Mycoplasma hyopneumoniae* with the porcine respiratory epithelium as observed by electron microscopy, *Infect. Immun.* **37**:1162–1164.

Tully, J. G., 1993, Biology of the Mollicutes, in: *Rapid Diagnosis of Mycoplasmas* (I. Kahane and A. Adoni, eds.), Plenum Press, New York.

Watson, H. L., McDaniel, L. S., Blalock, D. K., Fallon, M. T., and Cassell, G. H., 1988, Heterogeneity among strains and a high rate of variation within strains of a major surface antigen of *Mycoplasma pulmonis*, *Infect. Immun.* **56**:1356–1363.

Watson, H. L., Blalock, D. K., and Cassell, G. H., 1990, Variable antigens of *Ureaplasma urealyticum* containing both serovar-specific and serovar-crossreactive epitopes, *Infect. Immun.* **58**:3679–3688.

Yayoshi, M., Araake, M., Hayatsu, E., Kawakubo, Y., and Yoshoka, M., 1984, Characterization and pathogenicity of hemolysis mutants of *Mycoplasma pneumoniae*, *Microbiol. Immunol.* **28**:303–310.

Yoder, H. W., Jr., and Hofstad, M. S., 1964, Characterization of avian mycoplasma, *Avian Dis.* **8**:481–512.

Chapter 9

The Cytadhesins of *Mycoplasma pneumoniae* and *M. genitalium*

Joel B. Baseman

1. INTRODUCTION

Selective adhesion mechanisms permit microorganisms to colonize mucous membranes and other tissues and resist clearing mechanisms of the host. Although electrostatic and hydrophobic forces contribute to the adherence event, the specificity of the interaction is regulated by microbial surface components, termed adhesins, and complementary stereospecific structures associated with the host, termed receptors.

Mycoplasma pneumoniae and *M. genitalium* are human pathogens that share numerous features, such as a flask shape and tiplike organelle that mediates adherence (Baseman *et al.*, 1982b; Tully *et al.*, 1983). The importance of the cytadherence event for mycoplasmas is apparent when one considers the limited genome size of these prokaryotes. For example, the molecular sizes of the genomes of *M. pneumoniae* and *M. genitalium* are 500 and 400 MDa, respectively, the latter considered the smallest self-replicating biological system (Su and Baseman, 1990). Thus, mycoplasmas are biosynthetically deficient and dependent on the microenvironment (i.e., intimate contact with the host) for obtaining essential precursors such as nucleotides, fatty acids, sterols, and amino

Joel B. Baseman   Department of Microbiology, The University of Texas Health Science Center at San Antonio, San Antonio, Texas 78284.

Subcellular Biochemistry, Volume 20: Mycoplasma Cell Membranes, edited by Shlomo Rottem and Itzhak Kahane. Plenum Press, New York, 1993.

acids. This chapter describes the chemical and molecular basis of adherence of *M. pneumoniae* and *M. genitalium* and emphasizes the critical nature of this event in colonization and *in vivo* survival. Earlier reviews by Razin (1985; Razin and Yogev, 1988) will serve as excellent sources of background information and additional publications.

2. IDENTIFICATION OF PROTEIN ADHESINS OF *M. PNEUMONIAE*

The ability of *M. pneumoniae* to penetrate the network of host defenses such as ciliary motion and mucus blanket, and physically and intimately adhere to target cells is a prerequisite for successful colonization and subsequent infection. Figure 1 demonstrates the unique tip-oriented adherence of *M. pneumoniae* to tracheal epithelium and provides visual evidence for the existence of a highly

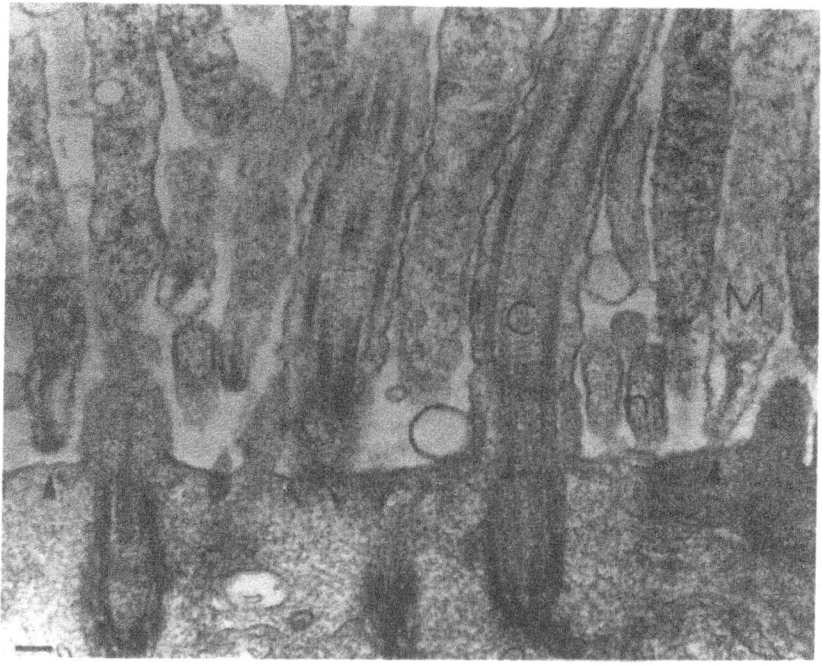

FIGURE 1. Transmission electron photomicrograph of a hamster trachea ring infected with *M. pneumoniae*. Note the orientation of the mycoplasmas via their specialized tiplike organelle which permits close association with the respiratory epithelium. × 50,000. M, mycoplasma; m, micro-villus; C, cilia.

polar attachment mechanism (Collier and Baseman, 1973; Hu *et al.*, 1977). Early studies showed that *M. pneumoniae* attached to respiratory epithelium rapidly and in a nearly linear fashion over several hours (Collier and Baseman, 1973; Powell *et al.*, 1976). Subsequently, tissue cytopathology and ciliostasis developed which correlated with the number of adherent mycoplasmas and the length of coincubation. Pretreatment of tracheal cells with neuraminidase or loss of mycoplasma viability significantly reduced mycoplasma adherence, indicating that the metabolic and chemical integrity of the host cell–mycoplasma interaction was critical (Collier and Baseman, 1973; Powell *et al.*, 1976; Hu *et al.*, 1976). In addition, brief trypsin treatment of *M. pneumoniae* markedly reduced mycoplasma adherence to host cells, suggesting that specific mycoplasma proteins might function as mediators of cytadherence (Hu *et al.*, 1977). This possibility was substantiated by gel electrophoretic analysis of trypsin-treated mycoplasmas, which revealed the selective loss of a 170-kDa surface protein, designated P1 (Hu *et al.*, 1977). Resynthesis of this protein by *M. pneumoniae* directly correlated with restoration of adherence capabilities.

How does protein P1 confer this unique polarity of attachment? The obvious close juxtaposition of *M. pneumoniae* to host cell surfaces necessitates that mechanisms exist to orient the tiplike terminal structure and to target appropriate host receptors. Immunoferritin electron microscopy was employed using monospecific antiserum directed against the P1 protein to determine the membrane distribution of P1. As seen in Figure 2, P1 molecules were densely clustered at the tip organelle in close association with a naplike mycoplasma structure (Baseman *et al.*, 1982b). Since less dense P1 foci were scattered randomly along the unit membrane, it appeared that the mobilization and clustering of P1 at the tip were a prerequisite for successful parasitism.

Functional support for the role of P1 in cytadherence was further reinforced by the demonstration that anti-P1 monoclonal and monospecific antibodies and anti-P1 Fab fragments blocked mycoplasma adherence to respiratory epithelium or erythrocytes without affecting the metabolic state of the mycoplasmas (Krause and Baseman, 1983). In the case of erythrocytes, neuraminidase pretreatment of erythrocytes or trypsin pretreatment of mycoplasmas markedly reduced hemadsorption, analogous to mycoplasma adherence to respiratory epithelium (Baseman *et al.*, 1982a). Antibodies raised to other mycoplasma surface molecules, such as non-adhesin-related proteins, did not diminish mycoplasma cytadherence (Morrison-Plummer *et al.*, 1986). Another experimental approach to further identify putative mycoplasma adhesins utilized an *in vitro* binding assay in which chemically stabilized, glutaraldehyde-fixed eukaryotic cells were incubated with detergent-solubilized ^{125}I-labeled *M. pneumoniae* proteins. Selective binding of the P1 protein in a receptor–ligand-mediated fashion was observed, further implicating this molecule in adherence (Krause and Baseman, 1982).

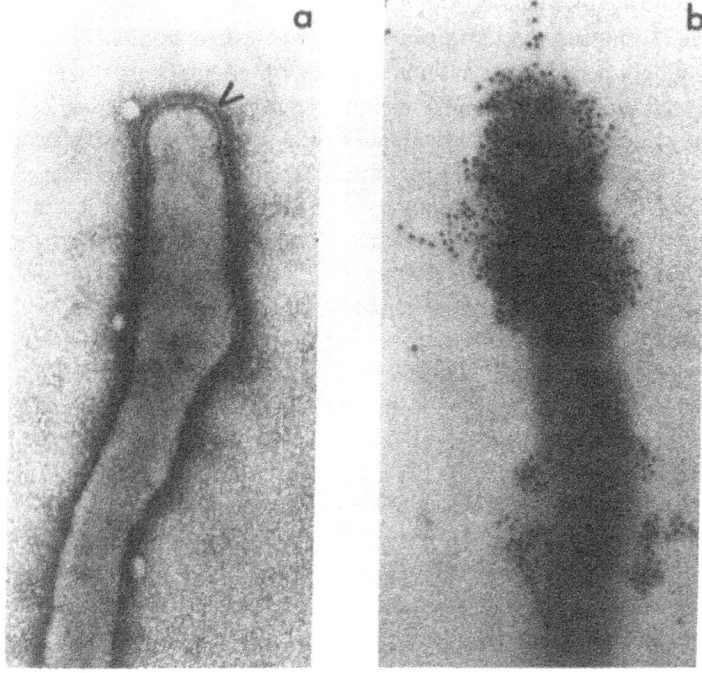

FIGURE 2. High-resolution electron microscopy of the truncated tip organelle of *M. pneumoniae* demonstrating (a) nap (arrow) and (b) P1 clusters at the tip region. (a) × 130,000; (b) × 150,000.

3. ISOLATION OF CYTADHERENCE-DEFICIENT *M. PNEUMONIAE* MUTANTS

Unfortunately, the existence of a noncytadhering mutant that possessed P1 suggested that P1 alone was insufficient for mediating attachment. Also, the possibility existed that other molecules might be required either to assist P1 in recognizing receptor(s), concentrating P1 at the tip, or maintaining the structural and functional integrity of the tip organelle. Spontaneously arising, noncytadhering mutants of *M. pneumoniae* arose at high frequency (10^{-2} to 10^{-3}) from the virulent, cytadhering parent strain (Krause *et al.*, 1982). Analysis of protein profiles by one- and two-dimensional acrylamide gel electrophoresis permitted mutant categorization into four classes (Table I; Krause *et al.*, 1982; Baseman *et al.*, 1987). Each of the mutant classes demonstrated greatly reduced adherence to both respiratory epithelium and erythrocytes, and these adherence-deficient mycoplasmas were avirulent based on decreased *in vivo* survival and lack of pathology in the hamster model of infection. To further clarify the putative role of these

Table I
Summary of the Protein Profiles of Spontaneous Noncytadhering Mutant Classes
of *Mycoplasma pneumoniae* and Homologous Class-Specific Cytadhering Revertants

Strain	Phenotype	Protein profiles								
		HMW1	HMW2	HMW3	HMW4	A	B	C	P1	P30
M129-B25C	Wild type	+	+	+	+	+	+	+	+	+
Class I	Mutant	±	−	−	−	+	+	+	+	+
	Revertant	+	+	+	+	+	+	+	+	+
Class II	Mutant	+	+	+	+	+	+	+	+	−
	Revertant	+	+	+	+	+	+	+	+	+
Class III	Mutant	+	+	+	+	−	−	−	+	+
	Revertant	+	+	+	+	+	+	+	+	+
Class IV	Mutant	+	+	+	+	−	−	−	−	+
	Revertant	+	+	+	+	+	+	+	+	+

^{a} +, protein present; ±, protein markedly deficient or a minor comigrating polypeptide present; −, protein absent. Designations: HMW1, 210 kDa; HMW2, 190 kDa; HMW3, 140 kDa; HMW4, 215 kDa; A, 72 kDa; B, 85 kDa; C, 37 kDa; P1, 170 kDa; P30, 30 kDa.

non-P1 proteins in the adherence event, spontaneously arising cytadherence-positive revertants were derived from each mutant class (Table I; Krause *et al.*, 1983). Reversion to the cytadhering phenotype was accompanied by resynthesis of the implicated proteins, recognition of host receptor(s), and restoration of virulence capabilities.

Ulstrastructural comparisons of the wild-type parent and noncytadhering mutants dramatically revealed the essential properties of these mycoplasma proteins in maintaining the structural and functional integrity of the tip organelle and promoting the distribution and localization of P1. For example, as seen in Figure 3, mutants exhibited abnormal tip morphology. What appears to be the normal naplike and truncated structure characteristic of the wild-type tip (Figure 2) is either partially or completely lacking among the mutants (Baseman *et al.*, 1982b). Additional studies have described a cytoskeletonlike network of filaments that remains following treatment of *M. pneumoniae* with the nonionic detergent Triton X-100 (Meng and Pfister, 1980). A subpopulation of P1 adhesin molecules appears associated with this Triton-insoluble cytoskeletal shell (Kahane *et al.*, 1985). These data, along with the earlier observations describing noncytadhering mutants of *M. pneumoniae* (Table I), reinforced the critical nature of these non-P1 proteins in the lateral movement and polar clustering of the P1 adhesin. Also, the non-P1 proteins may represent adhesins distinct from P1 that are capable of targeting similar or different populations of host cell receptors. Consistent with this hypothesis is the observation that monoclonal antibodies directed against the 30-kDa protein (designated P30), which is absent in

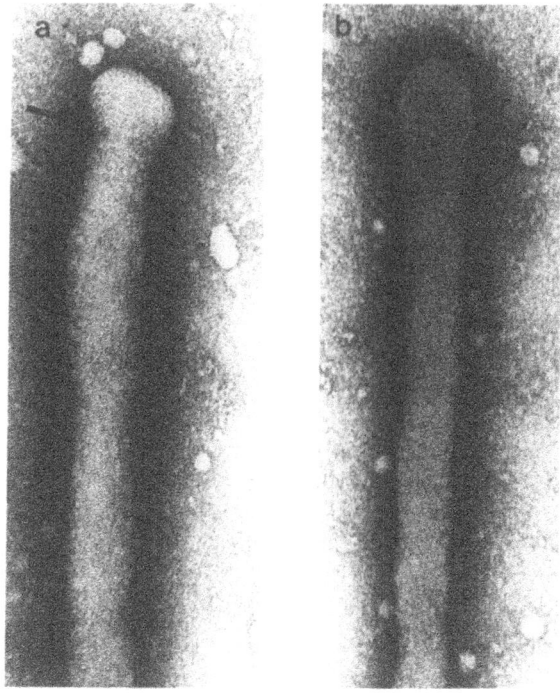

FIGURE 3. Negative staining of different noncytadhering mycoplasma mutant classes. (a) Mutant class III with truncated tip (arrow) but no nap. (b) Mutant class I with neither truncation nor nap. × 225,000.

class II noncytadhering *M. pneumoniae* mutants (Table I), block mycoplasma cytadherence (Morrison-Plummer *et al.*, 1986). In addition, P30 clusters at the tip organelle of wild-type *M. pneumoniae* in a fashion similar to P1 (Baseman *et al.*, 1987). Interestingly, P30 appears to be trypsin-resistant, in direct contrast to the high trypsin sensitivity of P1. This proposed relationship between P1 and P30 and the other mycoplasma proteins implicated in cytadherence is reminiscent of the close networking of specific proteins in maintaining the structural and functional integrity of erythrocytes and platelets (Fox and Boyles, 1988).

4. CHARACTERIZATION OF THE P1 ADHESIN GENE OF *M. PNEUMONIAE*

Initial attempts to clone *M. pneumoniae* DNA fragments encoding the P1 adhesin and to generate these proteins in an *E. coli* host failed because of the unusual property of mycoplasmas to use the universal stop codon UGA as a

Table II
Characterization of the P1 Gene of *Mycoplasma pneumoniae*

1. Determination of amino-terminal amino acid sequence of the purified P1 protein by gas-phase microsequencing
2. Generation of oligonucleotide probes (14-mer, corresponding to amino acids 1–5; 18-mer, corresponding to amino acids 7–12) complementary to mRNA
3. Identification of common DNA fragment (6 kb *Eco*RI) from *M. pneumoniae* genome by hybridization to both oligonucleotide probes
4. Cloning and sequencing of the P1 gene

tryptophan codon (Yamao *et al.*, 1985). Thus, *E. coli* reads this triplet codon as a termination signal, leading to the synthesis of truncated mycoplasma peptides (Treviño *et al.*, 1986). An alternate strategy was utilized that circumvented the need to express mycoplasma proteins in *E. coli*. Purification of the P1 adhesin was accomplished by anti-P1 monoclonal antibody affinity chromatography and preparative gel electrophoresis (Leith and Baseman, 1984; Su *et al.*, 1987). It was then feasible to proceed as illustrated in Table II and to sequence the 6-kb *Eco*RI fragment that had been identified by Southern blot analysis as hybridizing to both 14-mer and 18-mer oligonucleotide probes. An open reading frame of 4881 nucleotides was observed that correlated with the 18 amino acids determined by gas-phase sequencing of the purified P1 protein. However, these amino acids were not found at the amino-terminus of the open reading frame but rather started at position 60 of the deduced protein. These data, which were confirmed by Inamine *et al.* (1988a), suggested that P1 was synthesized as a precursor consistent with the observation that the extra 59 amino acids found at the amino-terminus appeared like a signal peptide (Oliver, 1985). Several interesting features of P1 are outlined in Table III and include the preferential usage of UGA to UGG codons, the proline-rich carboxy-terminus, the lack of intramolecular disulfide bonding, the location of the P1 structural gene in an operonlike organization (Inamine *et al.*, 1988b), and the substantial sequence homology with mammalian proteins. The latter is especially provocative and is consistent with the earlier discussion of the networking and trafficking of mycoplasma proteins in cytadherence and the importance of specific proteins in regulating membrane function (Fox and Boyles, 1988). Also, the shared homology between the P1 adhesin and matrix proteins in eukaryotes may correlate with previous observations of autoimmunelike mechanisms associated with mycoplasma infections (Biberfeld, 1971; Neimark, 1983; Wise and Watson, 1985).

Once the sequence of the P1 structural gene was determined, it was possible to examine the class IV noncytadhering mutant that lacked P1 (Table I) in an attempt to understand the control of P1 gene expression. A single base insertion of adenine into a stretch of seven adenines early in the P1 structural gene caused

Table III
Properties of the P1 Adhesin of *Mycoplasma pneumoniae*

	Number of amino acids	Estimated molecular weight
Precursor protein	1627	176,280
Mature protein	1560	169,750

Other features
  • Conventional transcription initiation sites about −35 and −10 upstream
  • No typical ribosomal binding site
  • 21 TGA (UGA) codons; 16 TGG (UGG) codons
  • Signal peptide of 59 amino acids
  • 13 of 26 amino acids at the carboxy-terminus are prolines
  • No cysteines
  • Sequence homology with mammalian cytoskeletal keratin and fibrinogen
  • Part of a multigene operon

a frameshift and the premature termination of the P1 sequence, explaining the absence of truncated P1-related peptides in this mutant class (Su *et al.*, 1989). Because the P1 gene exists in a multigene operon, the simultaneous loss of four proteins (P1, A, B, and C; see Table I) in class IV mutants might be explained by the location of these genes in the same operon. This possibility is consistent with a recent report by Sperker *et al.* (1991).

5. MULTICOPY NATURE AND SEQUENCE DIVERGENCY OF THE P1 GENE

Studies with numerous prokaryotic and eukaryotic pathogens indicate that genes encoding virulence-related proteins, such as adhesins, may exist as multiple gene families, providing a mechanism by which gene conversion regulates the expression of microbial adherence, colonization, and *in vivo* survival. Also, the occurrence of multiple copies of partial or incomplete gene sequences may result in high rates of mutation correlating with the frequency of spontaneously arising noncytadhering *M. pneumoniae* mutants (Krause *et al.*, 1982). Therefore, a repertoire of multiple P1-related gene sequences may exist to regulate the structural and functional properties of P1.

When the entire P1 structural gene was used to probe the *M. pneumoniae* genome in Southern blot analysis under stringent conditions, multiple bands were detected (Su *et al.*, 1988). Additional hybridization analysis of the P1 structural gene using 12 contiguous and nonoverlapping restriction fragments of the gene revealed that approximately one-third of the structural gene occurred in

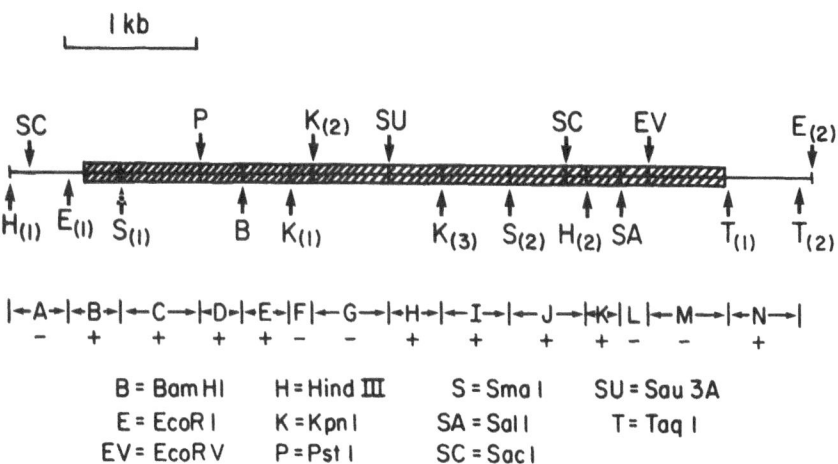

FIGURE 4. Restriction enzyme map of the P1 structural gene and surrounding regions. The boundary of each subclone from A to N is marked. Restriction enzyme sites that cut more than once are numbered starting from the 5' end. *Sau*3A and *Taq*I cut many times in the P1 gene, but only the sites used for subcloning purposes are shown. Hatched bars indicate the P1 structural gene. Numbers in parentheses indicate site numbers. Plus signs indicate the subclones that hybridized to between 3 and 8 multiple copies; minus signs indicate the subclones that hybridized to single copies.

single copies while approximately two-thirds occurred in multiple copies (Figure 4). Thus, the P1 gene consisted of a mosaic of DNA segments with distinct gene regions coding either for domains of the P1 protein involved in the recognition of host receptor(s) or for structural and functional properties, such as transmembrane and signal activities (Su *et al.*, 1988). How can the biofunctional domains of this complex structural gene be determined?

One experimental approach utilized anti-P1 monoclonal antibodies previously shown to be highly effective at blocking mycoplasma cytadherence (Morrison-Plummer *et al.*, 1986). A λgt11 recombinant DNA expression library of *M. pneumoniae* was constructed, and clones expressing P1 epitopes were selected by screening the library with anti-P1 monoclonal antibodies (Dallo *et al.*, 1988). Reactive clones were sequenced, and sequence comparisons revealed that each clone mapped near the COOH-terminal region of the P1 gene. Further analysis revealed that these sequences localized to single-copy regions of the gene, reinforcing the critical function of these sequences in mycoplasma adherence and virulence. The use of two anti-P1 adherence-blocking monoclonal antibodies that recognized distinct P1 protein epitopes permitted the identification of a 13-amino-acid cytadherence-related sequence (Dallo *et al.*, 1988). Another approach to define regions of the P1 protein that mediated cytadherence utilized adherence-blocking antibodies and overlapping synthetic peptides that correlated with three regions of the P1 adhesin: the $^{N}H_2$-terminal domain, a domain near the middle of the protein, and the COOH-terminal domain (Gerstenecker and Jacobs, 1990). This study not only reinforced the role of the previously identified single-copy COOH domain of P1 and its encoded 13-amino-acid peptide but also implicated other sites of the molecule that were located in multiple-copy regions (Su *et al.*, 1988). An interesting topological configuration of the P1 molecule was proposed that required a close spatial relationship between the three domains to form a native and functionally active adherence site (Gerstenecker and Jacobs, 1990). These data were particularly relevant because they clarified an earlier observation that described the abrogation of *M. pneumoniae* cytadherence following trypsin treatment of mycoplasmas (Hu *et al.*, 1977). Under these experimental conditions, several protease-generated, P1-derived peptides ranging from 85 to 22 kDa remained anchored to the *M. pneumoniae* membrane and immunoreactive with anti-P1 blocking antibodies (Baseman *et al.*, 1985). Taken together, these data clearly established the importance of topography and folding within the P1 molecule for recognition of host receptors and successful tissue colonization by *M. pneumoniae*.

The proposed surface-exposed multiloop model of the P1 adhesin and the evidence that cytadherence-associated P1 gene sequences appear to localize to single- and multiple-copy regions of the P1 structural gene suggested that recombination events could alter biological and structural properties of the adhesin. Examination of the P1 genes from different clinical isolates of *M. pneumoniae* revealed restriction length fragment polymorphisms permitting the classification of the clinical isolates into two groups (Su *et al.*, 1990b; Dallo *et al.*, 1990b). Recombination between the P1 structural gene and the P1-related multiple-copy sequences could generate new restriction sites assuming that the related gene fragments differed from the P1 gene in specific nucleotides. When the P1 structural genes from representative clinical isolates were cloned and sequenced, two major differences and several minor differences in the nucleotide sequences were detected between the two groups (Su *et al.*, 1990a). For example, one region of about 500 bp shared 72% homology and one region of about 700 bp shared 90% homology. These distinctions resulted in significant changes in the amino acid composition. The newly identified, divergent P1 cytadhesin consisted of 1635 amino acids with a calculated molecular weight of 176,893 (compare to Table III). As might be predicted, the single-copy regions of the P1 structural genes were essentially conserved between the two groups of clinical isolates. In contrast, previously identified multiple-copy regions displayed considerable sequence divergency. In order to further clarify the origin of the observed P1 gene diversity, oligonucleotide probes specific to the diverged regions of the P1 structural genes from the two groups of clinical isolates (Su *et al.*, 1990a,b) were used to clone and sequence multicopy P1-related DNA segments among selected clinical strains. Data indicated that individual oligonucleotide probes not only hybridized to multiple copies within each mycoplasma group but also hybridized to the genomes of the heterologous group. The existence in the clinical isolates of these repetitive sequences, which share homology with the P1 structural genes, is consistent with genome instability among *M. pneumoniae* (Krause *et al.*, 1982) and P1 gene variation as a result of recombination events (Su *et al.*, 1993). Since the P1 adhesin binding site is theorized to be comprised of multiple loops, one of which is located in a single-copy region and one in a multiple-copy region (Gerstenecker and Jacobs, 1990), the sequence differences observed among clinical isolates may influence mycoplasma affinity and tropism for target cells, antigenicity of P1 adhesin domains as previously suggested, and virulence potential (Su *et al.*, 1990a).

6. CHARACTERIZATION OF THE 30-kDa ADHESIN (P30) GENE OF *M. PNEUMONIAE*

Mutant analysis and immunological and ultrastructural studies also implicated the previously described P30 adhesin of *M. pneumoniae* in cytadherence (Baseman *et al.*, 1987). In order to clone and sequence the P30 gene, *M. pneumoniae* genomic libraries were constructed in the expression vector λgt11, and positive recombinant clones were identified using a pool of anti-P30 monoclonal antibodies (Dallo *et al.*, 1990a). The sequence data identified an open reading frame of 825 nucleotides and a protein of 275 amino acids with a molecular mass of approximately 30 kDa. Unexpectedly, the deduced P30 amino acid sequence revealed considerable homologies with the P1 adhesin and with various eukaryotic proteins such as vitronectin, collagen, and myosin, reminiscent of the sequence homologies of the P1 adhesin with mammalian matrix-related proteins (Su *et al.*, 1987; Dallo *et al.*, 1990a). Other relevant features of the P30 protein sequence were its high proline content (20.7%) and the presence of three types of repeat sequences at the carboxy end (Dallo *et al.*, 1990a). The function of these repeats is unclear although a spontaneous noncytadhering avirulent mutant of *M. pneumoniae* synthesizes a truncated "P30" protein (25 kDa) that differs from native P30 in the deletion of 8 of the 13 repeat sequences (unpublished data). Whatever structural and functional relationships exist between the P1 and P30 adhesin molecules, the membrane polarity of these proteins at the tip organelle and their involvement in receptor recognition are essential to *M. pneumoniae* cytadherence and virulence.

7. THE P1-ANALOGUE ADHESIN OF *M. GENITALIUM*

A new mycoplasma, *M. genitalium*, was isolated in 1980 from the urethra of nongonococcal urethritis patients (Tully *et al.*, 1983). This mycoplasma was flask shaped with a tiplike organelle similar to *M. pneumoniae*. Also, like *M. pneumoniae, M. genitalium* adhered to eukaryotic cells in a tip-mediated sialic acid-dependent fashion and shared extensive antigenic cross-reactivity with *M. pneumoniae* (Lind, 1982; Taylor-Robinson *et al.*, 1983; Tully *et al.*, 1983). Although only limited DNA homology (2 to 8%) was observed between *M. pneumoniae* and *M. genitalium* (Lind *et al.*, 1984; Yogev and Razin, 1986) and each displayed distinct DNA fingerprinting patterns (Baseman *et al.*, 1984), a search was initiated to identify putative adhesin molecules of *M. genitalium* using P1-related probes.

The application of immunoblotting techniques and anti-P1 monoclonal and monospecific antibodies provided the initial evidence that cross-reactive epitopes existed between the P1 adhesin of *M. pneumoniae* and a 140-kDa protein (P140) of *M. genitalium* (Morrison-Plummer *et al.*, 1987b). These proteins differed not only in mass but in amino acid sequence since several anti-P1 monoclonal antibodies that recognized distinct epitopes on P1 were incapable of binding to P140 (Clyde and Hu, 1986; Morrison-Plummer *et al.*, 1987b). Nonetheless, additional evidence clearly established the role of P140 in cytadherence. For example, the P140 protein of *M. genitalium* was shown to be surface located, clustered at the tip in close association with the nap layer as observed with *M. pneumoniae*, and strongly immunogenic in man and experimental animals like the P1 adhesin of *M. pneumoniae* (Hu *et al.*, 1987; Morrison-Plummer *et al*, 1987a,b). Most strikingly, a specific category of anti-P140 monoclonal antibodies was capable of binding to P1 and inhibiting *M. pneumoniae* cytadherence, suggesting similar biofunctional properties (Morrison-Plummer *et al.*, 1987b). This apparent link between the P1 and P140 adhesins of *M. pneumoniae* and *M. genitalium*, respectively, was dramatically reinforced by the discovery that *M. genitalium* could be coisolated with *M. pneumoniae* from throat specimens of patients hospitalized with pneumonia (Baseman *et al.*, 1988). The coexistence of these mycoplasmas in the respiratory tract parallels the biochemical and immunological similarities observed in their adhesins. Indeed, these adhesins may recognize identical or closely related host cell receptors, explaining the common tissue tropism of these mycoplasmas.

8. CHARACTERIZATION OF GENE AND PROTEIN SEQUENCE HOMOLOGIES BETWEEN THE P140 ADHESIN OF *M. GENITALIUM* AND THE P1 ADHESIN OF *M. PNEUMONIAE*

In order to identify the P140 gene, a λgt11 expression library of *M. genitalium* DNA was screened with a pool of anti-P140 monoclonal antibodies (Dallo *et al.*, 1989a). Positive recombinants were used to isolate DNA segments encoding the P140 gene, and all clones exhibited the same hybridization pattern with

Table IV
Similarities between *Mycoplasma pneumoniae*
and *M. genitalium* Adhesins

Preferential use of TGA for tryptophan
Extensive homologies at gene/protein levels
Single- versus multiple-copy nature of specific gene regions
Densely clustered at tip organelle
Signal peptide
Precursor protein processed to mature protein
No cysteines
Proline-rich carboxy-terminus
Isoelectric point at a basic pH
Strongly immunogenic in man and experimental animals
Protein sequence homologies with eukaryotic matrix components
Associated with operon–like organization

M. genitalium genomic DNA. Furthermore, the genetic relatedness between the P140 and P1 genes was established when 12 gene subclones spanning the entire P1 gene of *M. pneumoniae* were used to hybridize to *M. genitalium* DNA (Dallo *et al.*, 1989b). Out of the 12 subclones, 6 demonstrated the same hybridization pattern observed with the positive λgt11 recombinants while the remaining 6 subclones did not hybridize. Interestingly, the important single-copy regions of P1 that encoded the COOH-terminal cytadherence-mediating epitope(s) hybridized to *M. genitalium* genomic DNA, strengthening the biological and chemical relationships between the P1 and P140 adhesins.

Sequence analysis was performed on a restriction fragment from *M. genitalium* genomic DNA that contained the entire P140 gene (Dallo *et al.*, 1989a). A large open reading frame of 4335 nucleotides was detected which encoded a protein of 1445 amino acids with a calculated molecular mass of 159,668 Da. The deduced amino-terminus region of the P140 protein included positively charged amino acids followed by a stretch of hydrophobic amino acids, suggesting a signal sequence. Thus, the cleavage of this sequence as proposed earlier for

Table V
Differences between *M. pneumoniae* and *M. genitalium* Adhesins

G + C content of *M. genitalium* (39.9%) versus *M. pneumoniae* (53.5%) adhesin genes
Preferential use of A and T rather than G and C in two of three codon positions by *M. pneumoniae* (84 : 64)
Six of thirteen amino acids differ in analogue cytadherence binding site
Trypsin resistance of *M. genitalium* adhesin

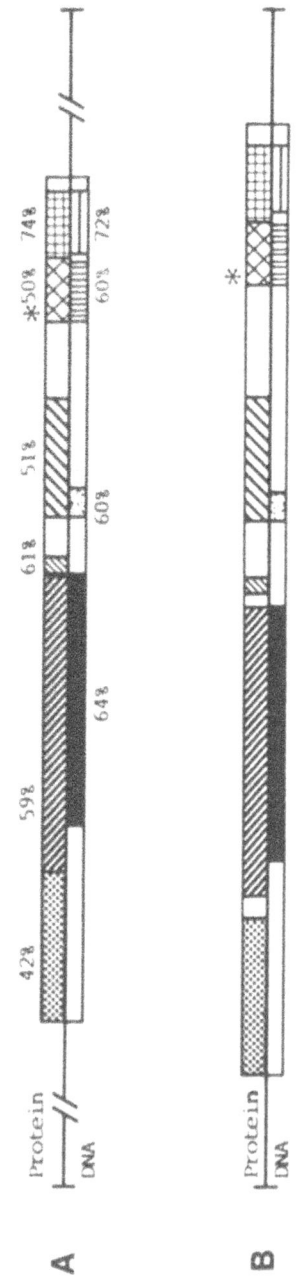

FIGURE 5. Regions of protein–gene homology between the adhesins of (A) *M. genitalium* and (B) *M. pneumoniae*. The percentages represent the degree of homology. The asterisks indicate the location of the 13-amino-acid cytadherence epitope of the *M. pneumoniae* P1 adhesin (-Gly-Ile-Val-Arg-Thr-Pro-Leu-Ala-Glu-Leu-Leu-Asp-Gly-) and the related epitope in the *M. genitalium* P140 adhesin (-Gly-Val-Val-Ser-Thr-Pro-Leu-Val-Asn-Leu-Ile-Asn-Gly-).

the P1 gene could correlate with a mature protein possessing a mass of 140 kDa. These data have been confirmed by Inamine *et al.* Figure 5 demonstrates the degree of gene (DNA) and protein homology observed between the adhesins of *M. pneumoniae* and *M. genitalium*, which was much higher than expected. Comparative properties of these adhesins appear in Tables IV and V.

The chemical and genetic relatedness of these adhesins clearly establishes the important evolutionary role of these virulence factors. Numerous flask-shaped mycoplasmas parasitic for man and animals have been reported (Kirchhoff *et al.*, 1984). The detection of gene and protein homologies associated with the adhesins of *M. pneumoniae* and *M. genitalium* and with the putative adhesins of other distinct mycoplasma species (Dallo and Baseman, 1990, and unpublished data) demonstrates the existence of a family of adhesin-related genes that is essential to the parasitic life cycle of these unique prokaryotes.

9. REFERENCES

Baseman, J. B., Banai, M., and Kahane, I., 1982, Sialic acid residues mediate *Mycoplasma pneumoniae* attachment to human and sheep erythrocytes, *Infect. Immun.* **38**:389–391.

Baseman, J. B., Cole, R. M., Krause, D. C., and Leith, D. K., 1982b, Molecular basis for cytadsorption of *Mycoplasma pneumoniae, J. Bacteriol.* **151**:1514–1522.

Baseman, J. B., Daly, K. L., Treviño, L. B., and Drouillard, D. L., 1984, Distinctions among pathogenic human mycoplasmas, *Isr. J. Med. Sci.* **20**:866–869.

Baseman, J. B., Drouillard, D. L., Leith, D. K., and Morrison-Plummer, J., 1985, Role of *Mycoplasma pneumoniae* adhesin P1 and accessory proteins in cytadsorption, in: *Molecular Basis of Oral Microbial Adhesin* (S. E. Mergenhagen and B. Rosan, eds.), American Society for Microbiology, Washington, D.C., pp. 18–23.

Baseman, J. B., Morrison-Plummer, J., Drouillard, D., Tryon, V. V., and Holt, S. C., 1987, Identification of a 32 kDa protein of *Mycoplasma pneumoniae* associated with hemadsorption, *Isr. J. Med. Sci.* **23**:474–479.

Baseman, J. B., Dallo, S. F., Tully, J. G., and Rose, D. L., 1988, Isolation and characterization of *Mycoplasma genitalium* strains from the human respiratory tract, *J. Clin. Microbiol.* **26**:2266–2269.

Biberfeld, G., 1971, Antibodies to brain and other tissues in cases of *Mycoplasma pneumoniae* infection, *Clin. Exp. Immunol.* **8**:319–333.

Clyde, W. A., Jr., and Hu, P. C., 1986, Antigenic determinants of the attachment protein of *Mycoplasma pneumoniae* shared by other pathogenic *Mycoplasma* species, *Infect. Immun.* **51**:690–692.

Collier, A. M., and Baseman, J. B., 1973, Organ culture techniques with mycoplasmas, *Ann. N.Y. Acad. Sci.* **225**:277–289.

Dallo, S. F., and Baseman, J. B., 1990, Cross-hybridization between the cytadhesin genes of *Mycoplasma pneumoniae* and *Mycoplasma genitalium* and genomic DNA of *Mycoplasma gallisepticum, Microb. Pathog.* **8**:371–375.

Dallo, S. F., Su, C. J., Horton, J. R., and Baseman, J. B., 1988, Identification of P1 gene domain containing epitope(s) mediating *Mycoplasma pneumoniae* cytadherence, *J. Exp. Med.* **167**:718–723.

Dallo, S. F., Chavoya, A., Su, C. J., and Baseman, J. B., 1989a, DNA and protein sequence homologies detected between the adhesins of *Mycoplasma genitalium* and *Mycoplasma pneumoniae*, *Infect. Immun.* **57**:1059–1065.

Dallo, S. F., Horton, J. R., Su, C. J., and Baseman, J. B., 1989b, Homologous regions shared by adhesin genes of *Mycoplasma pneumoniae* and *Mycoplasma genitalium*, *Microb. Pathog.* **6**:69–73.

Dallo, S. F., Chavoya, A., and Baseman, J. B., 1990a, Characterization of the gene for a 30-kilodalton adhesin-related protein of *Mycoplasma pneumoniae*, *Infect. Immun.* **58**:4163–4165.

Dallo, S. F., Horton, J. R., Su, C. J., and Baseman, J. B., 1990b, Restriction fragment length polymorphism in the cytadhesin P1 gene of human clinical isolates of *Mycoplasma pneumoniae*, *Infect. Immun.* **58**:2017–2020.

Fox, J. E. B., and Boyles, J. K., 1988, The membrane skeleton. A distinct structure that regulates the function of cells, *Bioessays* **8**:14–18.

Gerstenecker, B., and Jacobs, E., 1990, Topographical mapping of the P1-adhesin of *Mycoplasma pneumoniae* with adherence-inhibiting monoclonal antibodies, *J. Gen. Microbiol.* **136**:471–476.

Hu, P. C., Collier, A. M., and Baseman, J. B., 1976, Interaction of virulent *Mycoplasma pneumoniae* with hamster tracheal organ cultures, *Infect. Immun.* **14**:217–224.

Hu, P. C., Collier, A. M., and Baseman, J. B., 1977, Surface parasitism by *Mycoplasma pneumoniae* of respiratory epithelium, *J. Exp. Med.* **145**:1328–1343.

Hu, P. C., Schaper, U., Collier, A. M., Clyde, W. A., Jr., Horikawa, M., Huang, Y. S., and Barile, M. F., 1987, A *Mycoplasma genitalium* protein resembling the *Mycoplasma pneumoniae* attachment protein, *Infect. Immun.* **55**:1126–1131.

Inamine, J. M., Kenny, T. P., Loechel, S., Schaper, U., Huang, C. H., Bott, K. F., and Hu, P. C., 1988a, Nucleotide sequence of the P1 attachment-protein gene of *Mycoplasma pneumoniae*, *Gene* **64**:217–229.

Inamine, J. M., Loechel, S., and Hu, P. C., 1988b, Analysis of the nucleotide sequence of the P1 operon of *Mycoplasma pneumoniae*, *Gene* **73**:175–183.

Kahane, I., Tucker, S., Leith, D. K., Morrison-Plummer, J., and Baseman, J. B., 1985, Detection of the major adhesin P1 in Triton shells of virulent *Mycoplasma pneumoniae*, *Infect. Immun.* **50**:944–946.

Kirchhoff, H., Rowengarten, R., Lotz, W., Fischer, M., and Lopatta, D., 1984, Flask-shaped mycoplasmas: Properties and pathogenicity for man and animals, *Isr. J. Med. Sci.* **20**:848–853.

Krause, D. C., and Baseman, J. B., 1982, *Mycoplasma pneumoniae* proteins which selectively bind to host cells, *Infect. Immun.* **37**:382–386.

Krause, D. C., and Baseman, J. B., 1983, Inhibition of *Mycoplasma pneumoniae* hemadsorption and adherence to respiratory epithelium by antibodies to a membrane protein, *Infect. Immun.* **39**:1180–1186.

Krause, D. C., Leith, D. K., Wilson, R. M., and Baseman, J. B., 1982, Identification of *Mycoplasma pneumoniae* proteins associated with hemadsorption and virulence, *Infect. Immun.* **35**:809–817.

Krause, D. C., Leith, D. K., and Baseman, J. B., 1983, Reacquisition of specific proteins confers virulence in *Mycoplasma pneumoniae*, *Infect. Immun.* **39**:830–836.

Leith, D. K., and Baseman, J. B., 1984, Purification of a *Mycoplasma pneumoniae* adhesin by monoclonal antibody affinity chromatography, *J. Bacteriol.* **157**:678–680.

Lind, K., 1982, Serological cross-reactions between *Mycoplasma genitalium* and *M. pneumoniae*, *Lancet* **2**:1158–1159.

Lind, K., Lindhardt, O., Schütten, H. J., Blom, J., and Christiansen, C., 1984, Serological cross-reactions between *Mycoplasma genitalium* and *Mycoplasma pneumoniae*, *J. Clin. Microbiol.* **20**:1036–1043.

Meng, K. E., and Pfister, R. M., 1980, Intracellular structures of *Mycoplasma pneumoniae* revealed after membrane removal, *J. Bacteriol.* **144:**390–399.

Morrison-Plummer, J., Leith, D. K., and Baseman, J. B. 1986, Biological effects of anti-lipid and anti-protein monoclonal antibodies on *Mycoplasma pneumoniae, Infect. Immun.* **53:**398–403.

Morrison-Plummer, J., Jones, D. H., Daly, K., Tully, J. G., Taylor-Robinson, D., and Baseman, J. B., 1987a, Molecular characterization of *Mycoplasma genitalium* species-specific and cross-reactive determinants: Identification of an immunodominant protein of *M. genitalium, Isr. J. Med. Sci.* **23:**453–457.

Morrison-Plummer, J., Lazzell, A., and Baseman, J. B., 1987b, Shared epitopes between *Mycoplasma pneumoniae* major adhesin protein P1 and a 140-kilodalton protein of *Mycoplasma genitalium, Infect. Immun.* **55:**49–56.

Neimark, H., 1983, Mycoplasma and bacterial proteins resembling contractile proteins: A review, *Yale J. Biol. Med.* **56:**419–423.

Oliver, D., 1985, Protein secretion in *Escherichia coli, Annu. Rev. Microbiol.* **39:**615–648.

Powell, D. A., Hu, P. C., Wilson, M., Collier, A. M., and Baseman, J. B., 1976, Attachment of *Mycoplasma pneumoniae* to respiratory epithelium, *Infect. Immun.* **13:**959–966.

Razin, S., 1985, Mycoplasma adherence, in: *The Mycoplasmas* (S. Razin and M. F. Barile, eds.), Academic Press, New York, Vol. IV, pp. 161–202.

Razin, S., and Yogev, D., 1988, Molecular approaches to characterization of mycoplasmal adhesins, in: *Molecular Mechanisms of Microbial Adhesion* (E. Switalski, M. Hook, and E. Beachey, eds.), Springer-Verlag, Berlin, pp. 52–76.

Sperker, B., Hu, P. C., and Herrmann, R., 1991, Identification of gene products of the P1 operon of *Mycoplasma pneumoniae, Mol. Biol.* **5:**299–306.

Su, C. J., and Baseman, J. B., 1990, Genome size of *Mycoplasma genitalium, J. Bacteriol.* **172:**4705–4707.

Su, C. J., Tryon, V. V., and Baseman, J. B., 1987, Cloning and sequence analysis of cytadhesin gene (P1) from *Mycoplasma pneumoniae, Infect. Immun.* **55:**3023–3029.

Su, C. J., Chavoya, A., and Baseman, J. B., 1988, Regions of *Mycoplasma pneumoniae* cytadhesin P1 structural gene exist as multiple copies, *Infect. Immun.* **56:**3157–3161.

Su, C. J., Chavoya, A., and Baseman, J. B., 1989, Spontaneous mutation results in loss of the cytadhesin (P1) of *Mycoplasma pneumoniae, Infect. Immun.* **57:**3237–3239.

Su, C. J., Chavoya, A., Dallo, S. F., and Baseman, J. B., 1990a, Sequence divergency of the cytadhesin gene of *Mycoplasma pneumoniae, Infect. Immun.* **58:**2669–2674.

Su, C. J., Dallo, S. F., and Baseman, J. B., 1990b, Molecular distinctions among clinical isolates of *Mycoplasma pneumoniae, J. Clin. Microbiol.* **28:**1538–1540.

Su, C. J., Dallo, S. F., Chavoya, A., and Baseman, J. B., 1993, Possible origin of sequence divergency in the P1 cytadhesin gene of *Mycoplasma pneumoniae, Infect. Immun.* **61:**816–822.

Taylor-Robinson, D., Furr, F. M., and Tully, J. G., 1983, Serological cross-reaction between *Mycoplasma genitalium* and *M. pneumoniae, Lancet* **1:**527.

Treviño, L. B., Haldenwang, W. G., and Baseman, J. B., 1986, Expression of *Mycoplasma pneumoniae* antigens in *Escherichia coli, Infect. Immun.* **53:**129–134.

Tully, J. G., Taylor-Robinson, D., Cole, R. M., and Bove, J. M., 1983, *Mycoplasma genitalium*, a new species from the human urogenital tract, *Int. J. Syst. Bacteriol.* **33:**387–396.

Wise, K. S., and Watson, R. K., 1985, Antigenic mimicry of mammalian intermediate filaments by mycoplasmas, *Infect. Immun.* **48:**587–591.

Yamao, F., Muto, A., Kawauchi, Y., Iwami, M., Iwagami, S., Azumi, Y., and Osawa, S., 1985, UGA is read as tryptophan in *Mycoplasma capricolum, Proc. Natl. Acad. Sci. USA* **82:**2306–2309.

Yogev, D., and Razin, S., 1986, Common deoxyribonucleic acid sequences in *Mycoplasma genitalium* and *M. pneumoniae* genomes, *Int. J. Syst. Bacteriol.* **36:**426–430.

Chapter 10

Ion Pumps and Volume Regulation in Mycoplasma

Mitchell H. Shirvan and Shlomo Rottem

1. INTRODUCTION

Membrane-bound transport processes regulate the ionic environment within cells. These processes control the concentration of ions within a cell by moving ions across the cell membrane and, by doing so, generating electrochemical gradients. These gradients can be used for many essential functions such as volume regulation, regulation of intracellular pH, and transport of nutrients into the cell.

The Mollicutes are a class of microorganisms that do not have the peptidoglycan-based rigid cell wall. These organisms are bound by a single membrane, the plasma membrane, and must, therefore, rely on ion transport mechanisms for one of the most basic physiological functions, volume regulation. The three major families comprising the class Mollicutes are *Acholeplasmataceae*, *Mycoplasmataceae*, and *Spiroplasmataceae*, with the corresponding species *Acholeplasma*, *Mycoplasma*, and *Spiroplasma*. The trivial name mycoplasmas will be used in this review to denote any species included in the class

Mitchell H. Shirvan   Teva Pharmaceutical Industries, Jerusalem 91010, Israel.   **Shlomo Rottem**   Department of Membrane and Ultrastructure Research, The Hebrew University–Hadassah Medical School, Jerusalem 91010, Israel.

Subcellular Biochemistry, Volume 20: Mycoplasma Cell Membranes, edited by Shlomo Rottem and Itzhak Kahane. Plenum Press, New York, 1993.

Mollicutes, whereas the trivial names acholeplasmas and spiroplasmas will be used when reference is made to members of the specific genus.

Mycoplasmas are the more prevalent parasites of man, animals, plants, and insects. Having various natural habitats, these organisms must have different transport mechanisms to adapt to the different environments. For example, *Mycoplasma gallisepticum,* the causative agent of a respiratory disease in poultry, must contend with an environment which is rich in Na^+. The object of this review is to describe the membrane-bound ion-transport processes in mycoplasmas.

2. PRIMARY ION TRANSPORT SYSTEMS

A primary transport mechanism is a process whereby the energy for transport comes directly from the potential energy of cellular metabolites. The transport of ions across a sealed cell membrane, without a compensatory movement of ions, results in the generation of a chemical potential for that ion. For example, a H^+-ATPase would generate an electrochemical potential of H^+. This potential, also known as the proton-motive force ($\Delta\mu_{H+}$), consists of a chemical gradient of H^+ (ΔpH) and an electrical gradient (membrane potential, $\Delta\Psi$). In bacteria, H^+ (Harold, 1983) and Na^+ (Stewart *et al.,* 1985; Tokuda and Unemoto, 1981, 1982, 1983) transporting respiratory chains, and more recently Na^+-transporting decarboxylases (Dimroth, 1980, 1987) are examples of primary transport. Primary transport in prokaryotes, however, is best exemplified by ATPases which use the chemical energy from the hydrolysis of the terminal phosphate bond of ATP to drive the transport process. ATPases that transport ions have been classified into three major classes (Dean *et al.,* 1984; Pedersen and Carafoli, 1987): F, P, and V. V-class ATPases, which refer to ATPases found in vacuoles, as well as other intracellular organelles of eukaryotes, are the least characterized and have not been described in bacteria (Al-Awqati, 1986; Pedersen and Carafoli, 1987). P-class ATPases (or E_1 and E_2) operate through a phosphorylated intermediate and exist in two conformation states during their reaction cycle. They are exemplified by the K^+-transporting ATPases of *Escherichia coli* (Walderhaug *et al.,* 1986) and *Streptococcus faecalis* (Furst and Solioz, 1986). The F-class ATPases, also referred to as F_1F_0, have two major structural domains, a peripheral membrane portion (F_1) and an integral membrane portion (F_0). They have been found in various bacteria and are exemplified by the plasma membrane H^+-ATPase of *E. coli.* The F_1 moiety of this ATPase carries the catalytic activity and consists of two major subunits, α and β, and three minor subunits, γ, δ, and ϵ. The catalytic site of the enzyme is believed to be on the β subunit, which is very highly conserved in nature (Futai and Kanazawa, 1983; Pedersen and Carafoli, 1987). The F_0 moiety is integral to the membrane, and is composed of three subunits, a, b, c (Senior, 1990).

2.1. *Mycoplasma gallisepticum*

2.1.1. ATP-Dependent Na^+ Pump

Since the natural habitat of *M. gallisepticum* cells is a high Na^+ environment, they must extrude Na^+ from the cell for survival. The extrusion of $^{22}Na^+$ from *M. gallisepticum* cells, suspended in an isoosmotic medium containing 250 mM NaCl, was shown to be energy dependent (Shirvan *et al.*, 1989a). T. H. Wilson proposed (Linker and Wilson, 1985a) that this Na^+ efflux was mediated by a combination of a H^+-translocating ATPase and a Na^+/H^+ antiporter. It was therefore suggested that efflux is driven by the $\Delta\mu_{H^+}$. In support of this proposal, the addition of glucose to Na^+-loaded cells resulted in a concomitant Na^+ efflux and an internal acidification.

By contrast, Shirvan *et al.* (1989a) reported several lines of evidence indicating that a Na^+/H^+ antiporter is not present in *M. gallilsepticum* cells. The most significant results showed that $^{22}Na^+$ efflux from $^{22}Na^+$-loaded cells was not inhibited in the presence of the proton ionophore carbonyl cyanide *m*-chlorophenyl-hydrazone (CCCP), which collapses the $\Delta\mu_{H^+}$. These results were observed whether the Na^+ gradient was in the downhill or uphill direction, i.e., in cells suspended in a Na^+-free medium or in a medium in which the Na^+ concentration was much greater than the internal Na^+ concentration. Further, the generation of ΔpH by the ammonium chloride dilution procedure (Nakamura *et al.*, 1986) did not drive $^{22}Na^+$ uptake into cells (Shirvan *et al.*, 1989a), sealed membrane vesicles (Cirillo *et al.*, 1987; Shirvan *et al.*, 1989a), or reconstituted membrane vesicles (Shirvan *et al.*, 1989a). Transport studies, therefore, indicated that a Na^+/H^+-exchange activity is not present in these cells. Further support comes from physiological studies in which *M. gallisepticum* cells grew and remained viable in a high-salt-containing medium (124 mM NaCl) in the presence of CCCP (Shirvan *et al.*, 1989a). Consistent with these observations, no hybridization was observed between a plasmid containing the *E. coli* Na^+/H^+-antiporter gene and *M. gallisepticum* genomic DNA under conditions of low stringency, where a homology of approximately 50% could be detected. It was therefore proposed that Na^+ is transported from *M. gallisepticum* by an alternate mechanism not involving secondary proton transport (Shirvan *et al.*, 1989a).

Na^+ extrusion was further characterized by following the efflux of Na^+ from intact *M. gallisepticum* cells. The addition of glucose to the medium resulted in the generation of a $\Delta\Psi$, negative inside the cell (Shirvan *et al.*, 1989b). This negative potential drove the uptake of H^+ into the cell upon the addition of proton ionophores. The energy-dependent efflux was specific for Na^+ (Shirvan *et al.*, 1989b) and could not be replaced by K^+, Li^+, Rb^+, or Cs^+. Since the assay necessitates the generation of a $\Delta\Psi$ across the cell membrane for efflux to be visualized, only electrogenic transport is followed. The electrogenic nature of this Na^+ transport is consistent with $^{22}Na^+$ efflux studies in intact cells, where efflux was stimulated by conditions that reduce $\Delta\Psi$, i.e., high external K^+ in the presence of valinomycin. Efflux was also substantially reduced by conditions that increase $\Delta\Psi$, i.e., valinomycin alone, indicating that the membrane potential in the cell was reaching the reversal potential of the Na^+ pump (Shirvan *et al.*, 1989b). In addition, Na^+ efflux occurred in the presence of proton ionophores which collapse $\Delta\mu_{H^+}$, further indicating that Na^+ transport was not a secondary process coupled to H^+, but a primary electrogenic transport.

Further transport studies showed an almost total inhibition of Na^+efflux by the ATPase inhibitors dicyclohexylcarbodiimide (DCCD) and diethylstilbestrol (DES) (Shirvan *et al.*, 1989b), but little or no inhibition occurred in the presence of azide, *N*-ethylmaleimide (NEM), or vanadate. These results strongly suggest that the electrogenic pump responsible for Na^+ transport in *M. gallisepticum* is an ATPase.

2.1.2. ATP-Dependent H^+ Pump

The presence of an ATP-dependent H^+ pump in *M. gallisepticum* was suggested by the work of Rottem *et al.* (1981) and Scheifer and Schummer (1982), which showed that the $\Delta\mu_{H^+}$ across the plasma membrane was energy dependent. Indeed, it was proposed that most, if not all, prokaryotes are believed to possess a proton-translocating ATPase (Harold and Kakinuma, 1985; Razin, 1978) and are capable of generating a $\Delta\mu_{H^+}$. Linker and Wilson (1985b) showed that DCCD not only inhibited 95% of the total membrane-bound ATPase activity, but also collapsed both ΔpH and $\Delta\Psi$. Consistent results of Schummer *et al.* (1981) showed a significant reduction or inhibition of ΔpH by DCCD. In addition, when the pH of the suspension medium was varied over the external pH range of 6.0–8.5, the internal pH of the cells (pH_i) remained fairly constant, ranging from $pH_i = 6.8$ at an external pH of 6.0 ($\Delta pH_{mV} = 50$ mV) to $pH_i = 7.5$ at an external pH of 8.5 ($\Delta pH_{mV} = 25$ mV) (Shirvan *et al.*, 1989b). In the presence of the proton ionophores SF6847 and CCCP, at an external pH = 6.0, ΔpH_{mV} decreased by approximately 50%. However, at pH 8.5, where ΔpH is in an opposite direction to that observed at pH 6.0 (inside relatively acidic), the ΔpH_{mV} was slightly increased. These results may suggest an H^+-ATPase activity in the membrane, which is more active at the acidic pH range.

2.1.3. Inhibition Profile, pH Sensitivity, and Substrate Specificity of the Membrane-Bound ATPase Activity

The membrane-bound ATPase activity of *M. gallisepticum* was characterized by Linker and Wilson (1985b) and further studied by Shirvan *et al.*

(1989b). Membrane preparations selectively hydrolyzed purine nucleoside triphosphates and dATP. ADP, although not a substrate, inhibited ATP hydrolysis. The membranes exhibited an ATPase activity over a broad pH range, with an optimum activity between pH 7.0 and 7.5. ATPase activity showed an obligatory requirement for divalent cations, which was fulfilled by Mg^{2+} and to a lesser extent by Co^{2+}, Mn^{2+}, or Ca^{2+} (Linker and Wilson, 1985b). Na^+ had practically no effect on the ATPase activity at the acidic pH range (pH 5.0–6.5). At the neutral pH range (pH 7.0–7.5), the ATPase activity was moderately stimulated ($\sim$ 37%) (Shirvan et al., 1989b), whereas at the more alkaline pH range (pH 8.5–9.0), the ATPase activity was activated threefold by 10 mM NaCl. The ATPase activity of M. gallisepticum was markedly inhibited ($\sim$ 95%) by DCCD, though at concentrations somewhat higher than those required to inhibit known F-class ATPases. The enzyme was also very sensitive to DES, quercetin, and p-chloromercuribenzoate (pCMB) (Linker and Wilson, 1985b; Shirvan et al., 1987, 1989b). The activity was only moderately sensitive to azide or vanadate, and was insensitive to NEM, ouabain, oligomycin, or efrapeptin (Linker and Wilson, 1985b; Shirvan et al., 1989b) as well as to aurovertin (Rasmussen et al., 1992).

2.1.4. Immunological Studies Supporting the Presence of an F-Type ATPase

In support of the view that mycoplasmas possess an F-class ATPase, monospecific polyclonal antibodies against the β subunit of the E. coli F-ATPase were found to cross-react with a 52-kDa protein present in the cell membrane of M. gallisepticum (Zilberstein et al., 1986) as well as A. laidlawii, A. axanthum, A. granularum, M. capricolum, M. mobile, M. pneumoniae, M. hominis, S. citri, and Spiroplasma sp. strain BNR1 (Fischer et al., 1988; Rottem et al., 1987; Zilberstein et al., 1986).

In M. gallisepticum the protein was shown to be catalytically active, exhibiting up to 44% of the total membrane-bound ATPase activity. In contrast to other F-class ATPases (Fillingame, 1981), M. gallisepticum ATPase activity could not be released from the membrane by repeated washings with either low- or high-salt solutions, in the presence or absence of EDTA. These findings suggest that in mycoplasmas, the F-class ATPase undergoes structural modifications which allow for its integration into the cell membrane (Zilberstein et al., 1985). In addition, the lower sensitivity of the M. gallisepticum ATPase to DCCD and its insensitivity to oligomycin (Linker and Wilson, 1985b,c; Shirvan et al., 1986b) may also reflect the difference in the association of the mycoplasmal F-ATPase with the membrane, relative to other F-ATPases, or may be due to the presence of a second ATPase.

2.1.5.   Amino Acid Sequence of the *M. gallisepticum* ATPase Operon Deduced from Cloned Genes

Recently the H^+-ATPase operon of *M. gallisepticum* has been identified by heterogeneous hybridization (Rasmussen *et al.*, 1987, 1990) and the operon cloned and completely sequenced (Rasmussen *et al.*, 1992). Analysis of the operon shows that it contains a cluster of nine structural genes encoding eight subunits of the ATPase, as well as a gene (*I*) that precedes the *B* gene and codes for a hydrophobic polypeptide (Table I, Figure 1). The gene order is *I* (I subunit), *B* (a subunit), *E* (c subunit), *F* (b subunit), *H* (δ subunit), *A* (α subunit), *G* (γ subunit), *D* (β subunit), and *C* (ε subunit). The location of the polypeptide encoded by the *I* gene suggests that it may be analogous to the regulator *Unc* I gene of the *E. coli* ATP synthase operon (Gay and Walker, 1981). The α and β subunits of the *M. gallisepticum* ATPase are characterized by a high degree of identity with the corresponding subunits of *E. coli* (51 and 65%, respectively). The α subunit is the largest subunit in *E. coli* and is known to possess three functional regions: a membrane binding region at the NH_2-terminus (residues 1–15), a nucleotide binding domain (residues 170–177), and an α/β signal transmission region (residues 345–375) (Senior, 1990). In *M. gallisepticum* and *E. coli,* the amino acid sequence of this subunit at the functional regions showed a high degree of homology, suggesting that the α subunit of both organisms evolved from the same ancestor gene, and have a very similar three-dimensional structure.

The β subunit of *M. gallisepticum* contains the catalytic site for the ATPase

Table I
Characterization of the *M. gallisepticum* ATPase Subunits
Encoded by Genes of the *atp* Operon

atp genes		ATPase subunits			
Genes	% G + C	Subunit	Number of residues	Size (kDa)	% identity with *E. coli*
atp I	32.6	I	169	18.7	16.1 (19/120)[a]
atp B	37.8	a	297	33.3	22.6 (61/269)
atp E	38.3	c	96	10.0	25.6 (20/78)
atp F	32.4	b	198	22.7	21.3 (33/155)
atp H	32.0	δ	181	21.6	25.3 (44/174)
atp A	37.2	α	518	57.8	51.2 (256/500)
atp G	33.0	γ	289	33.1	27.9 (78/279)
atp D	38.0	β	470	51.5	65.2 (307/471)
atp C	35.8	ε	123	14.9	19.2 (25/130)

[a]Number of identical amino acids/No. of total amino acids.

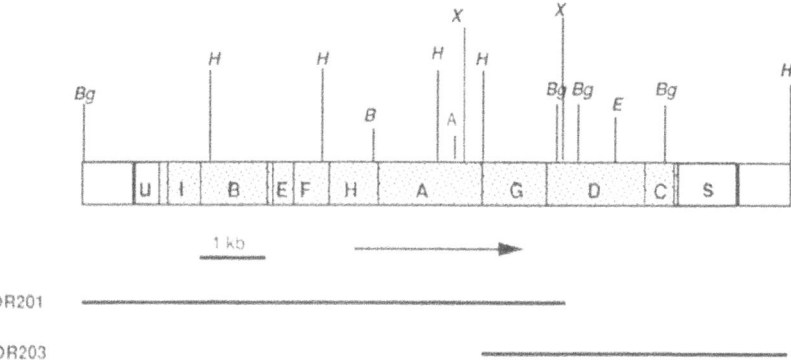

FIGURE 1. Physical map of the region in the *M. gallisepticum* genome encoding the *atp* operon. The *atp* genes are denoted by capital letters. U and S are two open reading frames flanking the *atp* operon. Dotted rectangles show the reading frames of the nine *atp* genes (lightly dotted) as well as U and S (heavily dotted). The following restriction sites are shown: A, *Ava*I; B, *Bam*HI; Bg, *Bgl*II; E, *Eco*RI; H, *Hind*III; X, *Xba*I. Arrow shows the direction of translation.

complex (Futai and Kanazawa, 1983) and its structure showed an unusually high degree of homology with the corresponding subunits of various microbial species, chloroplasts and mitochondria (Rasmussen *et al.*, 1992; Walker *et al.*, 1985), suggesting that the catalytic site is highly conserved in nature. However, an important difference has been reported between the mycoplasma β subunit and the β subunit from other F-class ATPases, which has been correlated with sensitivity to the ATPase inhibitor aurovertin. The antibiotic aurovertin inhibits the activity of F-class ATPases of various species. It appears that an arginine residue at position 398 of the *E. coli* F-ATPase, or an equivalent position on the β subunit from other species, is essential for ATPase activity to be sensitive to aurovertin (Senior, 1990). For example, in *E. coli*, a mutation which replaced this arginine residue with a histidine (R398 → H) resulted in an ATPase activity which was resistant to aurovertin. In *M. gallisepticum*, the residue equivalent to *E. coli* R398 is an asparagine residue (N406). As expected, the ATPase activity of *M. gallisepticum* was found to be almost completely resistant to high concentrations of aurovertin (Rasmussen *et al.*, 1992).

The a, b, c, γ, δ, and ε subunits showed a low degree of homology with the corresponding *E. coli* ATP synthase genes (22–28%). The amino acid sequences, as deduced from the nucleotide sequences, were also extensively evaluated by computer analysis with respect to secondary structure predictions, polarity profiles, and homologies with corresponding subunits from other organisms. It has been concluded from these previous studies that the δ and ε subunits are required for binding of F_1 to F_0 (Senior, 1990). Similar to that found in *E. coli*, the δ and ε subunits of *M. gallisepticum* have highly helical structures similar to those found in *E. coli*, and most probably form part of the stalk linking F_1 to F_0. Additional studies on the δ subunit of *M. gallisepticum* showed a very strong potential for amphipathic α-helices in the COOH-terminal region. It was suggested that such α-helices are interacting with a hydrophobic segment of the b subunit and, as a result, the F_1 portion could not be detached from the membrane by varying the osmolarity of the medium (Rasmussen *et al.*, 1992).

2.1.6. Does the *M. gallisepticum* Membrane Possess a Second ATPase Activity?

It is apparent from the immunochemical (Zilberstein *et al.*, 1986) and genetic studies (Rasmussen *et al.*, 1990) that *M. gallisepticum* membranes possess an F-class ATPase. Nonetheless, the properties of the membrane-bound ATPase activity are not consistent with any known ATPases, mainly with respect to Na^+ stimulation at the alkaline pHs, sensitivity to DCCD, and tight association with the membrane. Therefore, the possibility of a second ATPase cannot be excluded. Several lines of evidence were presented in support of the notion of a second ATPase activity. The first observation was that two pH optima of total membrane-bound ATPase activity were observed, in the presence and absence of Na^+ (Shirvan *et al.*, 1989b). In the absence of Na^+, no sharp pH optimum of total ATPase activity was recorded, with maximum activity recorded between pH 7.0 and 8.0. In the presence of Na^+ (10 mM), however, the pH optimum of the total membrane ATPase activity was shifted to a more alkaline range, pH 8.5–9.0. Additional support comes from the immunological studies of Zilberstein *et al.* (1986), whereby antibodies generated against the β subunit of the *E. coli* F-ATPase immunoprecipitated only 44% of the total *M. gallisepticum* ATPase activity. These finding are consistent with the proposal that there are two ATPase activities in the *M. gallisepticum* membrane (Shirvan *et al.*, 1989b). A further indication of two ATPases was reported by Shirvan *et al.* (1989b), who showed that polyclonal antibodies generated against the yeast plasma membrane H^+-ATPase, a P-class ATPase, cross-reacted with a 68-kDa protein present in *M. gallisepticum* membranes.

Structurally, all P-ATPases have been reported to have a peptide of approximately 70–100 kDa, which contains the phosphorylation and ATP binding sites (Pedersen and Carafoli, 1987). In *M. gallisepticum*, the protein that cross-reacted with the antibody to the yeast plasma membrane ATPase had a molecular mass that was consistent with these findings. In addition, plasma membrane P-class H^+-ATPases are inhibited by DCCD and DES (Pedersen and Carafoli, 1987), which can explain the sensitivity of total membrane-bound ATPase activity to these compounds. Vanadate, on the other hand, is a potent inhibitor of P-ATPases and totally inhibited the yeast plasma membrane ATPase (Pedersen and Carafoli, 1987). The ATPase activity of a membrane with a second ATPase of the P class would theoretically be only partially inhibited by vanadate. Indeed, total *M. gallisepticum* membrane-bound ATPase activity showed an incomplete inhibition of ATPase activity (Linker and Wilson, 1985b; Shirvan *et al.*, 1989b). Although the inhibition varied with different vanadate and membrane preparations (24–60%), the maximum inhibition of 60% (Linker and Wilson, 1985b) is consistent with the additional presence of a P-ATPase in *M. gallisepticum*. Analogously, two ATPases have been reported in *Spiroplasma citri*, one being an F-class ATPase and the other a vandate-sensitive ATPase, probably of the P class (Simoneau and Labarere, 1991). If two ATPases are present in *M. gallisepticum*, the question arises as to which ATPase activity is the electrogenic Na^+ pump in *M. gallisepticum*.

Ion-transporting ATPases have been generally considered to be of the P class (Dean *et al.*, 1984; Pedersen and Carafoli, 1987). Examples of P-ATPases are the Na^+/K^+-, Ca^{2+}-, and H^+-ATPases of the plasma membrane of eukaryotic cells and the K^+-transporting ATPases of *E. coli* and *S. faecalis* (Al-Awqati, 1986; Furst and Solioz, 1986; Pedersen and Carafoli, 1987). No H^+-ATPases of this class have been found in prokaryotes (Al-Awqati, 1986; Pedersen and Carafoli, 1987). Therefore, "conventional wisdom" would indicate that the P-ATPase is the electrogenic Na^+ pump in *M. gallisepticum*. Consistent with this view, active Na^+ extrusion from cells was inhibited by DCCD and DES, but not by NEM or azide (Shirvan *et al.*, 1989a,b). By contrast, vanadate had little or no effect on the Na^+ pump. The interpretation of this result, however, is not clear for several reasons. Vanadate inhibits ATPase activity by binding to the phosphorylation site of the ATPase from the cytoplasmic side of the membrane and is equilibrated across the membrane via the anion exchange system (Macara *et al.*, 1980). In *M. gallisepticum*, no such exchange system has been shown. The charged vanadate ion may, therefore, be excluded from its site of action in the cell, a condition which would not be observed when studying ATPase activity in membranes. Furthermore VO_4^{3-} ions in the cell could be reduced to vanadyl ions (V^{4+}), and thereby inactivate the active species (Furst and Solioz, 1986).

It must be considered that another ATPase is responsible for Na^+ transport out of the cell. Studies on the immunoprecipitation of *M. gallisepticum* membranes with antibodies against the beta subunit of the F-ATPase showed that the percentage of ATPase activity that was immunoprecipitated corresponds approximately to the percentage of Na^+-stimulated ATPase activity found in *M. gallisepticum* membranes at pH 8.3 (Shirvan *et al.*, 1989b; Zilberstein *et al.*, 1986). In addition, the pH optimum of membrane-bound ATPase activity, in the presence of Na^+, was approximately the same as that of ATPase activity on the beta subunit (D. Zilberstein, unpublished data). These studies suggest that the F-class ATPase in the *M. gallisepticum* membrane is in fact the electrogenic Na^+ pump. Consistent with this idea, an F-class ATPase that transports Na^+ has recently been reported in *Propionigenium modestum* (Laubinger and Dimroth, 1988).

This enzyme is strongly inhibited by DCCD, and has a very similar subunit pattern as the F_1 subunit in the *E. coli* enzyme. Differences between this ATPase and the ATPase activity in *M. gallisepticum*, however, are reflected in the sensitivity of the *P. modestum* ATPase to azide (Laubinger and Dimroth, 1988; Shirvan *et al.*, 1989b). In addition, the membrane-bound ATPase in *Acholeplasma laidlawii* has been proposed to be an F-ATPase which acts as a primary Na^+ pump (Mahajan *et al.*, 1988; Rottem *et al.*, 1983) (see below for discussion).

In conclusion, the results accumulated so far indicate the presence of two ATPases in *M. gallisepticum*, a Na^+-ATPase and a H^+-ATPase. The former is stimulated by Na^+ and has a pH optimum at 8.5–9.0, and some characteristics of an F-class ATPase. The latter is observed in the absence of Na^+, has a broad pH optimum, and is most probably a P-class ATPase. The observations are in contrast to the more typical F-class ATPases that generally transport H^+, and the P-class ATPases in bacteria that generally transport cations. Further, it cannot be ruled out that the F-ATPase of *M. gallisepticum* transports Na^+ in addition to H^+. It has recently been reported by Boyer (1988) that the specificity of H^+ pumps may be due to the molecules on the transporter that binds the form of H^+ that exists in aqueous solutions, namely the hydronium ion (H_3O^+). It was further suggested that the mechanism of binding of H_3O^+ may be quite analogous to that of Na^+, and achieved by an appropriate arrangement of the coordinating atoms. The structural arrangement of these atoms may be analogous to the cation binding site in the crown ether, dicyclohexyl-18-crown-6. By assuming such a mechanism, the ATPase may be able to transport Na^+ in addition to H^+. This dual relationship between Na^+ and H^+ has been found in other prokaryotes. The melibiose carrier of *E. coli* takes up melibiose into the cell in response to a $\Delta\mu_{H^+}$ or a $\Delta\mu_{Na^+}$, and only a small structural change, a replacement of Pro-122 for Ser, is necessary to lose the $\Delta\mu_{H^+}$ coupling and acquire a coupling to $\Delta\mu_{Li^+}$ (Boyer, 1988; Yatzyu *et al.*, 1985). Further, it has been proposed by Dibrov *et al.* (1988) that in *Vibrio alginolyticus*, ATP synthesis by the ATP synthase is driven by either $\Delta\mu_{H^+}$ or $\Delta\mu_{Na^+}$, depending on whether the external H^+ concentration is either high or low. Reconstitution experiments of purified forms of the ATPases in *M. gallisepticum*, as well as expression of the cloned F-ATPase, should provide more definitive answers to these questions.

2.2. *Acholeplasma laidlawii*

2.2.1. ATP-Dependent Na^+ Pump

A. laidlawii B cells, suspended in isoosmotic solutions of NaCl or KCl, swelled and slowly lysed in the absence of an energy source (Jinks *et al.*, 1978). Addition of glucose to the cell medium resulted in a reversal of the swelling process. It was proposed from these experiments that the membrane ATPase acts as an ion pump which actively extrudes cations, probably Na^+, from the cell. Na^+ transport from cells was directly demonstrated using ^{23}Na-NMR spectroscopy (Mahajan *et al.*, 1988). *A. laidlawii* extruded Na^+ from the cell against a Na^+ concentration gradient upon the addition of glucose to the suspension medium. Electrogenic Na^+ transport, however, has not been detected in these cells (Lelong and Rottem, unpublished data). The finding that Na^+ transport in this organism was influenced by the physical state of membrane lipids (Mahajan *et al.*, 1988) strongly suggested that transport was occurring via an intrinsic membrane protein.

2.2.2. Inhibition Profile and pH Sensitivity of Membrane-Bound ATPase Activity

Studies on *A. laidlawii* membranes supported the notion that Na^+ transport was occurring via the membrane-bound ATPase. ATP hydrolysis was stimulated three- to fourfold in the presence of 10 mM Na^+. Kinetic analysis showed that Na^+-stimulated activity was also influenced by the physical state of the membrane, apparently affecting V_{max} but not the K_m for ATP (Jinks *et al.*, 1978).

Studies on the total membrane-bound ATP activity in *A. laidlawii* showed that the pH optimum for the membrane-bound activity was similar to that found in *M. gallisepticum* membranes, under low-Na^+ conditions. The activity showed no sharp pH optimum, and was essentially constant between pH 7.0 and 8.0. The activity dropped off rapidly at pH values lower than 7.0 and above 8.0 (Jinks *et al.*, 1978; Rottem and Razin, 1966). Further, the activity showed a strict Mg^{2+} dependency, which could be partially replaced by Mn^{2+} or Co^{2+}, but not by Ca^{2+}. Inhibition profiles showed the membrane-bound ATPase activity to be strongly inhibited by DCCD, NEM, and *p*CMB, but not by azide or ouabain (Rottem *et al.*, 1983). Since ATPase activity was not inhibited by ouabain or stimulated by K^+, it would be very unlikely that a Na^+/K^+-ATPase, similar to the one found in animal cells, exists in *A. laidlawii*. Further, the lack of inhibition by azide would indicate that the classical bacterial F-ATPase, such as that found in *E. coli*, does not exist. In addition, the NEM concentration necessary to inhibit ATPase activity in *A. laidlawii* (Rottem and Razin, 1966) was much higher than the concentrations required to inhibit V-class ATPases (Dean *et al.*, 1984).

2.2.3. Structure of the Purified ATPase

Polyclonal antibodies against the β subunit of the *E. coli* plasma membrane ATPase cross-reacted with a 52-kDa protein in the membrane of *A. laidlawii* (Zilberstein *et al.*, 1986). These results suggest the presence of an F-class AT-

Pase in the *A. laidlawii* membrane. An ATPase activity was purified by Lewis and McElhaney (1983) as well as by Chen *et al.* (1984) and resolved into subunit structure resembling an F-ATPase containing five subunit bands, designated α, β, γ, δ, and ϵ. Ultracentrifugation studies showed that the purified enzyme, under low-salt conditions, behaves as a mixture containing a major species of ~ 600 kDa and a smaller species of ~ 100 kDa.

One of the possible differences between the *A. laidlawii* B ATPase and a classical F-ATPase may lie in the α subunit of this enzyme. Results from Lewis and McElhaney (1983), using photoactivable phospholipids in the presence of cross-linking agents, suggest that the α subunit of the purified ATPase is buried in the hydrophobic core of the lipid bilayers (George *et al.*, 1985; Rottem *et al.*, 1983). This is in contrast to typical F_1 moieties which do not penetrate the hydrophobic core of the lipid bilayer. This finding may also suggest why the F_1 moiety of this enzyme cannot be released from the membrane in low-ionic-strength buffers containing EGTA.

Further studies using cross-linking agents suggested that the γ and δ subunits are likely to be composed of two subunits, indicating a seven-subunit complex (Rottem *et al.*, 1983). It has not been reported, however, whether the β subunit of the purified ATPase cross-reacts with the antibody against the E. coli F-ATPase.

2.2.4. Inhibition Profile and Substrate Specificity of Purified ATPase

The properties of the purified enzyme were found to be similar to those of the membrane-bound ATP hydrolase activity, with respect to substrate inhibition. The purified enzyme was completely inhibited by 10 mM ATP, and in comparison with other nucleoside triphosphates, the hydrolysis of ATP occurred at a faster rate. Although the structure of the purified enzyme was similar to an F-class ATPase, the enzyme was not very sensitive to classical F-ATPase inhibitors of animal or bacterial cells. Most bacterial F-ATPases are inhibited by azide, whereas animal F-ATPases are inhibited by oligomycin. Both prokaryotic and eukaryotic F-ATPases have been shown to be sensitive to DCCD (Dean *et al.*, 1984; Pedersen and Carafoli, 1987). The most effective inhibitors of the purified enzyme were vanadate and leucinostatin. Each, however, inhibited the *A. laidlawii* enzyme only 50% at very high concentrations (50 μM). For example, inhibition of the Na^+, K^+-ATPase from dog kidney, which is a prototype ATPase containing a phosphorylated intermediate, was inhibited by vanadate at concentrations ≤ 0.1 μM (Dean *et al.*, 1984). A more modest inhibition of the *A. laidlawii* ATPase was observed with DCCD, which inhibited the activity by 22% at 50 μM. Oligomycin (100 μg/ml) inhibited the *A. laidlawii* ATPase activity by 15%. A similar inhibition was observed by efrapeptin at a concentration of 1 mg/ml (Rottem *et al.*, 1983). No inhibition was observed with azide, aurovertin, quercetin, or ouabain. This inhibition pattern of the purified enzyme differs from that found for the total ATPase activity found in *A. laidlawii* membranes. The significant reduction in DCCD inhibition of the purified enzyme compared with the membrane ATPase activity may be due to a reduced affinity of the DCCD binding protein for DCCD, or a loss of this protein during the purification procedure. If the inhibition profile of the purified enzyme does represent the actual inhibition pattern of the ATPase, the results may further suggest a fundamental difference between the *A. laidlawii* B membrane-bound Na^+-ATPase and a classical F-ATPase that is found in prokaryotes.

2.2.5. Reconstitution of Purified ATPase and Na^+ Transport Studies

Na^+ transport by this ATPase was observed by $^{22}Na^+$ tracer studies using proteoliposomes containing the purified enzyme and dimyristoylphosphatidylcholine (DMPC) lipids (Mahaja *et al.*, 1988). In this system, an energy-dependent $^{22}Na^+$ uptake into the proteoliposomes occurred, with a concomitant hydrolysis of ATP. At 37°C, approximately 1 mole of ATP was hydrolyzed per mole of Na^+ transported. Na^+ transport was inhibited approximately 50% by the ATPase inhibitors vanadate (500 µM) and leucinostatin (500 µM). The percent inhibition of Na^+ transport by these inhibitors was similar to the percent inhibition of ATPase activity that was observed with the purified enzyme. The concentrations used, however, were ten times greater than those reported for the purified enzyme system. The results are indicative of an active Na^+ efflux from *A. laidlawii* being mediated by an unusual F-class ATPase.

2.3. *Mycoplasma mycoides* subsp. *capri*

2.3.1. H^+-ATPase Activity

Generation of ΔpH and $\Delta \psi$ (inside negative above that of nonenergized cells) in *M. mycoides* subsp. *capri* was found to be energy dependent and the magnitude of the potential generated was dependent on the external glucose concentration (Benyoucef *et al.*, 1981b; Schummer *et al.*, 1981). Thus, the addition of glucose to deenergized cells suspended in a medium at pH 7.2 resulted in an alkalinization of pH_i by about 0.5 pH unit, and the generation of $\Delta \psi$ to approximately −85 mV (Benyoucef *et al.*, 1981a). In Na^+-depleted cells, conditions where a Na^+/H^+ antiporter would not be considered to be active, alkalinization of the cytosol occurred directly without any other transient changes in pH_i (Benyoucef *et al.*, 1982b). In addition, the generation of these potentials was related to the apparent rate of DCCD-sensitive H^+ extrusion by the cells (Benyoucef *et al.*, 1981b). Further, the generated ΔpH and $\Delta \psi$ were also sensitive to DCCD inhibition (Benyoucef *et al.*, 1981a; LeBlanc and Le Grimellec,

1979; Schummer *et al.*, 1981). From these results, a Mg^{2+}-dependent H^+-AT-Pase was proposed in membranes of *M. mycoides* subsp. *capri* (Benyoucef *et al.*, 1981a, 1982b).

2.3.2. ATP-Dependent Na^+ and K^+ Efflux

Intact *M. mycoides* subsp. *capri* cells, depleted of intracellular K^+, accumulated K^+ from the extracellular medium (Benyoucef *et al.*, 1982a). In the absence of extracellular Na^+, the level of K^+ accumulation was linearly related to the amplitude of the transmembrane electrical potential across the membrane ($\Delta\psi$). By contrast, in the presence of extracellular Na^+, K^+ accumulation in the cell exceeded the level that would be expected if $\Delta\psi$ alone was the driving force. The potassium-motive force ($\Delta\mu_{K^+}$) in fully energized cells suspended in a Na^+-containing medium was approximately 120 mV, whereas $\Delta\psi$ under these conditions was approximately 90 mV (inside negative). This Na^+ stimulation of K^+ uptake was shown to be dependent on $\Delta\psi$ and the Na^+ concentration (Benyoucef *et al.*, 1982a). The apparent K_m for Na^+ was approximately 1 mM and the maximal effect was observed at 10 mM Na^+. The presence of 10 mM Na^+ in the external medium led to an approximately threefold increase in intracellular K^+ accumulation. Since uptake of K^+ under these conditions consumed ATP and was not affected by the proton ionophore FCCP (Benyoucef *et al.*, 1982), it was proposed that K^+ accumulation was mediated by an ATP-dependent primary pump.

The kinetics of Na^+ efflux, from $^{22}Na^+$-loaded cells, significantly differed depending on whether cells were suspended in a Na^+-rich or Na^+-free medium (Benyoucef *et al.*, 1982b). Efflux into a Na^+-rich medium depended on the presence of an energy source. In addition, Na^+ efflux against a Na^+ concentration gradient required the additional presence of K^+. This dependency was specific for K^+, and could not be met by Rb^+. Collapsing the $\Delta\mu_{H^+}$ with FCCP did not affect the K^+-dependent Na^+ efflux, indicating that the requirement for K^+ was not the result of a possible coupling between a Na^+/H^+ antiporter and a K^+/H^+ antiporter. These results are consistent with a primary Na^+ pump directly associated with K^+. From the similarities in Na^+ and K^+ transport, it was proposed by Benyoucef *et al.* (1982b) that both active ion fluxes were executed by the same enzyme, a Na^+/K^+-dependent ATPase similar to the type found in higher organisms.

2.3.3. Inhibition Profile and Monovalent Cation Stimulation of ATPase Activity

The properties of the membrane-bound ATPase activity of *M. mycoides* subsp. *capri* were not consistent with the presence of a Na^+/K^+-dependent

ATPase. A Na^+/K^+-dependent ATPase would be expected to be stimulated by both Na^+ and K^+. Yet, at pH 7.0 ATPase activity was not stimulated by a high concentration (100 mM) of K^+, Li^+, or Rb^+. ATPase activity was stimulated by Na^+, although the K_m was relatively high, approximately 30 mM (Benyoucef *et al.*, 1982b). At high Na^+ concentration (100 mM) there was a twofold increase in ATPase activity. To explain why K^+ does not stimulate ATPase activity, as well as why such a high Na^+ concentration was necessary, it was proposed that during the membrane isolation procedure there was a loss of the Na^+/K^+ exchange protein from the membrane (Benyoucef *et al.*, 1982b).

The Na^+-stimulated ATPase activity was almost totally inhibited by DCCD (50 μM) and by 7-chloro-4-nitrobenzo-2-oxa-1,3-diazole (NBD, 100 μM). However, the activity was not affected by classical inhibitors of the Na^+/K^+-ATPase of animal cells, such as ouabain (500 μM) or vanadate (500 μM) (Benyoucef *et al.*, 1982b). The lack of inhibition by vanadate may further suggest that a P-class ATPase is not present in the *M. mycoides* subsp. *capri* membrane. The ATPase may be analogous to the inducible K^+-uptake system found in *S. faecalis* (Kakinuma and Igarashi, 1989), referred to as KtrII. The inhibition patterns of the ATPase as well as the lack of stimulation by K^+, however, indicate that this ATPase is different from Na^+/K^+-ATPases present in eukaryotic organisms. A model for the proposed ion transport mechanisms in *M. mycoides* subsp. *capri* is presented in Figure 2.

3. SECONDARY ION TRANSPORT SYSTEMS

3.1. Introduction

The second basic mechanism by which cations can be transported across a cell membrane are referred to as secondary transport processes. In these processes, the electrochemical potential of an ion is coupled to the movement of another molecule across the membrane. The most prevalent forms of secondary transport are those linked to $\Delta\mu_{H+}$.

3.2. $\Delta\mu_{H+}$-Driven Transport

Na^+/H^+ antiporters actively drives Na^+ across the cell membrane against its electrochemical potential. This is accomplished by coupling the movement of Na^+, in the opposite direction to that of H^+ moving down its electrochemical potential ($\Delta\mu_{H+}$). Through this mechanism, $\Delta\mu_{H+}$ actively drives the exchange of Na^+ for H^+. Most bacteria studied have this antiporter. In mycoplasmas, a Na^+/H^+ antiporter has been reported in *M. mycoides* subsp. *capri* (Benyoucef *et*

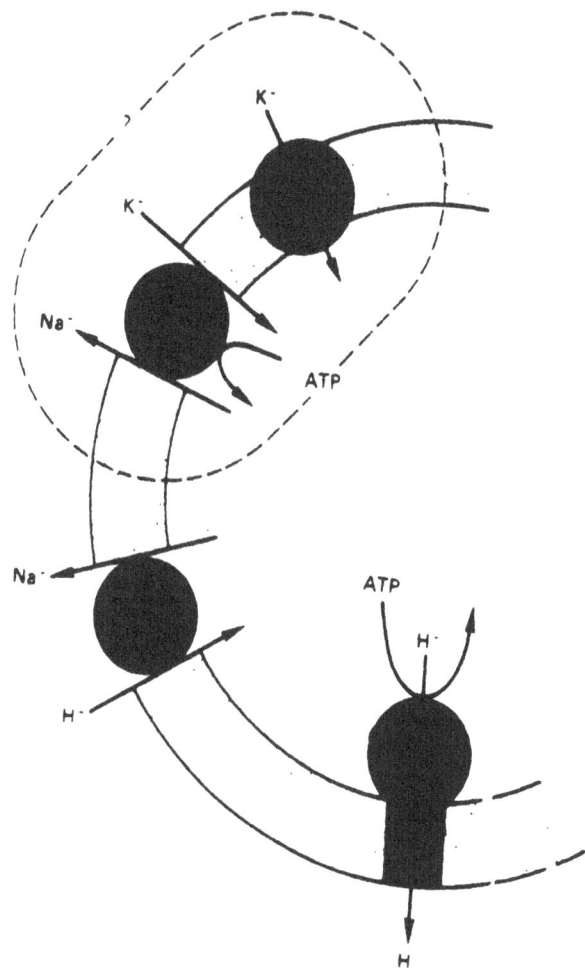

FIGURE 2. Proposed model for Na^+, K^+, and H^+ movements in *Mycoplasma mycoides* subsp. *capri*. Elements shown diagrammatically are the proton-translocating ATPase, a Na^+/H^+ antiport, a uniport system for K^+, and an ATP-consuming Na^+/K^+ exchange system (Na^+/K^+-dependent ATPase). The broken line indicates the possibility of a coupling between an active pump for Na^+ and a uniport system for K^+. (Modified from Benyoucef *et al.*, 1982b.)

al., 1982b), *A. laidlawii* (Lelong *et al.*, 1989), and recently in *S. floricola* (Shirazi and Rottem, unpublished data).

3.2.1. Na^+/H^+ Antiporter in *M. mycoides* subsp. *capri*

Evidence for a Na^+/H^+ antiporter in *M. mycoides* subsp. *capri* comes from studies in intact cells following the quenching/dequenching of the weak base fluorescent dye 9-aminoacridine (9-AA), and the efflux of $^{22}Na^+$. The addition of glucose to Na^+-loaded cells suspended in a Na^+-free medium resulted in a transient intracellular acidification and significantly enhanced Na^+ efflux from the cell (Benyoucef *et al.*, 1982b). These results were interpreted to indicate that H^+ were extruded from the cells by a H^+-ATPase, followed by a concomitant influx of H^+ and efflux of Na^+ via a Na^+/H^+ antiporter. The initial acidification was proposed to be due to a higher level of H^+ entering the cells from the antiporter, then being effluxed by the ATPase. As the Na^+/H^+ exchange reduced or stopped its activity, due to a depletion of intracellular Na^+ levels, the intracellular pH became more alkaline, since only the H^+-ATPase would then be mediating H^+ movements across the membrane (Benyoucef *et al.*, 1982b). Consistent with this proposal, no transient acidification was observed in Na^+-depleted cells. A Na^+/H^+ antiporter was, therefore, proposed to be active in *M. mycoides* subsp. *capri* cells, under the conditions of downhill Na^+ efflux.

Under conditions where there is an uphill Na^+ gradient (greater concentration of Na^+ in the suspension medium than in the cell) but no K^+ in the medium, an acidification of the medium was observed, although the alkalinization phase did not occur unless K^+ was present. These results were taken to indicate that there was a continuous Na^+/H^+ exchange activity occurring, and the intracellular Na^+ content was kept constant by a rapid recirculation of Na^+. Net Na^+ movement against a Na^+ concentration gradient has been proposed to occur by a primary mechanism (see primary Na^+ transport above) and not a Na^+/H^+ antiport mechanism as described in this section (Benyoucef *et al.*, 1982b).

Studies in intact *M. gallisepticum* cells, as well as *M. gallisepticum* sealed membrane vesicles and reconstituted membrane vesicles, showed that Na^+ movement across the membrane was not driven by $\Delta\mu_{H+}$ or inhibited by a collapse of $\Delta\mu_{H+}$ (Shirvan *et al.*, 1989a). Further, a collapse of $\Delta\mu_{H+}$ continued to allow cells to grow and remain viable in a high-salt medium. These results along with others discussed above indicate that a Na^+/H^+ antiporter is not present in *M. gallisepticum* cells. Yet, Na^+-loaded *M. gallisepticum* cells exhibit a transient acidification that is observed fluorometrically upon the addition of glucose to the Na^+-free cell medium (Linker and Wilson, 1985a). These experiments question whether the results obtained from the fluorometric assay described above for *M. mycoides* subsp. *capri* cells are sufficient to conclude that a Na^+/H^+ antiporter is active in the membrane.

3.2.2. H^+/cation Antiport Activity in *A. laidlawii*

In *A. laidlawii*, a H^+/cation antiport activity was reported in intact cells (Lelong *et al.*, 1989) using the ammonium chloride dilution procedure to generate a ΔpH (inside acid) across the cell membrane. In this procedure, NH_3 diffuses out of NH_4^+-loaded cells leaving behind H^+ which results, in the presence of high

Cl⁻, in the generation of only a ΔpH (inside acid) (Nakamura *et al.*, 1986). In the experiments, changes in the ΔpH were followed using the fluorescent dye acridine orange. The results showed that the presence of NaCl or KCl alone was not sufficient to change the pH_i, and that both ions together were necessary (Lelong *et al.*, 1989). The change in pH_i was independent of the order of addition of the ions, and LiCl could not replace NaCl, and RbCl could not replace KCl. In addition, the exchange was not affected by the presence of 100 mM tetraphenylphosphonium (TPP⁺), a condition which has been reported to collapse $\Delta\psi$; this suggests that the additive effect of K⁺ (in the presence of Na⁺) was not to collapse the transmembrane potential. By contrast, generation of a ΔpH in sealed membrane vesicles, by the same procedure, resulted in intracellular alkalinization by the addition of either NaCl or KCl. The addition of HCl, however, either alone or when added following or preceding the addition of NaCl, had no effect on pH_i. The exchange was inhibited by NEM (10 mM), but not by DCCD (25 μM) or amiloride, an inhibitor of Na⁺/H⁺ antiporters in eukaryotic systems. These results were taken to indicate a H⁺/cation antiport activity in *A. laidlawii*. Since the cations added to the cell suspensions were chloride salts, it could not be determined from these experiments whether the exchanger is a (Na⁺ + K⁺)/H⁺ antiporter or a (Na⁺ + K⁺ + Cl⁻)/H⁺ antiporter as found in eukaryotes. The K⁺ dependence of exchange activity in intact cells, compared with the activity in sealed membrane vesicles, may suggest that the transporter was modified during the membrane isolation procedure, possibly by proteolysis. This contention may be supported by the finding in *S. faecalis*, where everted membrane vesicles prepared in the absence of protease inhibitors have been proposed to undergo proteolysis.

3.3. $\Delta\mu_{Na^+}$-Driven Transport

3.3.1. Na⁺/Phosphate Symporter in *M. mycoides* subsp. *capri*

A Na⁺/phosphate symporter has been proposed in *M. mycoides* subsp. *capri* cells (Benyoucef *et al.*, 1982a,b). The evidence for this transporter has been derived mainly from studies on Na⁺-dependent K⁺ accumulation in cells. Phosphate uptake into deenergized cells was energy dependent, and stimulated by the presence of NaCl. Further, the Na⁺ stimulation of K⁺ transport required inorganic phosphate (P_i). These results were taken to indicate that for the stimulation by Na⁺ of the Na⁺/K⁺−dependent ATPase, Na⁺ must first enter the cell as sodium phosphate, probably via a Na⁺/phosphate symporter (Benyoucef *et al.*, 1982b).

3.3.2. Na⁺/Cation Exchange Mechanism in *M. gallisepticum*

A nonelectrogenic cation/Na⁺ exchange mechanism was suggested from ²²Na⁺ efflux studies in intact preloaded *M. gallisepticum* cells (Shirvan *et al.*,

1989a). Under conditions where the Na^+ concentration in the medium was greater than 150 mM, $^{22}Na^+$ efflux was markedly enhanced in the presence of glucose. Efflux into an isoosmotic NaCl solution (250 mM NaCl) resulted in both an energy-dependent efflux and an energy-independent efflux. The energy-independent process resulted in the efflux of only 14% of the tracer, while the energy-dependent process resulted in 51% of the $^{22}Na^+$ being released. By contrast, in a medium containing low Na^+ ($<$ 150 mM), $^{22}Na^+$ efflux was energy-independent, and resulted in $>$ 90% of the $^{22}Na^+$ being released. The energy-independent release did not occur in the absence of external Na^+, nor could the requirement be met by K^+ or Li^+. The affinity of this process for Na^+ was, however, low (K_m = 29 mM). The energy-independence of this process is consistent with an exchange mechanism. In addition, the efflux was resistant to inhibition by the proton ionophores CCCP (10 μM) and SF6847 (1 μM), indicating that the exchange is not with H^+, nor driven by $\Delta\mu_{H^+}$. Further, this exchange did not appear to be electrogenic, as the addition of glucose, which resulted in the generation of $\Delta\psi$ (inside negative), and CCCP, which would be active in charge compensation, had no effect on the rate of exchange. Moreover, sodium thiocyanate, under conditions which collapse $\Delta\psi$, had no effect on efflux (Shirvan *et al.*, 1989a). The exchange activity was, however, inhibited by DCCD. These results are consistent with a separate Na^+/cation exchange mechanism, or may suggest that the exchange, at low external Na^+, may be carried out by the Na^+-ATPase (see section on primary transport). Such an electroneutral exchange has been reported for the Na^+/K^+-ATPase of eukaryotes (Garrahan and Glynn, 1967) and the K^+-ATPase of *E. coli* (Epstein, 1985). The physiological significance of this process may be to provide a mechanism for the exchange of Na^+ and K^+ in the cell. Such a process would provide for a route by which Na^+ can reenter the cell, thus completing a Na^+ loop.

4. ELECTROCHEMICAL POTENTIALS IN MYCOPLASMAS

4.1. Introduction

4.1.1. Generation of $\Delta\mu_{H^+}$

In bacteria, electrochemical potentials provide the driving force for a variety of essential physiological functions, such as nutrient uptake, ATP synthesis, flagellum-driven movement, and pH regulation (Padan and Schuldiner, 1987; Skulachev, 1989). This concept was first proposed in the chemiosmotic theory by Mitchell (1966). According to this theory, H^+ are transported across a membrane, creating a concentration gradient of H^+ (ΔpH), and an electrical gradient from the transported charge on the proton (membrane potential, $\Delta\psi$). Together these two components comprise the electrochemical proton gradient, also referred to as the proton-motive force ($\Delta\mu_{H+}$), which is equal to the sum of $\Delta\psi$ (mV) and ΔpH (mV). Quantitatively, $\Delta\mu_{H+} = \Delta\psi - 2.3(RT/F)\Delta$pH, where the constants $2.3(RT/F)$ at 37°C (310 K) are equivalent to 59 mV (R is the universal gas constant, T is the absolute temperature, F is Faraday's constant). This potential, which is the driving force for H^+ to return across the membrane, can be "harnessed" and used to drive the above processes. Many of the processes that are driven are essential, such as pH_i regulation and nutrient uptake (Padan and Schuldiner, 1987). Under usual growth conditions, most bacteria require this potential for survival. Evidence supporting the use of this potential, directly or indirectly, came from the laboratory of F. Harold (Harold and Brunt, 1977; Heefner et al., 1980; Heefner and Harold, 1982). In these studies, S. faecalis grew only in the absence of $\Delta\mu_{H+}$ if conditions were maintained in the growth medium which allowed the cell to maintain the essential properties that were driven by $\Delta\mu_{H+}$. In S. faecalis, this meant buffering the growth medium so that the pH was approximately that of pH_i, maintaining a high K^+ and low Na^+ level in the medium which would be reflected intracellularly, and adding to the medium relatively high concentrations of amino acids so they could be taken up into the cell without a $\Delta\mu_{H+}$ driving force. One of the ways that $\Delta\mu_{H+}$ is associated with maintaining pH_i in prokaryotes is through the action of a Na^+/H^+ antiporter. In E. coli, the antiporter transports H^+ into the cell under alkaline conditions which help acidify the cytosol and assists in the maintenance of pH_i. In S. faecalis, the H^+-ATPase has been implicated in the regulation of pH_i, by the observation that the synthesis of the ATPase is enhanced with acidification of the cytosol (Kobayashi et al., 1986).

4.1.2. Na^+ versus H^+-Motive Force

Most bacteria in their natural environment generate a $\Delta\mu_{H+}$ which is sufficient to drive vital physiological functions. However, some bacteria live in extreme environments, and are not able to generate a sufficient potential. These bacteria have been shown to use an alternative chemical potential, most notably that of Na^+, $\Delta\mu_{Na+}$ (Skulachev, 1989). Examples of bacteria which generate and use $\Delta\mu_{Na+}$ (Lanyi, 1979; Skulachev, 1989) are: alkalophilic bacteria, such as B. firmus (Krulwich et al., 1988); anaerobic bacteria that ferment carboxylic acids, such as P. modestum (Laubinger and Dimroth, 1988); and marine bacteria that are both halophilic and alkalotolerant, such as Vibrio alginolyticus (Tokuda and Unemoto, 1981, 1982). These bacteria generate a $\Delta\mu_{Na+}$, through the action of a primary Na^+ pump. In addition, many of these bacteria also have a primary H^+ pump, and under the appropriate conditions are able to generate a $\Delta\mu_{H+}$ as well. Through this diversity the cells enhance their ability to adapt to various environments. For example, V. alginolyticus can generate both a $\Delta\mu_{H+}$ and a $\Delta\mu_{Na+}$, by primary transport systems, and use these potentials for various essential functions. Under environmental conditions of high salt and alkaline pH, *V. alginolyticus* generated a $\Delta\mu_{Na+}$ (Tokuda and Unemoto, 1981, 1982, 1983) which has been shown to drive several essential functions such as nutrient uptake (Kakinuma and Unemoto, 1985; Tokuda *et al.*, 1982) and the rotation of the flagella (Chernyak *et al.*, 1983). Yet, when the external medium was at the neutral or at the acidic range, a $\Delta\mu_{H+}$ coupling was observed. Another microorganism which has a primary mechanism for transporting both Na$^+$ and H$^+$, but does not readily live in "extreme" conditions, is *S. faecalis* (Kakinuma and Igarashi, 1989). It has been proposed that under relatively high Na$^+$ conditions (60 mM), these cells require the induction of the Na$^+$-ATPase for growth. The function of the Na$^+$-ATPase activity is to extrude Na$^+$, and to take up K$^+$ into the cell. Na$^+$, during this process, acts as a counterion for K$^+$ uptake. The role of the generated $\Delta\mu_{Na+}$, however, is not understood.

In the following sections we will describe the electrochemical gradients that are generated in several mycoplasmas, propose how they might be generated, and their association with membrane structure.

4.2. *Acholeplasma laidlawii*

4.2.1. Establishment and Measurement of Electrochemical Gradients

An energy-dependent generation of $\Delta\mu_{H+}$ was reported in *A. laidlawii* (Clementz *et al.*, 1986; Lelong *et al.*, 1989; Scheifer and Schummer, 1982). $\Delta\mu_{H+}$ was approximately -50 mV when determined in early log-phase cells by the intracellular/extracellular partition of the lipophilic cation TPP$^+$. Similar $\Delta\mu_{H+}$ values were obtained when cells suspended in a Tris-HCl buffer containing K$^+$, Na$^+$, and PO$_4^{2-}$, adjusted either to pH 7.4 (Clementz *et al.*, 1986) or 8.0–8.5 (Lelong *et al.*, 1989). A comparable $\Delta\mu_{H+}$ (-52 mV) was observed by following changes in the fluorescence of the potential-sensitive carbocyanine dye, 3,3'-dipropyl-2,2'-thiodicarbocyanine iodide, in a (2-[N-Morpholino] ethanesulfonic acid) (MES) buffered Ringer solution, at pH 7.0 (Scheifer and Schummer, 1982). The measurement of $\Delta\mu_{H+}$ at pH 7.4, by Clementz *et al.* (1986), showed that the magnitude of the potential was due almost entirely to the generation of $\Delta\psi$. By contrast, Scheifer and Schummer (1982) showed that at pH 7.0, the cells generated a $\Delta\psi$ of -28 mV and a low ΔpH. The ΔpH, measured by the distribution of the weak acid 5,5-dimethyloxazoline-2,4-dione (DMO), was ~ 24 mV (a ΔpH of 60 mV = 1 pH unit). This study showed that at pH$_o$ = 6.0, the ΔpH was 0.7 pH unit (pH$_i$ = 6.7). As the pH$_o$ increased up to pH$_o$ = 8.0, a linear decrease in ΔpH was detected. At pH$_o$ = 8.0, ΔpH = 0 (pH$_o$ = pH$_i$). However, studies by Lelong *et al.* (1989) showed no ΔpH in the pH range 6.5–7.5, and an increase in ΔpH, as pH$_o$ increased from ~ 7.8 to 8.5. In the pH range 8.0–8.5, a ΔpH of 0.2–0.6 unit was observed in this study.

The involvement of the Na^+/H^+ antiporter in the regulation of pH_i in other prokaryotes, such as *E. coli* (Bassilana *et al.*, 1984a,b; Padan and Schuldiner, 1986, 1987), suggests that the H^+/cation antiporter reported in *A. laidlawii* (Lelong *et al.*, 1989) may be involved in pH homeostasis in the alkaline pH range. If it is involved, it does not, however, appear to be the only mechanism, as a small acidification of the cytosol still occurs even after $\Delta\mu_{H^+}$ is collapsed with CCCP. It is possible that at least part of the ΔpH is driven by $\Delta\psi$. In this case, the negative potential inside the cells would act as a driving force for H^+ to enter and acidify the cytoplasm. This would require a "leakiness" of the membrane to H^+, either via specific transport mechanisms, or by a permeability of the membrane to H^+.

4.2.2. H^+ Permeability of the Membrane, Internal pH, and Growth

The permeability of membrane vesicles to H^+ has been associated with the level of carotenoids in the membrane (Lelong *et al.*, 1989). A relatively low H^+ permeability was observed with carotenoid-rich membrane preparations, obtained by growing the cells in the presence of acetate, whereas carotenoid-poor membranes were highly permeable to H^+ (Lelong *et al.*, 1989). It has been suggested that the permeability of the membrane to H^+ is associated with the effect of carotenoids on the packing of the hydrocarbon chains of membrane phospholipids (Lelong *et al.*, 1989). The ability of *A. laidlawii* to maintain a partial ΔpH in the presence of CCCP, in the pH range 8.0–8.5 (Lelong *et al.*, 1989), may suggest that $\Delta\psi$ is generated in these cells by a primary transport mechanism not associated with H^+. Consistently, CCCP (10 μM) did not affect cell growth (Clementz *et al.*, 1986).

Though the role of ΔpH in the physiology of *A. laidlawii* is unknown, pH homeostasis appears to be essential for cell growth. *A. laidlawii* cells grew best in a slightly alkaline medium (pH 8.5). Under this condition, energized cells maintained a ΔpH of approximately 0.6 pH unit ($pH_i = 7.9$). Growth of *A. laidlawii* slowed down or even ceased when the pH of the medium dropped to $\sim 7.2–7.4$. As no ΔpH was reported in this pH range, the results indicate that a $pH_i > 7.4$ is required for cell growth and that the cessation of growth is associated with pH_i dropping below this pH (Lelong *et al.*, 1989).

4.2.3.   Effect of Membrane Potential on Membrane Lipids

It has been proposed that the stability of a biological membrane is related to the packing of the lipid molecules, and that an "optimal packing" is required for a stable bilayer (Wieslander *et al.*, 1980, 1982). A stable membrane must possess lipids that form lamellar structures, as well as lipids that form mesophase structures, such as reverse hexagonal phase structures. Between the lamellar and the reverse hexagonal lipids a delicate balance must be kept.

Major constituents of *A. laidlawii* lipids that have been proposed to play a role in this process are the glycolipids (Wieslander *et al.*, 1980), mainly mono-glucosyldiglyceride (MGDG) and diglucosyldiglyceride (DGDG). MGDG tends to form reverse hexagonal phase structures, whereas DGDG tends to form lamellar structures. The cell must, therefore, regulate the relative amounts of these glycolipids in the membrane. A major regulatory signal appears to be $\Delta\psi$ (Clementz *et al.*, 1986). Hyperpolarization of the cells by the addition of valinomycin, in the absence of added K^+, resulted in a decrease in the MGDG/DGDG ratio. The addition of K^+ reverses the hyperpolarization of the cell, as well as the decrease in the lipid ratio, in a dose-dependent manner. Consistently, TPP^+ reduced $\Delta\psi$ and increased the MGDG/DGDG ratio. This regulation of membrane lipids by $\Delta\psi$ has been proposed to assist the cell to adapt to different environmental stimuli (Clementz *et al.*, 1986).

4.3. *Mycoplasma gallisepticum*

4.3.1. Establishment and Measurement of Electrochemical Gradients

M. gallisepticum cells also generate a substantial $\Delta\mu_{H^+}$, in the presence of an energy source, which is affected by the pH_o of the medium over the pH range 6.0–8.5 (Linker and Wilson, 1985a; Rottem and Razin, 1966; Scheifer and Schummer, 1982; Shirvan *et al.*, 1989b). The ΔpH, as determined by the intracellular/extracellular partition of either benzoic acid or methylamine, declined from 50 mV at $pH_o = 6.0$, to 0 as pH_o increased to pH 7.4. At $pH_o > 7.4$, ΔpH reversed direction to oppose $\Delta\psi$. At $pH_o = 8.5$, ΔpH was ~ 60 mV, which corresponds to a pH_i of 7.5 (Shirvan *et al.*, 1989b). Also, over the pH_o range 6.0–7.4, $\Delta\psi$, as determined by the partition of TPP^+, steadily increased from -45 mV to a maximum value of ~ -75 mV. $\Delta\psi$ at $pH_o > 7.4$ remained relatively constant (Shirvan *et al.*, 1989b). Analogous $\Delta\psi$ results were obtained in a separate study which compared the ratio of the concentrations (in/out) of TPP^+, Rh^+, and K^+ in the presence of valinomycin (Rotter *et al.*, 1981). Therefore, it appears that $\Delta\psi$ over the pH range 6.0–8.5 is insufficient to compensate for the loss of the pH gradient, and the $\Delta\mu_{H^+}$ declines from a maximum value of -112 mV at pH_o 7.0 to a minimum value of -21 mV at pH_o 8.5 (Shirvan *et al.*, 1989b). The $\Delta\mu_{H^+}$ was approximately the same in cells harvested at early exponential, mid-exponential, and stationary phases of growth (Rottem and Razin, 1966). The pH_i over this pH range remained fairly constant. Lower values of $\Delta\psi$ and ΔpH were reported in *M. gallisepticum* when measured by potential-sensitive cyanine dyes and DMO, respectively (Scheifer and Schummer, 1982). At pH_o 7.0, $\Delta\psi = -48$ mV and $\Delta pH = -24$ mV (or 0.4 pH unit). The ΔpH decreased linearly as the pH_o increased to 8.0, at which point $\Delta pH = 0$. At pH 7.0, $\Delta\mu_{H+}$ was -72 mV.

It was proposed that $\Delta\mu_{H+}$ is generated in *M. gallisepticum* by an electrogenic H^+-ATPase (Linker and Wilson, 1985a; Rottem *et al.*, 1981; Scheifer and Schummer, 1982). If this is the case, then permeabilizing the membrane to H^+ by proton ionophores which collapse $\Delta\psi_{H+}$ should result in the collapse of $\Delta\psi$. However, $\Delta\psi$ is resistant to the presence of CCCP and the proton ionophore SF6847, over external pH values which correspond to the physiological growth range (Shirvan *et al.*, 1989b). Yet, considering that $\Delta\psi$ is inhibited by DCCD (Linker and Wilson, 1985a; Shirvan *et al.*, 1989b) and that these cells have not yet been found to have other primary pumps except for ATPases, the results are compatible with the generation of $\Delta\psi$ by an electronic ATPase that transports ions besides H^+. The results are consistent with the generation of $\Delta\psi$ by the Na^+-ATPase discussed above. The collapse of $\Delta\psi$ by the addition of 100 mM KCl to the suspension medium (Shirvan *et al.*, 1989b) further suggests that there is an electrogenic K^+ uptake mechanism which may be associated with maintaining the level of $\Delta\psi$ in the cells. One such mechanism that has been proposed for *M. gallisepticum* is a K^+/H^+ symporter (discussed below).

The reduction of ΔpH by approximately half at pH 6.0, in the presence of proton ionophores (Shirvan *et al.*, 1989b), is indicative of a nonequilibrium distribution of H^+ at this pH range, and may suggest the involvement of H^+-ATPase activity as well as Na^+-ATPase activity in regulating ΔpH at the more acidic pH_o values. At the alkaline pH range, ΔpH is approximately equal to $\Delta\psi$, in the presence or absence of proton ionophores. These results may suggest that pH_i at the external alkaline pH range is regulated by $\Delta\psi$, which in turn is maintained by the Na^+-ATPase.

4.3.2. Support for the Presence of a H^+-ATPase

Support for a H^+-ATPase in the membrane, as well as a function for a localized pH gradient (acid outside), comes from fusion studies between mycoplasma cells and enveloped viruses of the Papamyxovirus group. Influenza virions fuse with erythrocyte membranes at pH 5.2, but not at pH 7.4 (Choppin and Scheid, 1980; Citovsky *et al.*, 1986; White *et al.*, 1983). Yet, this same virion fuses with *M. gallisepticum* at neutral and at high pH values (Citovsky *et al.*, 1986). The fusion at high pH values is inhibited by CCCP, without affecting the fusion at pH 5.2. By contrast, CCCP had no effect, at pH 7.4, on the fusion between *M. gallisepticum* and another virion of this group, Sendai virus, which normally has active fusogenic activity at this pH. These results were taken to indicate that a H^+-ATPase in *M. gallisepticum* was generating a localized acidic environment on the outer side of the membrane, and this localized low-pH environment was stimulating the fusion between influenza virus and the cells.

4.4. *Mycoplasma mycoides* subsp. *capri*

4.4.1. Establishment of Gradients

A comparison study of $\Delta\mu_{H+}$ in *M. mycoides* subsp. *capri*, *M. gallisepticum*, and *A. laidlawii* showed that *M. mycoides* subsp. *capri* and *M. gallisepticum* have the same $\Delta\mu_{H+}$ at $pH_o = 7.0$ ($\Delta\mu_{H+} = -72$ mV), whereas the $\Delta\mu_{H+}$ of *A. laidlawii* is lower ($\Delta\mu_{H+} = -52$ mV) (Scheifer and Schummer, 1982). In this study, $\Delta\psi$ of *M. mycoides* subsp. *capri*, as measured by potential-sensitive cyanine dyes, was -28 mV at pH_o 7.0. ΔpH at this pH_o was -24 mV, which corresponded to a pH_i that was 0.4 pH unit more alkaline than the extracellular medium. The ΔpH was determined using DMO. Consistent, but somewhat larger values for ΔpH were obtained using the weak acid butyrate and the weak base methylamine, employing flow dialysis and filtration techniques (Benyousef *et al.*, 1981a). At pH_o 7.2, the cytosol was 0.5 pH unit more alkaline than the extracellular medium. This corresponds to a ΔpH of -85 mV. Reduction of pH_o from 7.7 to 5.7 resulted in a decrease in pH_i, and an increase in ΔpH from 0.2 to 1.0 pH unit. The change in $\Delta\psi$ was approximately that of the change in ΔpH, decreasing from -90 mV to -60 mV. Since these changes largely compensated for each other, the resulting $\Delta\mu_{H+}$ remained almost constant over this pH_o range, $\Delta\mu_{H+} = -115$ mV (Benyousef *et al.*, 1981a). The inhibition by DCCD of $\Delta\psi$ (LeBlanc and Le Grimellec, 1979) and ΔpH (Benyousef *et al.*, 1981a) as well as the inhibition of ΔpH by CCCP and nigericin (Benyousef *et al.*, 1981a), is consistent with the generation of the $\Delta\mu_{H+}$ by the H^+-ATPase.

When Na^+ is absent from the suspension medium, K^+ accumulates within the cell to the level that is predicted from the amplitude of $\Delta\psi$ (Benyousef *et al.*, 1982). These results are consistent with K^+ uptake via a uniport mechanism. In the presence of Na^+ in the medium, K^+ accumulates to a level that is greater than that predicted from $\Delta\psi$. At pH_o 7.2, $\Delta\mu_{K+} = -120$ mV and $\Delta\psi = -90$ mV. Studies described in the ATPase section above indicate that K^+ is taken up into the cells by a Na^+/K^+-dependent ATPase that is modulated by $\Delta\psi$.

4.4.2. Effect on Membrane Structure

Energization of the cells shows very little effect on the physical state of the membrane lipids, as determined by fluorescence polarization and ESR experiments (Le Grimellec *et al.*, 1982). However, there appear to be conformational changes in membrane proteins. Under the conditions tested above, there was a significant increase in the degree of exposure of primary amino groups of membrane proteins to the aqueous medium. This effect was blocked by DCCD and FCCP, suggesting that the increased exposure is associated with the generation of $\Delta\mu_{H+}$ in these cells. It was further concluded that the conformational changes were associated primarily with the generation of a ΔpH across the membrane.

4.5. *Ureaplasma urealyticum*

Establishment of Gradient

In *U. urealyticum* it has been shown that the production of ATP requires the concomitant activity of its cytoplasmic urease and membrane-bound ATPase (Romano *et al.*, 1986). Furthermore, the production is drastically reduced by CCCP, indicating the importance of $\Delta\mu_{H+}$ in this process. It has been proposed by Masover and Hayflick (1973) that ATP may be generated through the formation of an ion gradient coupled to urea hydrolysis (Masover and Hayflick, 1973). This concept of how an electrochemical gradient is generated in these cells was extended by Romano *et al.*, who proposed that urea hydrolysis generates an NH_4^+ diffusion potential, and that the diffusion of NH_4^+ across the membrane generates a $\Delta\psi$.

5. VOLUME REGULATION

5.1. Introduction and Theory

Volume regulation occurs in cells as a result of the net flow of water between the intracellular membrane-bound compartment and the external medium. The direction of flow reflects the compartment in which the water would have the lowest free energy. Two factors which significantly affect the free energy of water are the intracellular colloid osmotic pressure and the number of osmotically active components in a cell. A colloid osmotic pressure results from the presence of nondiffusible macromolecules within a cell (Lewis and McElhaney, 1983; Wilson, 1954). In addition, as most of these macromolecules are polyanionic, they act to attract smaller permeable cations into the cell. This is referred to as the Gibbs–Donnan potential (Leaf, 1956, 1959; Wilson, 1954). This entrance of osmotically active components would be reinforced by the chemical potential of small diffusible ions that might exist between the medium and the cells. For example, a suspension of cells in 250 mM NaCl would have a $\Delta\mu_{Na+}$ directed into the cell, which, depending on the permeability of the membrane, would drive Na^+ into the cell. An increase in the intracellular ion concentration further acts to lower the free energy of water within the cell, by increasing its entropy. Therefore, water flows into the cell, resulting in an increase in its volume. In nonenergized cells, where you would not expect an active net transport of ions, water will continue to flow into the cell either until the hydrostatic pressure increases sufficiently to increase the activity of the water in that compartment (MacKnight and Leaf, 1978) or until the cell lyses. Most prokaryotic organisms protect themselves from osmotic lysis by the presence of a rigid peptidoglycan-based cell wall. Animal cells, on the other hand, depend mostly on a mechanism that regulates the osmotic balance between the intracellular space and external environment. This is accomplished mostly by the active extrusion of one or more ions in excess of their Donnan distribution. The resulting effect is that the extruded ion(s) acts in a similar manner as an extracellular impermeant ion, osmotically balancing the colloid osmotic pressure of intracellular macromolecules (Leaf, 1956, 1959; Wilson, 1954). Therefore, to prevent lysis, animal cells must have an active extrusion mechanism for ions. In mycoplasmas, the lack of a rigid peptidoglycan-based cell wall necessitates that these microorganisms also have a mechanism to extrude ions in order to prevent osmotic lysis. The permeability properties of the semipermeable membrane in these cells are also important in the volume regulatory process. The following section will describe these processes in *M. gallisepticum* and *A. laidlawii.*

5.2. *Mycoplasma gallisepticum*

5.2.1. Permeability of the Membrane to Small Ions

M. gallisepticum cells are more resistant to osmotic lysis than other mycoplasmas and do not behave as an ideal osmometer (Rottem and Razin, 1972; Rottem and Verkleij, 1982). This appears to be the case in nonenergized as well as energized cells, suggesting that the reason for this phenomenon is at the level of the cells' plasma membrane. The membrane of *M. gallisepticum,* as with other mycoplasmas, has been shown to have a limited permeability to small monovalent cations (Rottem *et al.,* 1983; Shirvan *et al.,* 1989a). Yet, the physical properties of this membrane appear to differ from those of many other mycoplasmas. *M. gallisepticum* possesses unusual segregated lipid domains in its membrane, due to the presence of disaturated phosphyatidylcholine (PC) (Rottem and Markowitz, 1979; Rottem and Verkleij, 1982). The high amount of disaturated PC induces the formation of segregated lipid domains: a cholesterol-poor domain which is in a relative solid state (solid region), and a cholesterol-rich lipid domain, which remains in a fluid state (fluid region). It was suggested that the disordered boundaries at the interface between the fluid and solid regions allow for the premeatin of small osmotic molecules. Under conditions of osmotic stress, internal osmotic compounds could efflux from the cell, reducing internal osmotic pressure, and act to buffer the cell against swelling. The reason why both energized and nonenergized *M. gallisepticum* cells are far more resistant to rapid swelling and lysis than other mycoplasmas has been explained by this theory (Rottem and Verkleij, 1982).

Although *M. gallisepticum* are more resistant to osmotic lysis than other mycoplasmas, the suspension of these cells in an isoosmotic medium of NaCl (250 mM), KCl (250 mM) or LiCl (250 mM) does result in the swelling and eventual lysis of these cells, when incubated at 37°C (Shirvan *et al.*, 1989a). Only a moderate swelling is observed when cells are incubated in solutions where the cation is impermeable. Almost no swelling is observed in Cl$^-$ salt solutions, in which the cation is impermeable, or in sucrose, which is also impermeable to the cell (Shirvan *et al.*, 1989a). The swelling of these cells in these high-osmotic-strength solutions is indicative of Na$^+$, K$^+$, and Li$^+$ entering the cells, increasing the intracellular osmotic pressure. The results further suggest that Cl$^-$ is entering the cell as an osmotic counterion, allowing Na$^+$ to enter as an electroneutral species. Interestingly, as the cells swell, the segregated lipid domains in the membrane do not allow a sufficient permeability for the ions that diffused into the cell to diffuse out of the cell and prevent the osmotic crisis. Since the entrance of Na$^+$ results in cellular swelling, volume regulation under the above condition depends on the active extrusion of Na$^+$.

5.2.2. Energy-Dependent Ion Transport

The prevention of swelling of *M. gallisepticum* cells, suspended in isoosmotic NaCl, is energy-dependent (Lewis and McElhaney, 1983; Razin, 1978; Rottem *et al.*, 1981; Shirvan *et al.*, 1989a). Conversely, swelling is stimulated by the ATPase inhibitor DCCD. It was originally proposed that the extrusion of Na$^+$ was through a secondary Na$^+$/H$^+$ antiporter, which was driven by $\Delta\mu_{H+}$ (Rottem *et al.*, 1981). However, later studies showed that collapsing $\Delta\mu_{H+}$ with the proton ionophore CCCP does not stimulate swelling (Lewis and McElhaney, 1983; Shirvan *et al.*, 1989a). Furthermore, a physiological cell volume is maintained and the cells continue to grow in a high-salt medium, in the presence of CCCP (Shirvan *et al.*, 1989a). These studies are consistent with the transport studies which show that, under conditions of osmotic stress, Na$^+$ is *not* extruded from cells via a Na$^+$/H$^+$ antiporter.

Cellular swelling and osmotic lysis is stimulated by the ATPase inhibitor DES (Shirvan *et al.*, 1987) as well as DCCD (Lewis and McElhaney, 1983; Rottem *et al.*, 1981) when cells are suspended in isoosmotic NaCl (250 mM) or solutions of other monovalent cations. These studies support the transport studies, which indicate that under conditions of osmotic stress, Na$^+$ is extruded from the cell by the proposed Na$^+$-ATPase. The data are consistent with volume regulation occurring by the extrusion of Na$^+$ via a primary electrogenic Na$^+$-

ATPase. As *M. gallisepticum* is found in lung epithelium, it must contend with a high-Na^+ medium. Volume regulation, when the cells are in their natural habitat, may, therefore, be analogous to the volume regulatory mechanism found in animal cells. The Na^+ pump in *M. gallisepticum,* however, is certainly different from the Na^+/K^+-ATPase found in animal cells.

Upon incubation of *M. gallisepticum* in isoosmotic KCl (250 mM), the cells also swell (Shirvan *et al.,* 1989a). The addition of an energy sources does not, however, protect the cells from swelling, as is the case when cells are incubated in isoosmotic NaCl. On the other hand, an almost complete protection is observed when glucose is added in the presence of CCCP. A transport mechanism which is consistent with these results is the constitutive K^+ transport systems found in *S. faecalis* and *E. coli,* referred to as KtrI (Bakker and Harold, 1980; Kakinuma and Harold, 1985) and Trk (Stewart *et al.,* 1985), respectively. Both mechanisms have been proposed to be K^+/H^+ symporters that are regulated by phosphorylation. In *S. faecalis,* Kakinuma and Harold (1985) suggest that in the presence of a substantial $\Delta\mu_{H+}$, the protonic potential supports K^+ accumulation via KtrI. In the absence of a large $\Delta\mu_{H+}$, KtrI functions chiefly in the direction of K^+ efflux. In the experiments with *M. gallisepticum,* the role of the proton ionophore may be both to collapse $\Delta\mu_{H+}$ and to allow the process to continue electroneutrally. To maintain K^+ at a steady-state level, two or more positive charges would have to exit the cell for each K^+ that diffused into the cell: one positive charge for the K^+, and at least one for the H^+. The role of the proton ionophore in this case would be to prevent the process from stopping due to the buildup of a charge gradient, by allowing the entrance of H^+ in response to the $\Delta\psi$. The role of glucose in this case would be to act as an energy source and allow the cell to generate ATP for the eventual phosphorylation of the transporter. It has been further suggested that phosphorylation of KtrI plays an important physiological function by preventing the depletion of internal K^+, in periods when no energy source is available, and therefore no $\Delta\mu_{H+}$ (Kakinuma and Harold, 1985). As with the electrogenic Na^+ pump, the K^+ transport mechanism in *M. gallisepticum* may act to help regulate electrochemical gradients as well as to act in the regulation of intracellular volume.

5.3. *Acholeplasma laidlawii*

5.3.1. Permeability of the Membrane to Small Ions

The plasma membrane surrounding *A. laidlawii* B cells serves as an effective permeability barrier to small monovalent cations, such as Na^+ (Mahajan *et al.,* 1988). Yet, incubation of *A. laidlawii* B in isoosmotic solutions of NaCl or KCl, in the absence of an energy source, results in the swelling and lysis of the cells (Jinks *et al.,* 1978) due to a greater rate of influx of Na^+ than efflux. The inward leakage of Na$^+$, as studied by ^{23}Na nuclear magnetic resonance, does not occur when the cells are incubated at low temperatures in isotonic glucose-free buffer (Mahajan *et al.*, 1988). Similar results are observed in merthiolate-poisoned cells incubated at 37°C, possibly suggesting that the leakage of Na$^+$ is through specific protein-mediated transport processes. Furthermore, in contrast to the behavior of dimyristoylphosphatidylcholine (DMPC) liposomes, which have an increased leakiness of the membrane to ions at low temperatures where gel–liquid-crystalline phase transitions occur, the permeability barrier of the *A. laidlawii* B membrane is maintained when membrane lipids are in the gel state as well as when gel and liquid-crystalline lipids coexist in the membrane (Mahajan *et al.*, 1988).

5.3.2. Energy-Dependent Ion Transport

The incubation of *A. laidlawii* B cells in an isoosmotic NaCl solution, in the presence of an energy source, results in the cells maintaining a physiological volume. The volume regulatory process is associated with a greater rate of active Na$^+$ extrusion from the cell than the rate of inward leakage of Na$^+$ (Jinks *et al.*, 1978; Mahajan *et al.*, 1988). Reconstitution studies of the purified Na$^+$-stimulated F-class ATPase isolated from *A. laidlawii* B (see ATPase section above), as well as studies in intact cells, show that Na$^+$ is extruded by this ATPase, and is therefore intricately involved with the volume regulatory process (Mahajan *et al.*, 1988).

6. CONCLUDING REMARKS

In this chapter we have reviewed the current understanding of ion transport, membrane potential, membrane-bound ATPases, and volume regulatory processes in mycoplasmas. A great deal has been learned over the last decade. The results have been consistent with our belief that although these organisms are the smallest and simplest self-replicating prokaryotes, their transport processes are analogous to those found throughout the spectrum of living cells; therefore, mycoplasmas serve as the simplest model to study the above essential processes. As much of the current physiological and biochemical work is now being complemented by studies at the level of the genome, many basic structure–function relationship questions should be answered in the next decade.

7. REFERENCES

Al-Awqati, Q., 1986, *Annu. Rev. Cell Biol.* **2**:179–199.
Bakker, E. P., and Harold, F. M., 1980, *J. Biol. Chem.* **255**:433–440.
Bassilana, M., Damianco, E., and Leblanc, G., 1984a, *Biochemistry* **23**:1015–1022.

Bassilana, M., Damianco, E., and Leblanc, G., 1984b, *Biochemistry* **23**:5288–5294.

Benyoucef, M., Rigaud, J.-L., and Leblanc, G., 1981a, *Eur. J. Biochem.* **113**:491–498.

Benyoucef, M., Rigaud, J.-L., and Leblanc, G., 1981b, *Eur. J. Biochem.* **113**:499–506.

Benyoucef, M., Rigaud, J.-L., and Leblanc, G., 1982a, *Biochem. J.* **208**:529–538.

Benyoucef, M., Rigaud, J.-L., and Leblanc, G., 1982b, *Biochem. J.* **208**:539–547.

Boyer, P. D., 1988, *Trends Biochem. Sci.* **13**:5–7.

Chen, J. W., Sun, Q., and Hwang, F., 1984, *Biochim. Biophys. Acta* **777**:151–154.

Chernyak, B. V., Dibrov, P. A., Glagolev, A. N., Sherman, M. Y., and Skulachev, V. P., 1983, *FEBS Lett.* **164**:38–42.

Choppin, P. W., and Scheid, A., 1980, *Rev. Infect. Dis.* **2**:40–58.

Cirillo, V. P., Katzenell, A., and Rottem, S., 1987, *Isr. J. Med. Sci.* **23**:380–383.

Citovsky, V., Rottem, S., Nussbaum, O., Laster, Y., Rott, R., and Loyter, A., 1987, *J. Biol. Chem.* **263**:461–467.

Clementz, T., Christiansson, A., and Wieslander, A., 1986, *Biochemistry* **25**:823–830.

Dean, G. E., Fishkes, H., Nelson, P. J., and Rudnick, G., 1984, *J. Biol. Chem.* **259**:9569–9574.

Dibrov, P. A., Skulachev, V. P., Sokolov, M. V., and Verkhovskaya, M., 1988, *FEBS Lett.* **233**:355–358.

Dimroth, P., 1980, *FEBS Lett.* **122**:234–236.

Dimroth, P., 1987, *Microbiol. Rev.* **51**:320–340.

Epstein, W., 1985, *Curr. Top. Membr. Transp.* **23**:153–175.

Fillingame, R. H., 1981, *Curr. Top. Bioenerg.* **11**:35–106.

Fischer, M., Shirvan, M. H., Platt, M. W., Kirchhoff, H., and Rottem, S., 1988, *J. Gen. Microbiol.* **134**:2385–2392.

Furst, P., and Solioz, M., 1986, *J. Biol. Chem.* **251**:4302–4308.

Futai, M., and Kanazawa, H., 1983, *Microbiol. Rev.* **47**:285–312.

Garrahan, P. J., and Glynn, I. M., 1967, *J. Physiol. (London)* **192**:159.

Gay, N. J., and Walker, J. E., 1981, *Nucleic Acid Res.* **9**:3919–3926.

George, R., Lewis, R., and McElhaney, R. N., 1985, *Biochim. Biophys. Acta* **821**:253–258.

Harold, F. M., 1983, *The Vital Force: A Study of Bioenergetics*, Freeman, San Francisco.

Harold, F. M., and Brunt, J. V., 1977, *Science* **197**:372–373.

Harold, F. M., and Kakinuma, Y., 1985, *Ann. N.Y. Acad. Sci.* **456**:375–383.

Heefner, D. L., and Harold, F. M., 1982, *Proc. Natl. Acad. Sci. USA* **79**:2798–2802.

Heefner, D. L., Kobayashi, H., and Harold, F. M., 1980, *J. Biol. Chem.* **255**:11403–11407.

Jinks, D. C., Silvius, J. R., and McElhaney, R. N., 1978, *J. Bacteriol.* **136**:1027–1036.

Kakinuma, Y., and Harold, F. M., 1985, *J. Biol. Chem.* **260**:2086–2091.

Kakinuma, Y., and Igarashi, K., 1989, *J. Bioenerg. Biomembr.* **21**:679–692.

Kakinuma, Y., and Unemoto, T., 1985, *J. Bacteriol.* **163**:1293–1295.

Kobayashi, H., Suzuki, T., and Unemoto, T., 1986, *J. Biol. Chem.* **261**:627–630.

Krulwich, T. A., Hicks, D. B., Seto-Young, D., and Guffanti, A. A., 1988, *Crit. Rev. Microbiol.* **16**:15–36.

Lanyi, J. K., 1979, *Biochim. Biophys. Acta* **559**:377–398.

Laubinger, W., and Dimroth, P., 1988, *Biochemistry* **27**:7531–7537.

Leaf, A., 1956, *Biochem. J.* **62**:241–248.

Leaf, A., 1959, *Ann. N.Y. Acad. Sci.* **72**:396–404.

Leblanc, G., and Le Grimellec, C., 1979, *Biochim. Biophys. Acta* **554**:168–179.

Le Grimellec, C., Lajqunesse, D., and Rigaud, J.-L., 1982, *Biochim. Biophys. Acta* **687**:281–290.

Lelong, I., Shirvan, M. H., and Rottem, S., 1989, *FEMS Lett.* **59**:71–76.

Lewis, R. N. A. H., and McElhaney, R. N., 1983, *Biochim. Biophys. Acta* **735**:113–122.

Linker, C., and Wilson, T. H., 1985a, *J. Bacteriol.* **163**:1243–1249.

Linker, C., and Wilson, T. H., 1985b, *J. Bacteriol.* **163**:1250–1257.

Linker, C., and Wilson, T. H., 1985c, *J. Bacteriol.* **163**:1258–1262.

Macara, I. G., Kustin, K., and Cantley, L. C., 1980, *Biochim. Biophys. Acta* **629**:95–106.

MacKnight, A. D., and Leaf, A., 1978, in: *Physiology of Membrane Disorders* (T. E. Andreoli, J. F. Hoffman, and D. D. Fanestil, eds.), Plenum Press, New York, pp. 315–334.

Mahajan, S., Lewis, R. N. A. H., George, R., Sykes, B. D., and McElhaney, R. N., 1988, *J. Bacteriol.* **170**:5739–5746.

Masover, G. K., and Hayflick, L., 1973, *Ann. N.Y. Acad. Sci.* **225**:118–130.

Mitchell, P., 1966, *Biol. Rev.* **41**:445–502.

Nakamura, T., Hsu, C.-M., and Rosen, B. P., 1986, *J. Biol. Chem.* **261**:678–683.

Padan, E., and Schuldiner, S., 1986, *Methods Enzymol.* **125**:337–352.

Padan, E., and Schuldiner, S., 1987, *J. Membr. Biol.* **95**:189–197.

Pedersen, P. L., and Carafoli, E., 1987, *Trends Biochem. Sci.* **12**:146–150.

Rasmussen, O. F., Shirvan, M. H., Rottem, S., and Christiansen, C., 1987, *Isr. J. Med. Sci.* **23**:393–397.

Rasmussen, O. F., Shirvan, M. H., Rottem, S., and Christiansen, C., 1990, *Zentralbl. Bakteriol.* **20**:619–621.

Rasmussen, O. F., Shirvan, M. H., Margalit, H., Christiansen, C., and Rottem, S., 1992, *Biochem. J.* **285**:881–888.

Razin, S., 1978, *Microbiol. Rev.* **42**:414–470.

Romano, N., La Licata, R., and Russo, A. D., 1986, *Pediatr. Infect. Dis. Suppl.* **5**:S308–S312.

Rottem, S., and Markowitz, O., 1979, *Biochemistry* **18**:2930–2935.

Rottem, S., and Razin, S., 1966, *J. Bacteriol.* **92**:714–722.

Rottem, S., and Razin, S., 1972, *J. Bacteriol.* **110**:699–705.

Rottem, S., and Verkleij, A. J., 1982, *J. Bacteriol.* **149**:338–345.

Rottem, S., Linker, C., and Wilson, T. H., 1981, *J. Bacteriol.* **145**:1299–1304.

Rottem, S., Shirvan, M. H., and Gross, Z., 1983, *Yale J. Biol. Med.* **56**:405–411.

Rottem, S., Shirvan, M. H., Barile, M. F., and Zilberstein, D., 1987, *Isr. J. Med. Sci.* **23**:389–392.

Scheifer, H.-G., and Schummer, U., 1982, *Rev. Infect. Dis. Suppl.* **4**:S65–S70.

Schummer, U., Scheifer, H.-G., and Gerhardt, U., 1981, *Curr. Microbiol.* **5**:371–374.

Senior, A. E., 1990, *Annu. Rev. Biophys. Chem.* **19**:7–41.

Shirvan, M. H., Schuldiner, S., and Rottem, S., 1987, *Isr. J. Med. Sci.* **23**:384–388.

Shirvan, M. H., Schuldiner, S., and Rottem, S., 1989a, *J. Bacteriol.* **171**:4410–4416.

Shirvan, M. H., Schuldiner, S., and Rottem, S., 1989b, *J. Bacteriol.* **171**:4417–4424.

Simoneau, P., and Labarere, J., 1991, **137**:179–185.

Skulachev, V. P., 1989, *FEBS Lett.* **250**:106–114.

Stewart, L. M. D., Bakker, E. P., and Booth, I. R., 1985, *J. Gen. Microbiol.* **131**:77–85.

Tokuda, H., and Unemoto, T., 1981, *Biochem. Biophys. Res. Commun.* **102**:265–271.

Tokuda, H., and Unemoto, T., 1982, *J. Biol. Chem.* **257**:10007–10014.

Tokuda, H., and Unemoto, T., 1983, *J. Bacteriol.* **156**:636–643.

Tokuda, H., Sugasawa, M., and Unemoto, T., 1982, *J. Biol. Chem.* **257**:788–794.

Walderhaug, M. O., Dosch, D. C., and Epstein, W., 1986, in: *Bacterial Ion Transport* (B. P. Rosen and S. Silver, eds.), Academic Press, New York.

Walker, J. E., Fearnley, I. M., Gay, N. J., Gibson, B. W., and Northrop, F. D., 1985, *J. Mol. Biol.* **184**:677–701.

White, J., Kielian, M., and Helenius, A., 1983, *Q. Rev. Biophys.* **16**:151–197

Wieslander, A., Christiansson, A., Rilfors, L., and Lindblom, G., 1980, *Biochemistry* **19**:3650–3655.

Wieslander, A., Christiansson, A., Rilfors, L., Khan, A., Johansson, L.B.-A., and Lindblom, G., 1982, *Rev. Infect. Dis. Suppl.* **4**:S43–S49.

Wilson, T. H., 1954, *Science* **120**:104–105.

Yatzyu, H., Shiota, S., Futai, M., and Tajima, A., 1985, *J. Bacteriol.* **162**:933–937.

Zilberstein, D., Shirvan, M. H., Barile, M., and Rottem, S., 1986, *J. Biol. Chem.* **261**:7109–7111.

Chapter 11

Transport Systems in Mycoplasmas

Vincent P. Cirillo

1. INTRODUCTION

Transport processes are classified on the basis of kinetics and energetics (Figure 1). Using D and L isomers of glucose as examples, simple, Fickian diffusion kinetics are characteristic of unmediated transport which does not discriminate between the D and L isomers. Unmediated transport represents transport through the lipid bilayer. Discrimination between D and L isomers and saturation kinetics are characteristic of mediated transport. Mediated transport is carried out by intrinsic membrane proteins and involves interaction between the solute and the proteins. The classification of mediated transport processes is based on biochemical and energetic criteria. In group translocation, the transported solute is derivatized by a membrane enzyme; in the group translocation of D-glucose, the sugar is transported and phosphorylated in a coordinated process. In carrier transport, the free solute, not a derivative, is transported. Carrier transport is further classified on the basis of energetic criteria. If the free sugar is transported energetically uphill (i.e., against its concentration gradient), it is classified as active transport; if the sugar is transported energetically downhill, it is classified as facilitated diffusion.

Vincent P. Cirillo   Department of Biochemistry and Cell Biology, State University of New York, Stony Brook, New York 11794.

Subcellular Biochemistry, Volume 20: Mycoplasma Cell Membranes, edited by Shlomo Rottem and Itzhak Kahane. Plenum Press, New York, 1993.

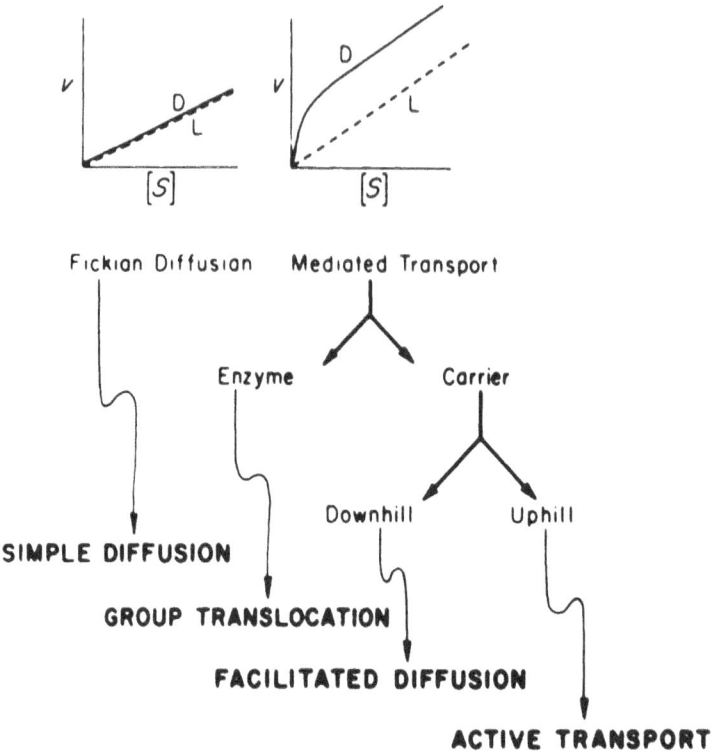

FIGURE 1. Classification of transport systems. Graphs show dependence of transport rate, v, on substrate concentration, [S], for a hypothetical pair of isomers, D and L. Mediated transport is classified on the basis of biochemical and energetic characteristics. (From Cirillo, 1979.)

Active transport processes are further classified on the basis of whether they are directly or indirectly linked to energy-yielding reactions (Figure 2). If the linkage is direct, they are classified as primary active transport process. These include ion pumps energized by electron transport or ATP hydrolysis and solute transport coupled directly to the hydrolysis of ATP (Figure 2). If the linkage is indirect, such as by coupling to the electrochemical gradients or membrane potentials created by the primary transport processes, they are classified as secondary active transport processes (Figure 2). The primary ion pumps shown in Figure 2 transport protons. While these are the most common primary ion pumps in bacteria, primary sodium pumps also occur in bacteria driven by electron transport chains, ATP hydrolysis, and unique decarboxylases (Skulachev, 1988). A primary sodium pump has been described for *M. gallisepticum* which is implicated in volume regulation (Shirvan *et al.*, 1989) and described in detail in Chapter 10 of this volume.

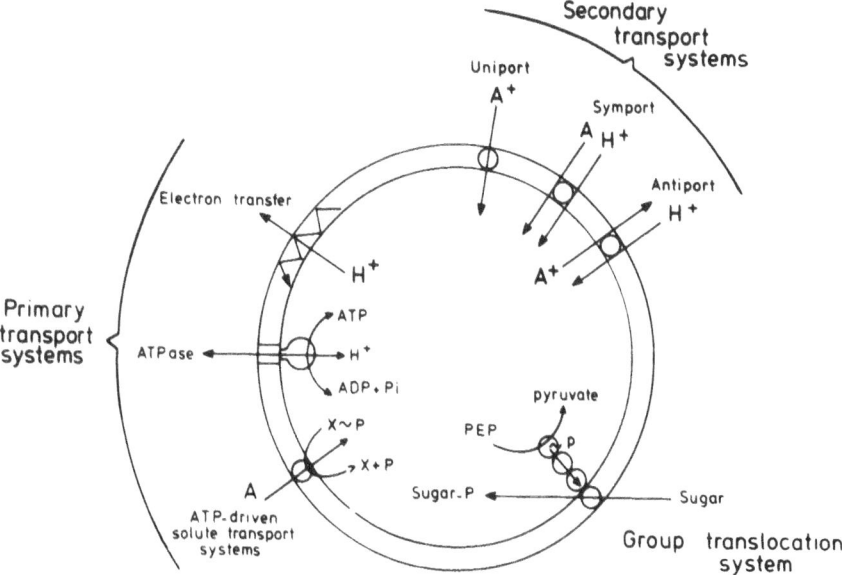

FIGURE 2. Differentiation among primary and secondary active transport and group translocation processes in bacteria. (From Konings *et al.*, 1987.)

2. THE PERMEABILITY BARRIER

2.1. The Barrier Function of the Lipid Bilayer

Deducing the nature of the permeability barrier from the characteristics of unmediated transport of homologous series of solutes has a long history going back to Overton at the end of the 19th century (Davson, 1989). These early permeability studies, together with physical and chemical studies of cell membranes and cell membrane lipids, led to the classical Davson–Danielli model of the cell membrane in which the membrane was described to be composed of a lipid bilayer in association with proteins (Davson, 1989). The lipid bilayer was identified as the principal permeability barrier to charged and polar molecules (Davson and Danielli, 1943). The current Singer–Nicholson (1972) fluid-mosaic model of the cell membrane differs from the original Davson–Danielli model principally with respect to the structure and arrangement of "intrinsic" and "extrinsic" membrane proteins.

2.2. Unmediated Transport of Polyols

Studies of the permeability of intact mycoplasma cells and liposomes derived from the membrane lipids have confirmed and extended our understanding

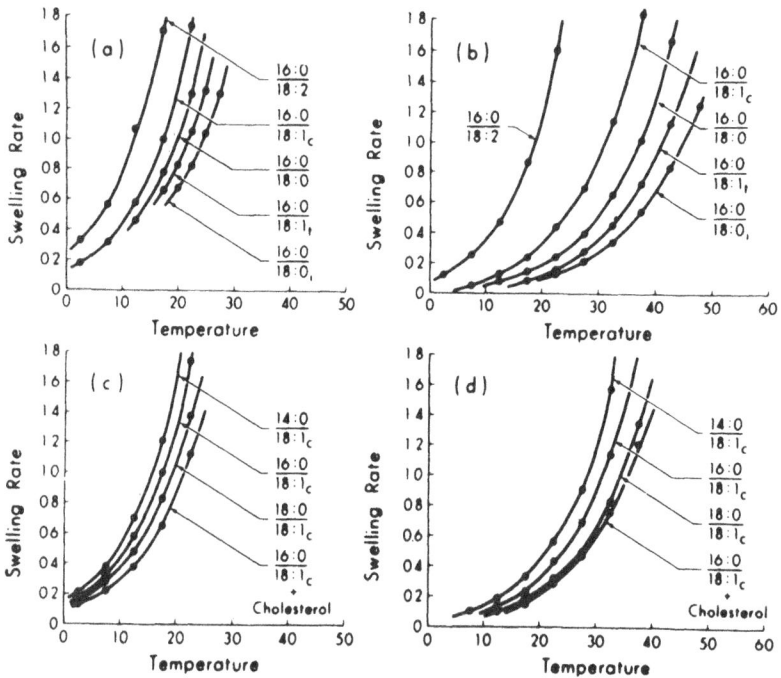

FIGURE 3. Initial swelling rates as a function of temperature for *A. laidlawii* cells grown in the presence of different combinations of fatty acids and cholesterol (a, c) and liposomes prepared from the total lipids of the same cells (b, d). Swelling rates were determined from the changes in absorbance, *A*, as a function of time, $d(1/A)/dt$. (Modified from McElhaney *et al.*, 1973.)

of the barrier function of the lipid bilayer. Mycoplasmas are natural fatty acid auxotrophs; their membrane lipid composition is determined by the fatty acid composition of the growth medium (McElhaney and Tourtellotte, 1969). McElhaney *et al.* (1973) measured the rate of uptake of a homologous series of polyols (glycol, glycerol, and erythritol) by intact *A. laidlawii* cells and liposomes prepared from the membrane lipids. The rate of uptake was determined from the rate of swelling of cells or liposomes suspended in isoosmotic (i.e., 400–500 mosmole) solutions under which conditions the cells and liposomes behave as ideal osmometers. In Figure 3, the rate of swelling of cells and liposomes of different fatty acid compositions are compared. There is a striking similarity between the permeability of the cells and the liposomes, confirming the barrier function of the membrane lipids. Analysis of the permeability of polyols in membranes of different fatty acid compositions allowed two conclusions about the process of unmediated, polyol transport in both cells and liposomes: (1) the permeability to a given polyol increases with an increase in the concentration of unsaturated fatty acids, but the activation energy is not affected by the fatty acid composition. This shows that the lower permeability in membranes with saturated fatty acids is due to a greater entropic resistance of polyol penetration into membranes with saturated fatty acids. In simple terms, this means that penetration of polyols causes less packing of the lipid side chains in membranes with unsaturated fatty acids than in membranes with saturated fatty acids. (2) The permeability decreases with an increase in the number of hydroxyl groups on the molecule. This is due primarily to the energy required to break the hydrogen bonds between the hydroxyl groups and water.

3. SUGAR TRANSPORT

Because mycoplasmas have a limited biosynthetic capacity, they are auxotrophic for many nutrients and, therefore, require a large variety of transport systems. Initial studies of solute transport in mycoplasmas established that mycoplasmas carry out both active transport and group translocation processes similar to, and as complex as, those in other bacteria (Rottem and Razin, 1966; Razin *et al.*, 1968; Cirillo, 1979). The complexity was a disappointment since it was hoped that mycoplasmas, with their reduced genome size, might have reduced the type and complexity of their transport processes (Cirillo, 1979).

Different mechanisms of sugar transport are observed in *Acholeplasma* species on the one hand and in *Mycoplasma* and *Spiroplasma* species on the other. In *A. laidlawii*, glucose is transported by secondary active transport coupled to the proton gradient. In fermentative species of *Mycoplasma* and *Spiroplasma*, sugar is transported by group translocation. The latter will be described in a later section.

3.1. Primary Ion Pumps and Secondary Active Transport

Secondary active transport coupled to ion electrochemical gradients created by a primary cation pump was first proposed by Crane (1962) to explain the dependence on the sodium pump of the active transport of sugars in the intestine. It was proposed that the dependence was indirect; the sodium pump (the Na^+/K^+-dependent ATPase) creates an electrochemical sodium ion gradient and the downhill transport of the sodium ion is linked, via a common carrier, to the uphill transport of the sugar. Mitchell incorporated the concept of ion-coupled transport in the chemiosmotic hypothesis and predicted that electrochemical proton gradients (i.e., the proton-motive force) created by proton pumps could be used, in addition to synthesizing ATP, to drive secondary active transport (Mitchell, 1961, 1966). The predicted proton-linked active transport processes were soon discovered in organelles, prokaryotes, and plant cells (Harold, 1986). Gradient coupling mechanisms are widely distributed among both eukaryotes and prokaryotes including mycoplasmas (Cirillo, 1979). The types and mechanisms of the primary cation pumps which create electrochemical ion gradients in mycoplasmas are described in detail in Chapter 10.

Secondary Active Transport in A. laidlawii

The conclusion that glucose transport in A. laidlawii is a carrier-mediated, secondary active transport process coupled to the proton-motive force is based on the following observations. (1) The rate of glucose uptake by cells is 100-fold greater than in liposomes (Read and McElhaney, 1975). (2) Cells take up D-glucose and its nonmetabolized analogue, 3-O-methylglucose (3-O-MG) in preference to L-glucose (Tarshis et al., 1976a,b). (3) Uptake shows saturation kinetics for D-glucose in whole cells and 3-O-MG in membrane vesicles with apparent K_m's of 21.2 and 4.6 μM, respectively (Tarshis et al., 1976a; Panchenko et al., 1975). (4) Transport is inhibited by the sugar transport inhibitors phloretin and phlorizin (Smith, 1969; Read and McElhaney, 1975). (5) Unmodified 3-O-MG is accumulated against a concentration gradient in whole cells and both free glucose and 3-O-MG are accumulated against a concentration gradient in membrane vesicles (Fedetov et al., 1975b; Tarshis et al., 1976ba). (6) Active transport is inhibited by inhibitors which dissipate the proton-motive force, dinitrophenol (DNP) and carbonyl cyanide m-chlorophenyl hydrazone (CCCP) (Fedetov et al., 1975a; Tarshis et al., 1976a,b). (7) Counterflow occurs between intracellular unlabelled 3-O-MG and extracellular ^{3}H-labeled 3-O-MG in CCCP-inhibited cells (Tarshis et al., 1976a).

Although mycoplasmas possess a primary sodium pump in addition to a proton pump (Shirvan et al., 1989, 1990), no secondary active transport coupled to the sodium gradient has yet been described.

3.2. Group Translocation

In a group translocation process, membrane transport and the first enzymatic step of metabolism is a single process carried out by a membrane-bound enzyme. For glucose, this would represent transport and sugar phosphorylation. This was such a compelling idea that it dominated the sugar transport field for at least two decades before being abandoned for most cells when significant differences were observed between the substrate specificity of transport and that of phosphorylating enzymes and when nonmetabolized analogues were shown to be transported without being phosphorylated. However, when "the last nail was about to be put into the coffin" of the sugar group translocation process, one of those surprises occurred which makes the scientific enterprise so exciting. Saul Roseman and his colleagues (Kundig et al., 1964), without looking for group translocation, discovered the PEP-dependent: sugar phosphotransferase system (PTS) in E. coli. The PTS involves a phosphorylation cascade between PEP and the sugar. PTSs were found to be widely distributed among facultative and obligate anaerobic bacteria (Romano *et al.*, 1979; Saier, 1985). The PTS has been found in a number of fermentative species of Mycoplasma and Spiroplasma (Van Demark and Plackett, 1972; Cirillo and Razin, 1973; Cirillo, 1979; Tarshis, 1991).

3.2.1. The PEP-Dependent : Sugar Phosphotransferase System

The PTS is a phosphorylation cascade which transfers the phosphoryl group from PEP to the transported sugar. It consists of both soluble and membrane-bound proteins. In the form originally described in *E. coli*, it consisted of two general energy-coupling proteins, Enzyme I and HPr, which are soluble, are successively phosphorylated on histidine residues and are common to all sugars which are transported by the PTS, and two, sugar-specific permease proteins that may be membrane-bound (IIA and IIB) or may consist of one soluble (III) and one membrane-bound (IIB') protein. The reaction sequence for glucose can be represented by the following equations:

$$\text{PEP} + \text{Enzyme I} \longrightarrow \text{Enzyme I}-P + \text{pyruvate} \qquad (1)$$

$$\text{Enzyme I}-P + \text{HPr} \xrightarrow[\text{Enzyme II}^{glc}\text{A/II}^{glc}\text{B}]{} \text{HPr}-P + \text{Enzyme I} \qquad (2)$$

$$\text{HPr}-P + \text{Glucose} \xrightarrow[\text{Enzyme III}^{glc}/\text{II}^{glc}\text{B'}]{} \text{HPr} + \text{Glucose}-P \qquad (3)$$

Rottem and Razin (1969) reported that *M. gallisepticum* accumulates α-methyl-D-glucopyranoside (α-MG). In bacteria, this substrate is phosphory-lated by the glucose-specific PTS (Romano *et al.*, 1970; Kornberg, 1976); its uptake is often a clue to the presence of a PTS. The presence of a PTS in *M. gallisepticum* was subsequently reported by Cirillo and Razin (1973); they also reported a PTS in *M. mycoides* subsp. *mycoides* and *M. mycoides* subsp. *capri*. A PTS was independently reported by Van Demark and Plackett (1972) in *Mycoplasma* sp. strain Y. The PTS has since been found in *Mycoplasma capricolum* (Jaffor Ullah and Cirillo, 1976, 1977) and in all species of *Spiroplasma* tested (Table I; Cirillo, 1979; Tarshis, 1991). It is absent in nonfermentative species of the genus *Mycoplasma* and in all species of *Acholeplasma* tested and in *Thermoplasma acidophilum* (Cirillo and Razin, 1973; Jaffor Ullah and Cirillo, 1976, 1977; Cirillo, 1979).

3.2.2. *M. Capricolum* PTS Components

3.2.2.1. Complementation between *M. capricolum* and *E. coli* PTS Components. Only the PTS of *M. capricolum* has been studied in detail. It contains PTS components which complement *E. coli* HPr, Enzyme I, and En-

Table I
Distribution of the PEP : Sugar Phosphotransferase System in Mollicutes

Present	Absent
Mycoplasma gallisepticum[a]	*Mycoplasma bovigenitalium*[c]
M. sp. strain Y[b]	*M. hominis*[c]
M. mycoides subsp. *mycoides*[c]	*M. agalactiae*[c]
M. mycoides subsp. *capri*[c]	*M. flocculare*[c]
M. capricolum[d]	*M. hypopneumoniae*[e]
Sprioplasma citri Morocco[e,g]	*Acholeplasma laidlawii*[c,f]
S. citri W[g]	*A. granularum*[c]
S. citri B[g]	*A. axanthum* 410[c]
S. citri 74[g]	*A. axanthum* 743[c]
S. floricola BNR1[g]	*Thermoplasma acidophilum*[e]
S. meliferum BC3[g]	
S. meliferum WIMP[g]	
S. apis B31[g]	
S. apis PPS1[g]	

[a]Rottem and Razin (1969).
[b]Van Demark and Plackett (1972).
[c]Cirillo and Razin (1973).
[d]Jaffor Ullah and Cirillo (1976, 1977).
[e]Cirillo (1979).
[f]Tarshis *et al.* (1976a,b).
[g]Tarshis (1991).

zyme II (Tables II and III; Jaffor Ullah and Cirillo, 1976). By measuring the PTS activity of mixtures of membranes as a source of Enzyme II and the soluble fraction of cell extracts (containing HPr and Enzyme I) as a source of HPr–*P*, it was found that *M. capricolum* Hpr–*P* is as good a substrate for *E. coli* Enzyme II as the *E. coli* HPr–*P* but the *E. coli* HPr–*P* is only about 10% as effective a substrate for the *M. capricolum* Enzyme II as *M. capricolum* HPr–P (Table II). Mixing experiments using purified Hpr and Enzyme I (Table III) showed that

Table II
In vitro Complementation between *M. Capricolum* and *E. coli*[a]

Membranes (Enzyme II)	Soluble fraction (Enzyme I and HPr)	α-MG phosphorylated/min (nmole)
Mycoplasma	*Mycoplasma*	0.97
Mycoplasma	*E. coli*	0.1
E. coli	*Mycoplasma*	0.8
E. coli	*E. coli*	0.78

[a]Data from Jaffor Ullah and Cirillo (1976).

Table III
In vitro Complementation Using Isolated PTS Components of *M. Capricolum* and *E. coli*[a]

Expt No.	Enzyme II	Enzyme I	HPr	α-MG phosphorylated/min (nmole)
1	*Mycoplasma*	*Mycoplasma*	*Mycoplasma*	0.69
2	*Mycoplasma*	*E. coli*	*E. coli*	0.06
3	*E. coli*	*Mycoplasma*	*Mycoplasma*	0.44
4	*E. coli*	*E. coli*	*E. coli*	0.43
5	*Mycoplasma*	*E. coli*	*Mycoplasma*	0.12
6	*E. coli*	*E. coli*	*Mycoplasma*	0.15
7	*Mycoplasma*	*Mycoplasma*	*E. coli*	0.02
8	*E. coli*	*Mycoplasma*	*E. coli*	0.09

[a]Data from Jaffor Ullah and Cirillo (1976).

Enzyme I of each species phosphorylates the heterologous HPr at about 20% of the activity that it shows for its homologous HPr (Expt 1 versus 5 and Expt 4 versus 8).

3.2.2.2. *M. capricolum* HPr and Enzyme I. The *M. capricolum* HPr and Enzyme I have been purified to homogeneity and characterized by Jaffor Ullah and Cirillo (1976, 1977). The HPr is very similar to that of *E. coli* and *S. aureus* in its molecular weight (9506 versus 9537 and 8630, respectively) and in its heat stability. However, it does not cross-react with antibodies raised against the *E. coli* HPr. It contains a single histidine which is necessary for its activity and is presumably the residue which is phosphorylated by Enzyme I. A fructose-specific HPr, HPr[fru], has been identified in extracts of fructose-grown, α-MG-resistant mutants (Mugharbil and Cirillo, 1978). It is ten times more active with the fructose-specific Enzyme II than HPr, but has not been purified or further characterized.

The *M. capricolum* Enzyme I is a tetramer with a molecular weight of 220,000 with an $\alpha_2\beta\tau$ subunit structure; the molecular weights of the subunits are 44,500, 62,000, and 64,500, respectively. By contrast, Enzyme I of *E. coli* and *S. aureus* are monomers with molecular weights between 70,000 and 90,000.

3.2.2.3. *M. capricolum* "Enzyme II." The Enzyme II activities have been identified only in crude membrane fractions. Constitutive activity for glucose and inducible activity for fructose are observed in *M. capricolum* (Mugharbil and Cirillo, 1978) and constitutive activities for glucose, mannose, and fructose in *M. mycoides* subsp. *capri* (Cirillo and Razin, 1973). The Enzyme II activity of *M. capricolum* was presumed to be comparable to the IIA/IIB type of *E. coli* (Cirillo, 1979); however, see below.

The Enzyme II nomenclature used above to describe the sugar-specific permease components of the PTS has recently been revised. The original nomenclature was based on permeases which consisted of two polypeptide chains designated IIA and IIB, if they were both membrane-bound, or III and IIB', if one was soluble and one was membrane-bound, respectively. Recent determination of the structure and amino acid sequences of some 20 permeases has shown that the PTS permeases can have a wide range of structures (Saier and Reizer, 1990). Although many permeases consist of two proteins, as originally described, others may consist of only one polypeptide or up to three proteins; however, irrespective of the number of polypeptides, the aggregate molecular weight is usually about 68,000, or about 630 amino acids. The permeases consist of three functional domains: a hydrophilic, Enzyme III-like domain which possesses the first phosphorylation site, a second hydrophilic protein which possesses the second phosphorylation site, and a hydrophobic, transmembrane domain which binds and transports the sugar substrate. These three domains are designated domains IIA, IIB, and IIC, respectively. The three domains may all occur on a single polypeptide chain, as in the case of the *E. coli* mannitol PTS, or may be separated on three different polypeptides, as in the case of the *E. coli* cellobiose PTS (Reizer *et al.*, 1990). Even more complex rearrangements occur in the fructose-specific enzymes. The complex rearrangements of the order and distribution of these domains on different numbers of polypeptides in different PTSs suggest that there has been a great deal of interdomain shuffling, splicing, fusion, deletion, and duplication in the evolution of the PTSs. Given this new information, the earlier presumption that the Enzyme II activity of *M. capricolum* was comparable to the IIA/IIB type of *E. coli* must be reconsidered. It will be interesting to determine the actual structure of the mycoplasma PTSs and to determine the extent to which they show evidence of domain shuffling and the other rearrangements described above.

3.2.3. The Regulatory Role of the PTS and Its Components

The PTSs serve both to transport their sugar substrates and to regulate the function and synthesis of non-PTS transporters. In the presence of a PTS sugar, the synthesis of inducible, non-PTS permeases is inhibited (repression) and the transport activity of inducible, non-PTS transporters is inhibited (inducer exclusion). In *E. coli* and *S. typhimurium*, these regulatory functions are mediated by the degree of phosphorylation of Enzyme III^{glc} (labeled RPr in Figure 4; Saier, 1977). In the phosphorylated form, Enzyme III^{glc} activates adenylate cyclase, enabling the synthesis of inducible, non-PTS permeases, whereas in its dephospho form, it inactivates the carrier function of non-PTS permeases. Since, in the presence of a PTS sugar, Enzyme III^{glc} will be in the dephospho form, adenylate cyclase is not activated and non-PTS permeases are inactivated. Thus,

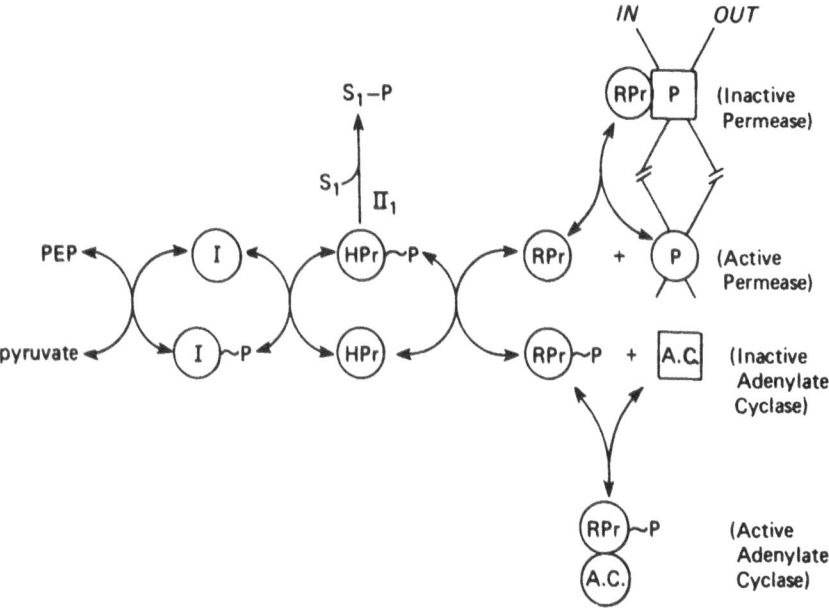

FIGURE 4. The PTS phosphorylaton cascade and the proposed regulation of non-PTS permeases and adenylate cyclase by the phosphorylated and dephosphorylated forms of Enzyme III^glc represented here as RPr. (From Saier, 1977.)

the degree of phorphorylation of III^glc is a sensor for the presence of PTS sugars which are used in preference to inducible, non-PTS sugars. The applicability of this model to mycoplasma was demonstrated for *M. capricolum* in which the intracellular level of cyclic AMP is decreased in the presence of the PTS substrates glucose and fructose (Mugharbil and Cirillo, 1978). The direct role of the PTS in the reduction of cyclic AMP was demonstrated by the loss of glucose sensitivity in mutants with a defective, glucose-specific Enzyme II with continued sensitivity to fructose for which there is a separate, inducible fructose specific Enzyme II.

The PTS may also regulate cellular metabolism by activating or inhibiting the activity of metabolic enzymes. Thus, the enzyme glycerol kinase is regulated by the PTS in both gram-negative and gram-positive cells. In *E. coli* and *S. typhimurium* the enzyme is regulated in a manner similar to the non-PTS permeases (Figure 4), namely the enzyme is inhibited by allosteric interaction with the dephospho form of Enzyme III^glc (Postma *et al.*, 1984; Novotny *et al.*, 1985). In *Streptococcus faecalis* the enzyme is activated tenfold by an Enzyme I and HPr-mediated phosphorylation (Deutscher and Sauerwald, 1986).

At the end of an earlier review, I commented, "While we marvel at the many [regulatory] functions of the PTS in bacteria and . . . in mycoplasmas, it must not be forgotten that many organisms (i.e., the acholeplasmas among the *Mollicutes*) do not have this system. This naturally raises the question of what mechanisms are used by these other organisms to serve these functions" (Cirillo, 1979). It seems that even in organisms lacking a functional PTS, PTS components may serve a regulatory role.

HPr and Enzyme I in the Absence of PTS Activity. Romano made the interesting discovery that although the heterofermentative lactic acid bacilli *Lactobacillus brevis* and *L. buchneri* lack a functional PTS, they possess an HPr (Romano *et al.*, 1987; Reizer *et al.*, 1988). In these cells, HPr is phosphorylated by the ATP-dependent Hpr(Ser) kinase which phosphorylates a serine residue instead of a histidine residue, as is the case for PEP-dependent, Enzyme I-mediated phosphorylation. The product, designated HPr[Ser(P)], is dephosphorylated by an HPr[(Ser–P)] phosphatase. The allosteric modulation by regulatory metabolites on the opposing protein kinase and phosphatase was taken to suggest a regulatory role for HPr (Reizer *et al.*, 1988); in lactic acid bacilli, the HPr is involved in the phenomenon of inducer exclusion (Romano *et al.*, 1987), i.e., the active expulsion of previously accumulated β-galactosides upon the addition of glucose. The possibility that HPr could exist in organisms which do not show PTS activity led Rottem and Reizer (unpublished data) to look for PTS components in *A. laidlawii*. Using *in vitro* complementation tests similar to those described in Table III, they found both HPr and Enzyme I activities as well as an ATP-dependent HPr kinase in the absence of a functional PTS system. In cells lacking a PTS, HPr and HPr kinase are likely to have a regulatory function. One possible regulatory mechanism would be phosphorylation of non-PTS proteins or enzymes. It will be recalled that in *Streptococcus faecalis,* Enzyme I and HPr phosphorylate the non-PTS enzyme, dihydroxyacetone and glycerol kinase, with a tenfold increase in kinase activity (Deutscher and Sauerwald, 1986). In *Acholeplasma* species, HPr, phosphorylated either by the PEP-dependent Enzyme I or the ATP-dependent kinase, is likely to have a regulatory function, possibly as a phosphoryl donor to a non-PTS enzyme. What that regulatory function is in *Acholeplasma* remains for future investigations to answer. The PTS continues to be full of surprises!

4. AMINO ACID TRANSPORT

Until the recent reconstitution into proteoliposomes of the arginine transport system from a *Spiroplasma* (Rottem and Shirazi, 1990), the only published work on amino acid transport in mycoplasma was that carried out by Razin *et al.* in 1968 on *l*-histidine transport in *M. fermentans* and *l*-methionine in *M. hominis*. The salient features of the latter transport processes are summarized in Table IV.

Table IV
Characteristics of Amino Acid Transport in Mycoplasmas[a]

	Mycoplasma fermentans	*Mycoplasma hominis*
Substrate	*l*-histidine	*l*-methionine
Kinetics		
K_m	80 μM	30 μM
V_{max}	17 nmole/mg/min	0.04 nmole/mg/min
Competitive inhibitors	*l*-arginine, *l*-ornithine	None (15 amino acids tested)
Accumulation ratio (C_i/C_o)	ca. 1000	30[*] (with chloramphenicol)
Energy sources		
Glucose	No stimulation	No stimulation
Na acetate	No stimulation	2× stimulation
Arginine	No stimulation	8× stimulation
Na acetate plus arginine	No stimulation	12× stimulation
Inhibitors		
*p*CMB	ca. 100% inhibition	ca. 100% inhibition
Na arsenate	ca. 40% inhibition	ca. 40% inhibition
K azide	ca. 25% inhibition	ca. 25% inhibition
NEM	ca. 25% inhibition	ca. 25% inhibition
IAA	ca. 20% inhibition	ca. 25% inhibition
KCN	ca. 20% inhibition	ca. 20% inhibition
NaF	No inhibition	No inhibition
DNP	No inhibition	No inhibition
Ouabain	No inhibition	No inhibition
Tl acetate	No inhibition	No inhibition
EDTA	No inhibition	ca. 80% inhibition
Efflux		
0°C	No efflux	Not done
37°C	$t_{1/2}$ ca. 15 min	Not done
Amino acid exchange 37°C	$t_{1/2}$ ca. 3 min	Not done

[a]Data from Razin *et al.* (1968).

4.1. Histidine Transport in *M. fermentans* Cells

A number of features of *l*-histidine transport in *M. fermentans* command special attention: (1) the absence of stimulation by exogenous energy sources, (2) the very high accumulation ratio (i.e., 1000/1), (3) the high V_{max}, (4) the relatively low efflux rate relative to the exchange rate, and (5) the relative insensitivity to the uncoupler DNP. Except for the lack of stimulation by exogenous energy sources, these features are characteristic of primary, ATP-driven solute transport systems (A in Figure 2) which occur in both gram-negative (Ames, 1990) and gram-positive bacteria (Konings *et al.*, 1987). Particularly relevant in this regard is the high accumulation ratio and insensitivity to the uncoupler. Konings *et al.* (1987) speculate that such amino acid-accumulating systems could supplement ion transport systems in osmoregulation. These early, provocative data should be confirmed and extended since they raise interesting and important questions. The nature of the endogenous substrate is another question that needs to be reinvestigated.

4.2. Methionine Transport in *M. hominis* Cells

The *l*-methionine transport system of *M. hominis* shares many of the features of the *l*-histidine system of *M. fermentans* but there are also significant differences. (1) There is a high affinity for its substrate, but, compared with the *l*-histidine system, it has a very low capacity. (2) The substrate specificity is very high; none of the 15 amino acids tested are competitive inhibitors. (3) Susceptibility to inhibitors is the same as for the *l*-histidine system except for its sensitivity to EDTA. (4) The accumulation ratio of 30 in the presence of chloramphenicol is modest compared with 1000 for *l*-histidine in *M. fermentans*. (5) The stimulation of transport by exogenous energy sources is a clear difference between the two systems. The stimulation by arginine suggests that energy from the dihydrolase pathway is involved (Smith, 1965). Strict comparisons between these two systems must be made with caution since only a single and a different amino acid transport system was studied in each organism. A more complete survey is clearly needed.

4.3. Reconstitution of *Spiroplasma* Arginine Transport in Proteoliposomes

Arginine is a major energy source for many mycoplasmas as seen above for *l*-methionine transport into *M. hominis* cells (Table IV). The arginine transport system from *S. melliferum* BC3 has been reconstituted in proteoliposomes.

Sealed proteoliposomes can be prepared by fusing purified intrinsic membrane proteins or membrane fractions with preformed liposomes by a freeze–thaw technique (Kasahara and Hinkle, 1977; Pick, 1981; Cirillo *et al.*, 1987). The liposomes are prepared by sonication of a mixtures of crude soybean phospholipids (Asolectin) and cholesterol. Fusion of the liposomes and the membranes is effected by freezing in a dry ice/acetone mixture or in liquid nitrogen followed by thawing at room temperature and briefly sonicating the thawed preparation. When ornithine-loaded proteoliposomes were diluted into arginine-containing medium, there was a rapid (22 nmole/mg protein per min), one-to-one exchange between the external arginine and the internal ornithine (Rottem and Shirazi, 1990). The exchange was insensitive to the uncouplers CCCP (at 10 μM) or SF 6847 (0.4 μM). The authors conclude that the driving force for arginine uptake *in vivo* is supplied by the concentration gradient of ornithine formed during arginine metabolism "thus preserving the energy obtained by the dihydrolase pathway."

5. FATTY ACID TRANSPORT

Since mycoplasmas are natural fatty acid auxotrophs, they must be able to take up fatty acids from the growth medium. The fatty acids are utilized primarily for the synthesis of membrane phospholipids and are not used as an energy source (Dahl et al., 1981; Dahl, 1988). It is surprising that the mechanism of fatty acid transport in mycoplasmas has received so little attention. In the earlier of the two papers cited above, Dahl et al. (1981) showed that the sterol content of the membranes affects the apparent K_m for oleate uptake. In cells grown in the presence of lanosterol (10 mg/ml), the K_m for oleate uptake was 16 μM; the addition of 0.5 mg/ml of cholesterol to the lanosterol medium decreased the K_m for oleate uptake to 3 μM which is the value found for cells grown in the presence of high cholesterol (10 mg/ml). Cholesterol does not affect the apparent K_m for palmitate uptake which is 2 μM both in lanosterol- and in cholesterol-grown cells. No attempt was made in this study to determine the mechanism of fatty acid uptake. In the more recent study, Dahl (1988) investigated fatty acid uptake in more detail. Fatty acid transport was found to be an energy-dependent process showing saturation kinetics and is inhibited by sulfhydryl reagents (e.g., IAA, NEM, and pCMB). Fatty acid uptake and esterification into phospholipids, mainly phosphatidylglycerol and cardiolipin, are tightly coupled; uptake only occurs under conditions which allow fatty acid esterification, namely the presence in the medium of glucose, glycerol, and potassium. Under these conditions, the free fatty acid content in the cells is less than 1% of the total radioactivity taken up after a 60-min incubation with radiolabeled fatty acid. Fatty acid uptake is inhibited by DCCD and valinomycin. The sensitivity of the uptake process to the latter inhibitors suggests that uptake depends on the membrane potential. However, since uptake is tightly coupled to phosphatidylglycerol synthesis and possibly to potassium transport, it is difficult to determine which step is membrane potential dependent. The tight coupling between fatty acid transport and esterification suggests that fatty acid uptake might be a group translocation process. Such a process is considered possible, but not proven, for fatty acid uptake in E. coli (Nunn, 1986). To test this hypothesis, Dahl looked for thiokinase (fatty acyl CoA synthase) activity in M. capricolum membranes. Surprisingly, no thiokinase activity was found; instead, a membrane-bound activity was detected which catalyzes the formation of fatty acyl-hydroxymate. This enzymatic activity is independent of CoA. Its role in fatty acid uptake is not understood. Also not understood is the requirement for potassium in fatty acid uptake. Thus, our understanding of the mechanism of uptake of these important molecules remains in an unsatisfactory state.

6. CONCLUSIONS

During what might be called the classical period, between 1960 and 1980, transport processes were studied in mycoplasmas hoping to find simplified versions of the complex systems found in other cells. It is now clear that, as far as transport processes are concerned, mycoplasmas are as complex as other bacteria. The future must focus on the unanswered questions raised by both the classical and the recent studies on mycoplasmas. Is the histidine transport system of *M. fermentans* an ATP-driven, primary active transport process? What role does amino acid transport play in osmoregulation in mycoplasmas? How much domain shuffling has occurred in the evolution of the mycoplasma PTS permeases? What are the functions of HPr and Enzyme I in *A. laidlawii?* Are fatty acids transported by group translocation? What is the nature and significance of the membrane enzyme which catalyzes the CoA-independent formation of fatty acyl-hydroxymates? With the new techniques of membranology and molecular biology, the future promises a rich harvest of answers—and new questions.

7. REFERENCES

Ames, G. F.-L., 1990, Energy coupling in bacterial periplasmic permeases, *J. Bacteriol.* **172:**4133–4137.

Cirillo, V. P., 1979, Transport systems, in: *The Mycoplasmas,* Volume 1 (M. F. Barile and S. Razin, eds.), Academic Press, New York, pp. 323–349.

Cirillo, V. P., and Razin, S., 1973, Distribution of a phosphoenolpyruvate dependent sugar phosphotransferase system in mycoplasmas, *J. Bacteriol.* **113:**212–217.

Cirillo, V. P., Katzenell, A., and Rottem, S., 1987, Sealed vesicles prepared by fusing *Mycoplasma gallisepticum* membranes and preformed lipid vesicles, *Isr. J. Med. Sci.* **23:**380–383.

Crane, R. K., 1962, Hypothesis for mechanism of intestinal active transport of sugars. *Fed. Proc.* **21:**891–895.

Dahl, J., 1988, Uptake of fatty acids by *Mycoplasma capricolum, J. Bacteriol.* **170:**2022–2026.

Dahl, J. S., Dahl, C. E., and Bloch, K., 1981, Effect of cholesterol on macromolecular synthesis and fatty acid uptake by *Mycoplasma capricolum, J. Biol. Chem.* **256:**87–91.

Davson, H., 1989. Biological membranes as selective barriers to diffusion of molecules, in: *Membrane Transport: People and Ideas* (D. C. Tosteson, ed.), Academic Press, New York, pp. 1–49.

Davson, H., and Danielli, J. F., 1943, *The Permeability of Natural Membranes,* Cambridge University Press, London.

Deutscher, J., and Sauerwald, H., 1986, Stimulation of dihydroxyacetone and glycerol kinase in *Streptococcus faecalis* by phosphoenolpyruvate-dependent phosphorylation catalyzed by enzyme I and HPr of the phosphotransferase system, *J. Bacteriol.* **166:**829–836.

Fedetov, N. S., Panchenko, L. F., and Tarshis, M. A. 1975a, Transport properties of membrane vesicles from *Acholeplasma laidlawii* I, *Folia Microbiol. (Prague)* **20**:470–479.

Fedetov, N. S., Panchenko, L. F., and Tarshis, M. A., 1975b, Transport properties of membrane vesicles from *Acholeplasma laidlawii* III, *Folia Microbiol. (Prague)* **20**:488–495.

Harold, F. M., 1986, *The Vital Force: A Study of Bioenergetics*, Freeman, San Francisco.

Jaffor Ullah, A. H., and Cirillo, V. P., 1976, *Mycoplasma* phosphoenolpyruvate-dependent sugar phosphotransferase system: Purification and characterization of the phosphocarrier protein, *J. Bacteriol.* **127**:1298–1306.

Jaffor Ullah, A. H., and Cirillo, V. P., 1977, *Mycoplasma* phosphoenolpyruvate-dependent sugar phosphotransferase system: Purification and characterization of enzyme I, *J. Bacteriol.* **131**:988–996.

Kasahara, M., and Hinkle, P. C., 1977, Reconstitution and purification of the D-glucose transporter from human erythrocytes, *J. Biol. Chem.* **252**:7384–7390.

Konings, W. N., de Vrij, W., Driessen, A. J. M., and Poolman, B., 1987, Primary and secondary transport in gram negative bacteria, in: *Sugar Transport and Metabolism in Gram-Positive Bacteria* (J. Reizer and A. Peterkofsky, eds.), Halsted Press, New York, pp. 270–294.

Kornberg, H. L., 1976, Genetics in the study of carbohydrate transport by bacteria, *J. Gen. Microbiol.* **96**:1–16.

Kundig, W., Gosh, S., and Roseman, S., 1964, Phosphate bound to histidine in a protein as an intermediate in a novel phosphotransferase system, *Proc. Natl. Acad. Sci. USA* **52**:1067–1074.

McElhaney, R. N., and Tourtellotte, M., 1969, Mycoplasma membrane lipids: Variations in fatty acid composition, *Science* **164**:433–434.

McElhaney, R. N., de Gier, J., van Deenen, L. L. M., and van der Neut-Kok, E. C. M., 1973, The effects of alterations in fatty acid composition and cholesterol content on the nonelectrolyte permeability of *Acholeplasma laidlawii* cells and derived liposomes, *Biochim. Biophys. Acta* **298**:500–512.

Mitchell, P., 1961, Coupling of phosphorylation to electron and hydrogen transfer by a chemiosmotic type of mechanism, *Nature* **191**:144–148.

Mitchell, P., 1966, Chemiosmotic coupling in oxidative and photosynthetic phosphorylation, *Biol. Rev.* **1**:445–502.

Mugharbil, U., and Cirillo, V. P., 1978, Mycoplasma phosphoenolpyruvate-dependent sugar phosphotransferase system: Glucose-negative mutant and regulation of intracellular cyclic AMP, *J. Bacteriol.* **133**:203–209.

Novotny, M. J., Fredericksen, W. L., Waygood, E. B., and Saier, M. H., Jr., 1985, Allosteric regulation of glycerol kinase by enzyme II^{glc} of the phosphotransferase system in *Escherichia coli* and *Salmonella typhimurium*, *J. Bacteriol.* **162**:810–816.

Nunn, W. D., 1986, A molecular view of fatty acid catabolism in *Escherichia coli*, *Microbiol. Rev.* **50**:179–192.

Panchenko, L. F., Fedotov, N. S., and Tarshis, M. A., 1975, Transport properties of membrane vesicles from *Acholeplasma laidlawii* II, *Folia Microbiol. (Prague)* **20**:480–487.

Pick, U., 1981, Liposomes with large trapping capacity prepared by freezing and thawing of sonicated phospholipid mixtures, *Arch. Biochem. Biophys.* **212**:186–194.

Postma, P. W., Epstein, W., Schuitema, A. R. J., and Nelson, S. O., 1984, Interaction between III^{glc} of the phosphoenolpyruvate : sugar phosphotransferase system and glycerol kinase of *Salmonella typhimurium*, *J. Bacteriol.* **158**:351–353.

Razin, S., Gottfried, L., and Rottem, S., 1968, Amino acid transport in *Mycoplasma*, *J. Bacteriol.* **5**:1685–1691.

Read, B. D., and McElhaney, R. N., 1975, Glucose transport in *Acholeplasma laidlawii* B: Dependence on the fluidity and physical state of membrane lipids, *J. Bacteriol.* **123**:47–55.

Reizer, J., Saier, M. M., Deutscher, J., Grenier, F., Thompson, J., and Hengstenberg, W., 1988, The phosphoenolpyruvate–sugar phosphotransferase system in gram-positive bacteria: Properties, mechanism and regulation, *Crit. Rev. Microbiol.* **15**:2977–3038.

Reizer, J., Reizer, A., and Saier, M. H., Jr., 1990, The cellobiose permease of *Escherichia coli* consists of three proteins and is homologous to the lactose permease of *Staphylococcus aureus*, *Res. Microbiol.* **141**:1061–1067.

Romano, A. H., Eberhard, S. J., Dingle, S. L., and MacDowell, T. D., 1970, Distribution of the phosphoenolpyruvate: glucose phosphotransferase system in bacteria, *J. Bacteriol.* **104**:808–813.

Romano, A. H., Trifone, J. D., and Brustolan, M., 1979, Distribution of the phosphoenolpyruvate: glucose transport system in fermentative bacteria, *J. Bacteriol.* **139**:93–97.

Romano, A. H., Brino, G., Peterkovsky, A., and Reizer, J., 1987, Regulation of β-galactoside transport and accumulation in heterofermentative lactic acid bacteria, *J. Bacteriol.* **169**:5589–5596.

Rottem, S., and Razin, S., 1966, Adenosine triphosphatase activity of Mycoplasma membranes, *J. Bacteriol.* **92**:714–722.

Rottem, S., and Razin, S., 1969, Sugar transport in *Mycoplasma gallisepticum*, *J. Bacteriol.* **97**:787–792.

Rottem, S., and Shirazi, I., 1990, An arginine-ornithine exchange system in spiroplasmas, *IOM Lett.* **1**:102–103.

Saier, M. H., Jr., 1977, Bacterial phosphoenolpyruvate: sugar phosphotransferase systems: Structural, functional, and evolutionary interrelationships, *Bacteriol. Rev.* **41**:856–871.

Saier, M. H., Jr., 1985, *Mechanisms and Regulation of Carbohydrate Transport in Bacteria*, Academic Press, New York.

Saier, M. H., Jr., and Reizer, J., 1990, Domain shuffling during evolution of the proteins of the bacterial phosphotransferase system, *Res. Microbiol.* **141**:1033–1038.

Shirvan, M. H., Schuldiner, S., and Rottem, S., 1989, Volume regulation in *Mycoplasma gallisepticum:* Evidence that Na$^+$ is extruded via a primary Na$^+$ pump, *J. Bacteriol.* **171**:4417–4424.

Shirvan, M. H., Schuldiner, S., and Rottem, S., 1990, Role of Na$^+$ cycle in cell volume regulation of *Mycoplasma gallisepticum*, *J. Bacteriol.* **171**:4410–4416.

Singer, S. J., and Nicolson, G. L., 1972, The fluid mosaic model of the structure of cell membranes, *Science* **175**:720–731.

Skulachev, V. P., 1988, *Membrane Bioenergetics*, Springer-Verlag, Berlin.

Smith, P. F., 1965, Amino acid metabolism by pleuropneumonia-like organisms, *J. Bacteriol.* **70**:552–556.

Smith, P. F., 1969, The role of lipids in membrane transport in *Mycoplasma laidlawii*, *Lipids* **4**:331–336.

Tarshis, M., 1991, Spiroplasma cells utilize carbohydrates via the phosphoenolpyruvate-dependent sugar phosphotransferase system, *Can. J. Microbiol.* **37**:477–479.

Tarshis, M. A., Bekkouzjin, A. G., Ladygina, V. G., and Panchenko, L. F., 1976a, Properties of the 3-O-methyl-D-glucose transport system in *Acholeplasma laidlawii*, *J. Bacteriol.* **125**:1–7.

Tarshis, M. A., Bekkouzjin, A. G., and Ladygina, V. G., 1976b, On the possible role of respiratory activity of *Acholeplasma laidlawii* cells in sugar transport, *Arch. Microbiol.* **109**:295–299.

Van Demark, P. J., and Plackett, P., 1972, Evidence for a phosphoenolpyruvate-dependent sugar phosphotransferase in *Mycoplasma* strain Y, *J. Bacteriol.* **111**:454–458.

Index

The manufacturer's authorised representative in the EU is Springer
Nature Customer Service Centre GmbH, Europaplatz 3, 69115 Heidelberg,
Germany. If you have any concerns regarding our products, please
contact ProductSafety@springernature.com

Printed and bound by CPI Group (UK) Ltd, Croydon, CR0 4YY

23/04/2026

02095623-0001